AF505689

SOLVING POLYNOMIAL EQUATION SYSTEMS

Volume III: Algebraic Solving

This third volume of four finishes the program begun in Volume I by describing all the most important techniques, mainly based on Gröbner bases, which allow one to manipulate the roots of an equation rather than just compute them.

The book begins with the "standard" solutions (the Gianni–Kalkbrener Theorem, Stetter Algorithm, the Cardinal–Mourrain result) and then moves on to more innovative methods (Lazard triangular sets, Rouillier's Rational Univariate Representation, the TERA Kronecker package). The author also looks at classical results, such as Macaulay's matrix, and provides a historical survey of elimination, from Bézout to Cayley.

This comprehensive treatment in four volumes is a contribution to algorithmic commutative algebra that will be essential reading for algebraists and algebraic geometers.

Encyclopedia of Mathematics and Its Applications

This series is devoted to significant topics or themes that have wide application in mathematics or mathematical science and for which a detailed development of the abstract theory is less important than a thorough and concrete exploration of the implications and applications.

Books in the **Encyclopedia of Mathematics and Its Applications** cover their subjects comprehensively. Less important results may be summarized as exercises at the ends of chapters. For technicalities, readers can be referred to the bibliography, which is expected to be comprehensive. As a result, volumes are encyclopedic references or manageable guides to major subjects.

All the titles listed below can be obtained from good booksellers or from Cambridge University Press. For a complete series listing visit www.cambridge.org/mathematics.

ENCYCLOPEDIA OF MATHEMATICS AND ITS APPLICATIONS

Solving Polynomial Equation Systems

Volume III: Algebraic Solving

TEO MORA

University of Genoa

CAMBRIDGE
UNIVERSITY PRESS

CAMBRIDGE
UNIVERSITY PRESS

University Printing House, Cambridge CB2 8BS, United Kingdom

Cambridge University Press is part of the University of Cambridge.

It furthers the University's mission by disseminating knowledge in the pursuit of
education, learning and research at the highest international levels of excellence.

www.cambridge.org
Information on this title: www.cambridge.org/9780521811552

First published 2015

A catalogue record for this publication is available from the British Library

ISBN – Volume I   978-0-521-81154-5 Hardback
ISBN – Volume II   978-0-521-81156-9 Hardback
ISBN – Volume III 978-0-521-81155-2 Hardback
ISBN – Volume IV 978-1-107-10963-6 Hardback

Joachim of Fiore's *Age of the Holy Spirit* offers a Hegelian synthesis between the Old and New Testament.

A. Buendia, *The Long March of the Red Brigades through Conquest,*
War, Famine and Death

God is my witness that I would sooner free your mind from mistakes than see me released from prison.

Martinek Húska Loquis

The computational effort required to implement this approach turned out to be orders of magnitude less than the effort which would be required by the direct techniques of decoding by exhaustive search. Using new techniques which are introduced in this book, it is now possible to build algebraic decoders which are orders of magnitude simpler than any that have previously been considered.

There is frequently a conflict between proofs which some people consider conceptually "simple" and proofs which lead to simple instrumentation. In this book I have attempted to provide the proofs which lead to the simplest implementations.

E.R. Berlekamp, *Algebraic Coding Theory*

Solomon Gandz in the final section of his introduction to the Mensuration of al-Khwarizmi wrote: "Euclid and his geometry [. . .] is entirely ignored by him when he writes on geometry. On the contrary, in the preface to his *Algebra*, Algorithm distinctly emphasizes his purpose of writing a popular treatise that, in contradiction to Greek theoretical mathematics, will serve the practical ends and needs of the people in their affairs of inheritance and legacies, in their law suits, in trade and commerces, in the surveying of lands and in the digging of canals. Hence, Algorithm appears to us not as a pupil of the Greeks but, quite to the contary, as the antagonist of [. . .] the Greek school, as the representative of the native popular science. At the Academy of Bagdad Algorithm represented rather the reaction against the introduction of Greek mathematics. His *Algebra* impresses us as a protest rather against the Euclid translation and against the whole trend of the reception of the Greek science."

Is it too much to read this quotation as a parable, interpreting Greeks as French and Euclid as Bourbaki?

R.F. Ree, *The Foundational Crisis, a Crisis of Computability?*

Contents

Preface

La gloria di colui che tutto move
per l'universo penetra, e risplende
in una parte piú e meno altrove.

My HOPE that this *SPES* series reaches completion has supported me over many years. These years have been devoted both to fixing the details of the operative scheme based on Spear's Theorem, which allows one to set a Buchberger Theory over each effective associative ring and of which I have been aware since my 1988 preprint "Seven variations on standard bases" and to satisfy my *horror vacui* by including all the relevant results of which I have been aware.

My *horror vacui* had the negative aspect of making the planned third book grow too much, forcing me to split it into two separate volumes. As a consequence the structure I planned 12 years ago and which anticipated a Hegelian (or Dante-like) trilogy, whose central focus was the Gröbnerian technology discussed in Volume II, was quite deformed and the result appears as a (Wagner-like?) tetralogy.

This volume contains Part six, *Algebraic Solving*, and is where I complete the task set out in Part one by discussing all the recent approaches. These are mainly based on the results discussed in Volume II, which allow one to effectively manipulate the roots of a polynomial equation system, thus fulfilling the aim of "solving" as set out in Volume I according to the Kronecker–Duval Philosophy: Trinks' Algorithm, the Gianni–Kalkbrener Theorem, the Stetter Algorithm, Dixon's resultant, the Cardinal–Mourrain Algorithm, Lazard's Solver, Rouillier's Rational Univariate Representation, the TERA Kronecker package.

Macaulay's Matrix and u-resultant, a historical tour of elimination from Bézout to Dixon, who was the last student of Cayley, the Lagrange resolvent and the investigation of it performed by Valibuze and Arnaudies are also covered.

Setting

1. Let k be an infinite, perfect field, where, if $p := \mathrm{char}(k) \neq 0$, it is possible to extract pth roots, and let k be the algebraic closure of k and $\Omega(k)$ the universal field over k.

Let us fix an integer value n and consider the polynomial ring

$$\mathcal{P} := k[X_1, \ldots, X_n]$$

and its k-basis

$$\mathcal{T} := \{X_1^{a_1} \cdots X_n^{a_n} : (a_1, \ldots, a_n) \in \mathbb{N}^n\}.$$

For each $\delta \in \mathbb{N}$ we will also set $\mathcal{T}_\delta := \{t \in \mathcal{T} : \deg(t) = \delta\}$.

2. We also fix an integer value $r \leq n$, set $d := n - r$ and consider

the field $K := k(V_1, \ldots, V_d)$,
its algebraic closure K and its universal field $\Omega(K) = \Omega(k)$,
the polynomial ring $\mathcal{Q} := K[Z_1, \ldots, Z_r]$ and
its K-basis $\mathcal{W} := \{Z_1^{a_1} \cdots Z_r^{a_r} : (a_1, \ldots, a_r) \in \mathbb{N}^r\}$.

All the notation introduced in the previous volumes will be applied also in this setting, with the proviso that everywhere $n, k, \mathcal{P}, \mathcal{T}$ are substituted by, respectively $r, K, \mathcal{Q}, \mathcal{W}$.

3. Each polynomial $f \in k[X_1, \ldots, X_n]$ is a unique linear combination,

$$f = \sum_{t \in \mathcal{T}} c(f, t)t,$$

of the terms $t \in \mathcal{T}$ with coefficients $c(f, t)$ in k and can be uniquely decomposed as $f = \sum_{\delta=0}^{d} f_\delta$, by setting

$$f_\delta := \sum_{t \in \mathcal{T}_\delta} c(f, t)t \quad \text{for each } \delta \in \mathbb{N},$$

where each f_δ is homogeneous, $\deg(f_\delta) = \delta$ and $f_d \neq 0$, so that $d = \deg(f)$.

4. Since, for each i, $1 \le i \le n$,

$$\mathcal{P} = k[X_1, \ldots, X_{i-1}, X_{i+1}, \ldots, X_n][X_i],$$

each polynomial $f \in \mathcal{P}$ can be uniquely expressed as

$$f = \sum_{j=0}^{D} h_j(X_1, \ldots, X_{i-1}, X_{i+1}, \ldots, X_n)X_i^j \quad h_D \ne 0,$$

and

$$\deg_{X_i}(f) := \deg_i(f) := D$$

denotes its degree in the variable X_i.

In particular, for $i = n$, we have

$$f = \sum_{j=0}^{D} h_j(X_1, \ldots, X_{n-1})X_n^j, \quad h_D \ne 0, \quad D = \deg_n(f);$$

the *leading polynomial* of f is $\mathrm{Lp}(f) := h_D$, and its *trailing polynomial* is $\mathrm{Tp}(f) := h_0$.

5. Given a finite basis $F := \{f_1, \ldots, f_u\} \subset \mathcal{P}$, we denote as

$$\mathbb{I}(F) := (F) := \left\{ \sum_{i=1}^{u} h_i f_i : h_i \in \mathcal{P} \right\} \subset \mathcal{P}$$

the ideal generated by F, and as

$$\mathcal{Z}(F) := \{\mathbf{a} \in \mathsf{k}^n : f(\mathbf{a}) = 0, \text{ for all } f \in F\} \subset \mathsf{k}^n$$

the algebraic variety consisting of each common root of all polynomials in F.

6. The support

$$\mathrm{supp}(f) := \{t \in \mathcal{T} : c(f, t) \ne 0\}$$

of f being finite, once a term ordering[1] $<$ on $\mathcal{T}$ is fixed, f has a unique representation as an ordered linear combination of terms:

$$f = \sum_{i=1}^{s} c(f, t_i)t_i : c(f, t_i) \in k \setminus \{0\}, \quad t_i \in \mathcal{T}, \quad t_1 > \cdots > t_s.$$

The *maximal term* of f is $\mathbf{T}(f) := t_1$, its *leading cofficient* is $\mathrm{lc}(f) := c(f, t_1)$ and its *maximal monomial* is $\mathbf{M}(f) := c(f, t_1)t_1$.

7. For any set $F \subset \mathcal{P}$ we write

- $\mathbf{T}_<\{F\} := \{\mathbf{T}(f) : f \in F\}$,
- $\mathbf{T}_<(F) := \{\tau\mathbf{T}(f) : \tau \in \mathcal{T}, f \in F\}$,

---

[1] A well-ordering $<$ on $\mathcal{T}$ will be called a term ordering if it is a semigroup ordering.

- $\mathbf{N}_<(F) := \mathcal{T} \setminus \mathbf{T}_<(F)$,
- $k[\mathbf{N}_<(F)] := \mathrm{Span}_k(\mathbf{N}_<(F))$

and we will usually omit the dependence on $<$ if there is no ambiguity.

8. Let $<$ be a term ordering on $\mathcal{T}$, $\mathsf{I} \subset \mathcal{P}$ an ideal and $\mathsf{A} := \mathcal{P}/\mathsf{I}$. Since $\mathsf{A} \cong k[\mathbf{N}_<(\mathsf{I})]$, there is, for each $f \in \mathcal{P}$, a unique

$$g := \mathrm{Can}(f, \mathsf{I}, <) = \sum_{t \in \mathbf{N}_<(\mathsf{I})} \gamma(f, t, <)t,$$

the *canonical form*, such that

$$g \in k[\mathbf{N}(\mathsf{I})] \quad \text{and} \quad f - g \in \mathsf{I}.$$

9. For an ideal $\mathsf{I} \subset \mathcal{P}$,

$$\mathsf{I} := \cap_{i=1}^{t} \mathfrak{q}_i$$

denotes an irredundant primary representation in $\mathcal{P}$; $d := \dim(\mathsf{I})$ its dimension and $r := r(\mathsf{I}) := n - d$ its rank; for each i, $\mathfrak{p}_i := \sqrt{\mathfrak{q}_i}$ is the associated prime.

10. For such an ideal I we will re-enumerate and re-label the variables as follows:

$$\{X_1, \ldots, X_n\} = \{V_1, \ldots, V_d, Z_1, \ldots, Z_r\},$$

so that

$$\mathsf{I} \cap k[V_1, \ldots, V_d] = (0), \quad d := \dim(\mathsf{I}),$$

and we will wlog assume that the primaries are ordered so that, for a suitable value $1 \leq \mathsf{r} \leq \mathsf{t}$,

$$\mathfrak{q}_i \cap k[V_1, \ldots, V_d] = (0), \quad \dim(\mathfrak{q}_i) = d \iff i \leq \mathsf{r}$$

so that the ideal

$$\mathsf{J} := \mathsf{I}\, k(V_1, \ldots, V_d)[Z_1, \ldots, Z_r] = \mathsf{I}\mathcal{Q}$$

is zero-dimensional and has, in $\mathcal{Q}$, the irredundant primary representation

$$\mathsf{J} := \cap_{i=1}^{\mathsf{r}} \mathfrak{q}_i \mathcal{Q}.$$

11. In general, when dealing with a zero-dimensional ideal, instead of

$$\mathsf{I} \subset \mathcal{P} = k[\mathcal{T}] = k[X_1, \ldots, X_n]$$

we prefer to use the notation

$$\mathsf{J} \subset \mathcal{Q} = K[\mathcal{W}] = K[Z_1, \ldots, Z_r].$$

12. For such a zero-dimensional ideal J, with a slight abuse of notation we will still set $\mathsf{A} := \mathcal{Q}/\mathsf{J}$ and denote as q_i its primary components in $\mathcal{Q}$; we also assume that

$$s := \deg(\mathsf{J}) = \dim(\mathsf{A})$$

and we denote, for each $f \in \mathcal{Q}$, $[f] \in \mathsf{A}$, its residue class modulo J and as Φ_f the endomorphism

$$\Phi_f : \mathsf{A} \to \mathsf{A}, \quad [g] \mapsto [fg].$$

13. In terms of a K-basis $\mathbf{q} = \{[q_1], \dots, [q_s]\}$ of A such that $\mathsf{A} = \mathrm{Span}_K(\mathbf{q})$, for each $g \in \mathcal{Q}$ the *Gröbner description of g* (Definition 29.3.3,) is the unique (row) vector

$$\mathbf{Rep}(g, \mathbf{q}) := (\gamma(g, q_1, \mathbf{q}), \dots, \gamma(g, q_s, \mathbf{q})) \in K^s,$$

which satisfies

$$[g] = \sum_j \gamma(g, q_j, \mathbf{q})[q_j].$$

14. A *Gröbner representation* (Definition 29.3.3) of J (or, better, of the algebra A) is the assignment of

- a K-linearly independent set $\mathbf{q} = \{[q_1], \dots, [q_s]\}$,
- the set $\mathcal{M} = \mathcal{M}(\mathbf{q}) := \left\{\left(a_{lj}^{(h)}\right) \in K^{s^2}, 1 \le h \le r\right\}$ of r square matrices,
- s^3 values $\gamma_{ij}^{(l)} \in K$,

which satisfy

(1) $\mathcal{Q}/\mathsf{J} \cong \mathrm{Span}_K(\mathbf{q})$,

(2) $[Z_h q_l] = \sum_j a_{lj}^{(h)}[q_j]$ for each $l, j, h, 1 \le l, j \le s, 1 \le h \le r$,

(3) $[q_i q_j] = \sum_l \gamma_{ij}^{(l)}[q_l]$ for each $l, j, h, 1 \le i, j, l \le s$.

A Gröbner representation is called a *linear representation* iff $\mathbf{q} = \mathsf{N}_<(\mathsf{J})$ w.r.t. a term ordering $<$.

15. For the zero-dimensional ideal $\mathsf{J} \subset \mathcal{Q}$ with irredundant primary representation $\mathsf{J} = \bigcap_{i=1}^r \mathsf{q}_i$ in $\mathcal{Q}$, we set, for each $i, 1 \le i \le \mathsf{r}$,

- $\mathsf{m}_i = \sqrt{\mathsf{q}_i}$, the associated maximal prime,
- $K_i := \mathcal{Q}/\mathsf{m}_i$, $K \subset K_i \subset \mathsf{K}$,
- $\mathcal{Q}_i := K_i[Z_1, \dots, Z_r]$,
- the irredundant primary representations $\mathsf{q}_i = \cap_{j=1}^{r_i} \mathsf{q}_{ij}$ and $\mathsf{m}_i = \cap_{j=1}^{r_i} \mathsf{m}_{ij}$ in $\mathcal{Q}_i$,
- the roots $\mathsf{b}_{ij} := (b_1^{(ij)}, \dots, b_r^{(ij)}) \in K_i^r \subset \mathsf{K}^r, 1 \le j \le r_i$,
- $d_{ij} := \mathrm{mult}(\mathsf{b}_{ij}, \mathsf{J}) = \deg(\mathsf{q}_{ij})$ for each $j, 1 \le j \le r_i$,

which satisfy:

(1) $\mathsf{m}_{ij} = (Z_1 - b_1^{(ij)}, \dots, Z_r - b_r^{(ij)})$,

(2) the $\mathsf{b}_{ij}, 1 \le j \le r_i$, are K-conjugate for each i,

(3) up to a renumeration, $\sqrt{\mathfrak{q}_{ij}} = \mathfrak{m}_{ij}$,

(4) $\mathfrak{m}_i = \mathfrak{m}_{ij} \cap \mathcal{Q}$,

(5) $\mathfrak{q}_i = \mathfrak{q}_{ij} \cap \mathcal{Q}$,

(6) for each $j, l, 1 \le j, l \le r_i$, $d_{ij} = d_{il} =: d_i$,

(7) $r_i = \deg(\mathfrak{m}_i) = [K_i : K]$,

(8) $\deg(\mathfrak{q}_i) = d_i r_i$,

(9) $J = \cap_{i=1}^{r} \cap_{j=1}^{r_i} \mathfrak{q}_{ij}$, $\sqrt{J} = \cap_{i=1}^{r} \cap_{j=1}^{r_i} \mathfrak{m}_{ij}$, are the irredundant primary representations in $K[Z_1, \ldots, Z_r]$,

(10) $\mathcal{Z}(J) = \{\mathfrak{b}_{ij} : 1 \le i \le r, 1 \le j \le r_j\}$,

(11) $\sum_{i=1}^{r} d_i r_i = s$.

16. With the notation above the ideal J has $s := \sum_{i=1}^{r} r_i$ roots; we will also denote this set of roots as

$$\mathcal{Z}(J) = \{\alpha_1, \ldots, \alpha_s\} \subset K^r, \quad \alpha_i = (a_1^{(i)}, \ldots, a_r^{(i)}).$$

For each such root α_i we write

- $\mathfrak{m}_{\alpha_i} = (Z_1 - a_1^{(i)}, \ldots, Z_r - a_r^{(i)})$,
- $\mathfrak{q}_i$ as the $\mathfrak{m}_{\alpha_i}$-primary component of J, so that $J = \cap_{i=1}^{s} \mathfrak{q}_i$ in $K \otimes_K \mathcal{Q}$,
- $s_i := \mathrm{mult}(\alpha_i, J) = \deg(\mathfrak{q}_i)$ as the multiplicity in J of α_i, so that $s = \sum_{i=1}^{s} s_i$.

17. A linear form $Y := \sum_{h=1}^{r} c_h Z_h$ is said to be an *allgemeine coordinate* for the zero-dimensional ideal J (Definition 34.4.7) iff

(a) there are polynomials $g_i \in K[Y], 0 \le i \le n$, g_0 monic, $\deg(g_i) < \deg(g_0)$, such that

$$G := (g_0(Y), Z_1 - g_1(Y), Z_2 - g_2(Y), \ldots, Z_r - g_r(Y))$$

is the reduced Gröbner basis of the ideal

$$J^+ := J + \left(Y - \sum_h c_h Z_h \right) \subset K[Y, Z_1, \ldots, Z_r]$$

w.r.t. the lex ordering induced by $Y < Z_1 < \cdots < Z_r$;

with the present notation this condition implies, among other things, that (Corollary 34.4.6)

(b) $\mathcal{Q}/J \cong K[Y]/g_0(Y)$

(c) for each $i, 1 \le i \le s$, $\beta_i := \sum_{h=1}^{r} c_h a_h^{(i)}$ is a root of g_0 with multiplicity s_i and

(d) $a_j^{(i)} = g_j(\beta_i)$ for each $i, 1 \le i \le s$, and each $j, 1 \le j \le r$,

(e) $g_0(Y) = \prod_{i=1}^{r} (Y - \beta_i)^{s_i}$,

(f) $f \in J \iff \mathbf{Rem}(f(g_1(Y), \ldots, g_r(Y)), g_0(Y)) = 0$.

Moreover, there is a Zarisky open set $U \subset K^n$ such that $Y := \sum_{h=1}^{r} c_h Z_h$ is an *allgemeine* coordinate for J iff $(c_1, \ldots, c_r) \in U$.

18. Given the polynomial ring $\mathcal{P} := k[X_1, \ldots, X_n]$ and its monomial k-basis $\mathcal{T}$, we introduce n futher variables $Y_1, \ldots, Y_n$ and denote

- $\mathcal{P}_Y := k[Y_1, \ldots, Y_n]$ and $\mathcal{T}_Y$ its corresponding monomial k-basis,
- $\mathcal{P}_\otimes := \mathcal{P} \otimes \mathcal{P}_Y = k[X_1, \ldots, X_n, Y_1, \ldots, Y_n]$, and $\mathcal{T}_\otimes$ its corresponding monomial k-basis $\mathcal{T}_\otimes := \{\tau \otimes \omega : \tau \in \mathcal{T}, \omega \in \mathcal{T}_Y\}$,
- for each $i, 0 \leq i \leq n$, we use the notation $h(\mathsf{X}_i)$ to denote the polynomial

$$h(\mathsf{X}_i) := h(Y_1, \ldots, Y_i, X_{i+1}, \ldots, X_n) \quad \text{for each } h(X_1, \ldots, X_n) \in \mathcal{P};$$

  in particular $h(\mathsf{X}_0) = h(X_1, \ldots, X_n)$ and $h(\mathsf{X}_n) = h(Y_1, \ldots, Y_n)$,
- for an ideal $\mathsf{I} = \mathbb{I}(f_1, \ldots, f_s) \subset \mathcal{P}$, with a slight abuse of notation we denote also as I the ideal in $\mathcal{P}_Y$ generated by $\{f_1(Y_1, \ldots, Y_n), \ldots, f_n(Y_1, \ldots, Y_n)\}$ and $\mathsf{A} := \mathcal{P}_Y / \mathsf{I}$; thus we have

$$\mathsf{A} \otimes_k \mathsf{A} = \mathcal{P}_\otimes / \mathbb{I}(f_i(X_1, \ldots, X_n), f_i(Y_1, \ldots, Y_n), 1 \leq i \leq n);$$

- finally we denote $\mathsf{I}_X := \mathsf{I} \otimes \mathcal{P}_Y \subset \mathcal{P}_\otimes$ and $\mathsf{I}_Y := \mathcal{P} \otimes \mathsf{I} \subset \mathcal{P}_\otimes$.

PART SIX

Algebraic Solving

And I beheld when he opened the sixth seal, and, lo, there was a great earthquake; and the sun become black as sackcloth of hair, and the moon became as blood;

And the stars of heaven fell unto the earth, even as a fig tree casteth her untimely figs, when she is shaken of a mighty wind.

Revelation (Authorized Version)

The things depending from Sun: spirituality, gold, carbuncle, heliotrope, lion, phoenix, seal.

E.C. Agrippa, De Occulta Phylosophia

So, according to the wil and meaning of *Fra.* C. R. C., we his brethren request again all the learned in Europe who shall read (sent forth in five languages) this our *Fama* and *Confessio*, that it would please them with good deliberation to ponder this our offer, and to examine most nearly and sharply their arts, and behold the present time with all diligence, and to declare their minde, either *communicato consilio*, or *singulatim* by print.

And although at this time we make no mention either of our names or meetings, yet nevertheless every one's opinion shall assuredly come to our hands, in what language so ever it be, nor any body shal fail, whoso gives but his name, to speak with some of us, either by word of mouth, or else, if there be some lett, in writing. And this we say for a truth, that whosoever shal earnestly, and from his heart, bear affection unto us, it shal be beneficial to him in goods, body, and soul; but he that is false-hearted, or onely greedy of riches, the same first of all shal not be able in any manner of wise to hurt us, but bring himself to utter ruine and destruction. Also our building, although one hundred thousand people had very near seen and beheld the same, shal for ever remain untouched, undestroyed, and hidden to the wicked world.

Sub umbra alarum tuarum, Jehova.

Fama Fraternitatis

39

Trinks

The first paper applying Buchberger's Algorithm being Trinks' proposal of an algorithm for solving polynomial equation systems, Trinks' Algorithm is the natural choice for opening this section on algebraic solving.

Trinks' Algorithm is essentially an effective reformulation of Gröbner's proof of Hilbert's Nullstellensatz (Section 20.3): given a zero-dimensional ideal $J \subset Q := K[Z_1, \ldots, Z_r]$, then Trinks' Algorithm, for each root $\alpha \in \mathsf{K}^{i-1}$ of $J \cap K[Z_1, \ldots, Z_{i-1}]$, computes iteratively and solves $\gcd(h(\alpha, Z_i) : h \in G_i) \in \mathsf{K}[Z_i]$ where G_i denotes a basis of the ideal $J \cap K[Z_1, \ldots, Z_i]$; the rôle of Gröbner bases consists in allowing the computation of such a basis of the elimination ideals.

The main improvement to Trinks' Algorithm, apart from the use of FGLM (see Sections 29.1 and 29.2) in order to efficiently deduce the needed lex Gröbner basis of J, is the Gianni–Kalkbrener proposal to use their Theorem 34.6.3; the evaluation at α of all polynomials in G_i and the computation of their gcd is thus reduced to the evaluation at α of the leading polynomials of some elements in G_i and of the first element whose leading polynomial does not vanish at α.

After recalling in Section 39.1 the basic tools provided by Gröbner bases w.r.t. solving, I present Trinks' Algorithm, in Section 39.2, and the Gianni–Kalkbrener Algorithm, in Section 39.3, and conclude with some comments which aim to read these algorithms in the setting of Kronecker–Duval Philosophy (Section 39.4). Finally, in Section 39.5, I discuss a solver from 1913 which already explicitly applies the main property of the lex term ordering and anticipates Macaulay's Lemma.

39.1 Recalling Gröbner

Let us consider

- an infinite, perfect field k,[1] where, if $p := \mathrm{char}(k) \neq 0$, it is possible to extract pth roots,
- the algebraic closure k of k,

---

[1] While the techniques discussed here apply in this general setting, we are mainly thinking of the case $k = \mathbb{Q}$, $\mathsf{k} := \mathbb{C}$; nevertheless, technically we need to (and we can) solve over $\mathbb{Q}(V_1, \ldots, V_d)$.

- the universal field $\Omega(k)$ over k (Definition 9.4.1),
- the polynomial ring $\mathcal{P} := k[X_1, \ldots, X_n]$,
- its k-basis $\mathcal{T} := \{X_1^{a_1} \cdots X_n^{a_n} : (a_1, \ldots, a_n) \in \mathbb{N}^n\}$,
- an ideal $\mathsf{I} := (F) := \mathbb{I}(F) := \left\{\sum_{i=1}^u h_i f_i : h_i \in \mathcal{P}\right\} \subset \mathcal{P}$ given by[2]
- a finite basis $F := \{f_1, \ldots, f_u\} \subset \mathcal{P}$,
- the algebraic affine variety $\mathcal{Z}(\mathsf{I}) := \{\mathsf{a} \in \mathsf{k}^n : f(\mathsf{a}) = 0, \text{ for each } f \in F\} \subset \mathsf{k}^n$.

Each polynomial $f \in k[X_1, \ldots, X_n]$ is therefore a unique linear combination

$$f = \sum_{t \in \mathcal{T}} c(f, t) t$$

of the terms $t \in \mathcal{T}$ with coefficients $c(f, t)$ in k; the support

$$\text{supp}(f) := \{t \in \mathcal{T} : c(f, t) \neq 0\}$$

of f being finite, once a term ordering $<$ on $\mathcal{T}$ is fixed,[3] f has a unique representation as an ordered linear combination of terms:

$$f = \sum_{i=1}^s c(f, t_i) t_i : c(f, t_i) \in k \setminus \{0\}, \quad t_i \in \mathcal{T}, \quad t_1 > \cdots > t_s;$$

the *maximal term* of f is $\mathbf{T}(f) := t_1$, its *leading cofficient* is $\text{lc}(f) := c(f, t_1)$ and its *maximal monomial* is $\mathbf{M}(f) := c(f, t_1) t_1$.

For any set $F \subset \mathcal{P}$ we denote

- $\mathbf{T}_<\{F\} := \{\mathbf{T}(f) : f \in F\}$,
- $\mathbf{T}_<(F) := \{\tau \mathbf{T}(f) : \tau \in \mathcal{T}, f \in F\}$,
- $\mathbf{N}_<(F) := \mathcal{T} \setminus \mathbf{T}_<(F)$,
- $k[\mathbf{N}_<(F)] := \text{Span}_k(\mathbf{N}_<(F))$;

we will usually omit the dependence on $<$ if there is no ambiguity. Recall that

Definition 39.1.1 (Buchberger). *A subset $G \subset \mathsf{I}$ will be called a* Gröbner basis *of I w.r.t. $<$ (Definition 22.2.1) if $\mathbf{T}(G) = \mathbf{T}\{\mathsf{I}\}$, that is, $\mathbf{T}\{G\}$ generates the monomial ideal $\mathbf{T}(\mathsf{I}) = \mathbf{T}\{\mathsf{I}\}$.*

For each $f \in \mathcal{P}$ the canonical form *of f w.r.t. I (Definition 22.2.13) is the unique polynomial*

$$g := \text{Can}(f, \mathsf{I}, <) = \sum_{t \in \mathbf{N}(\mathsf{I})} \gamma(f, t, <) t \in k[\mathbf{N}(\mathsf{I})]$$

such that $f - g \in \mathsf{I}$. $\qquad\qquad$ ⊡

Let us fix any term ordering $<$ on $\mathcal{T}$ and compute a Gröbner basis $G \subset \mathsf{I}$ of I w.r.t. $<$.

---

[2] Throughout the book I will use the notation $\mathbb{I}(F) \subset \mathcal{R}$ to denote the ideal generated by the basis F in the ring $\mathcal{R}$; when there is no ambiguity $\mathcal{R}$ will be not specified.

[3] Recall that (cf. Definition 22.1.2) a well-ordering $<$ on $\mathcal{T}$ that is also a semigroup ordering, that is, if

$$t_1 < t_2 \implies t t_1 < t t_2 \quad \text{for each } t, t_1, t_2 \in \mathcal{T}$$

is called a term ordering.

Then the following hold (cf. Remark 27.12.4):

- $\mathcal{Z}(\mathsf{I}) = \emptyset \iff 1 \in \mathsf{I} \iff 1 \in G$;
- $\mathcal{Z}(\mathsf{I})$ is infinite iff $\mathbf{N}(\mathsf{I})$ is infinite $\iff$ there exists i such that, for each $d \in \mathbb{N}$, $X_i^d \notin \mathbf{T}(G) = \mathbf{T}(\mathsf{I})$;
- $\mathcal{Z}(\mathsf{I})$ is finite iff $\mathbf{N}(\mathsf{I})$ is finite $\iff$ for each i there exists $d_i \in \mathbb{N} : X_i^{d_i} \in \mathbf{T}\{G\} \subset \mathbf{T}(\mathsf{I})$; moreover, in this case and under the assumption that I is radical, we have $\#\mathcal{Z}(\mathsf{I}) = \#\mathbf{N}(\mathsf{I})$.

The Kredel–Weispfenning Algorithm (Corollary 27.11.9) allows us to deduce from $\mathbf{T}(\mathsf{I})$ the dimension $d := \dim(\mathsf{I})$, the rank $r := n - d := r(\mathsf{I})$ of I and a maximal set of independent variables (Definition 27.11.4) $\{X_{i_1}, \ldots, X_{i_d}\}$ such that $\mathsf{I} \cap k[X_{i_1}, \ldots, X_{i_d}] = (0)$.

Then, we can re-enumerate and relabel the variables as follows:

$$\{X_1, \ldots, X_n\} = \{V_1, \ldots, V_d, Z_1, \ldots, Z_r\}, \quad \{X_{i_1}, \ldots, X_{i_d}\} = \{V_1, \ldots, V_d\},$$

so that

$$\mathsf{I} \cap k[V_1, \ldots, V_d] = (0),$$

and consider

    the field $K := k(V_1, \ldots, V_d)$,
    its algebraic closure $\mathbf{K}$
    and its universal field $\Omega(K) = \Omega(k)$,
    the polynomial ring $\mathcal{Q} := K[Z_1, \ldots, Z_r]$,
    its K-basis $\mathcal{W} := \{Z_1^{a_1} \cdots Z_r^{a_r} : (a_1, \ldots, a_r) \in \mathbb{N}^r\}$,
    the zero-dimensional ideal $\mathsf{J} := \mathsf{I}^e := \mathsf{I}K[Z_1, \ldots, Z_r]^4$
    and the unmixed ideal $\mathsf{J}^c := \mathsf{J} \cap \mathcal{P}$.

Then, if $\mathsf{I} = \cap_{i=1}^t \mathfrak{q}_i$ denotes any irredundant primary representation in $\mathcal{P}$ and we wlog assume that the primaries are ordered in such a way that, for a suitable value $1 \leq \mathsf{r} \leq \mathsf{t}$,

    $\{X_{i_1}, \ldots, X_{i_d}\}$ is a maximal set of independent variables for $\mathfrak{q}_i \iff i \leq \mathsf{r}$

then Corollary 27.5.19 grants that

$$\mathsf{J} := \mathsf{I}^e = \bigcap_{i=1}^{\mathsf{r}} \mathfrak{q}_i^e = \bigcap_{i=1}^{\mathsf{r}} \mathfrak{q}_i \mathcal{Q}$$

is an irredundant primary representation in $\mathcal{Q}$ and

$$\mathsf{J}^c := \mathsf{I}^{ec} = \bigcap_{i=1}^{\mathsf{r}} \mathfrak{q}_i \subset \mathcal{P}$$

is an irredundant primary representation.

---

[4] For this notation see Section 27.5.

Moreover, the Gianna–Trager–Zacharias (GTZ), Alonso, Raimondo, Giusti and Heintz (ARGH) and Caborah, Conti and Traverso (CCT) schemes (Chapter 35) allow us to compute unmixed ideals $\mathfrak{a}_j \subset \mathcal{P}$, giving a decomposition

$$\sqrt{\mathsf{I}} = \sqrt{\mathsf{J}^c} \cap \left(\bigcap_j \sqrt{\mathfrak{a}_j} \right).$$

Thus solving the ideal $\mathsf{I} \subset \mathcal{P}$ is reduced, via the Gröbner technique, to solving each unmixed GTZ, ARGH or CCT component and, in turn, solving each such component is reduced to solving the related zero-dimensional extension ideal.

39.2 Trinks' Algorithm

Thus we are reduced to consider a zero-dimensional ideal

$$\mathsf{J} \subset \mathcal{Q} := K[Z_1, \dots, Z_r],$$

which we assume to be given via a Gröbner basis $G_{\prec}$ w.r.t. the lexicographical ordering $\prec$ induced on $\mathcal{W}$ by $Z_1 \prec Z_2 \prec \cdots \prec Z_r$:

$$Z_1^{a_1} \cdots Z_r^{a_r} \prec Z_1^{b_1} \cdots Z_r^{b_r} \iff \text{there exist } j : a_j < b_j \text{ and } a_i = b_i \text{ for } i > j.$$

Then, if we denote, for i, $1 \le i < r$,

$\mathsf{J}_i := \mathsf{J} \cap K[Z_1, \dots, Z_i]$,
$\pi_i : \mathsf{K}^r \to \mathsf{K}^i$ as the canonical projection $\pi_i(a_1, \dots, a_r) = (a_1, \dots, a_i)$,
$G_i := G_{\prec} \cap K[Z_1, \dots, Z_i]$,

we have, for each i,

(1) $\mathcal{Z}(\mathsf{J}_i) = \pi_i(\mathcal{Z}(\mathsf{J})) = \{(a_1, \dots, a_i) : (a_1, \dots, a_r) \in \mathcal{Z}(\mathsf{J})\}$,
(2) G_i is the reduced lexicographical Gröbner basis of J_i (Corollary 26.2.4).

In particular, there is a unique monic polynomial $f(Z_1) \in K[Z_1]$ such that

$$\mathsf{J}_1 = \mathbb{I}(f) \quad \text{and} \quad \{f\} = G_{\prec} \cap K[Z_1].$$

For each $\alpha := (a_1, \dots, a_{i-1}) \in \mathsf{K}^{i-1}$, denote as $\Phi_\alpha : K[Z_1, \dots, Z_i] \to \mathsf{K}[T]$ the projection defined by

$$\Phi_\alpha(f) = f(a_1, \dots, a_{i-1}, T) \quad \text{for each } f \in K[Z_1, \dots, Z_i].$$

Theorem 39.2.1 (Trinks). *Let $\alpha := (a_1, \dots, a_{i-1}) \in \mathcal{Z}(\mathsf{J}_{i-1})$ and let $f \in \mathsf{K}[T]$ be a generator of the principal ideal $\Phi_\alpha(\mathsf{J}_i) \subset \mathsf{K}[T]$. Then, for each $b \in \mathsf{K}$,*

$$(a_1, \dots, a_{i-1}, b) \in \mathcal{Z}(\mathsf{J}_i) \iff f(b) = 0.$$

Proof. Let $h(Z_1, \dots, Z_i) \in \mathsf{J}_i$ be any polynomial such that

$$f(T) = \Phi_\alpha(h) = h(a_1, \dots, a_{i-1}, T).$$

$\mathsf{Z} := \mathbf{Solve}(F, L)$
where
    $F := \{f_1, \ldots, f_u\} \subset \mathcal{Q} := K[Z_1, \ldots, Z_r]$,
    $L \supset K$ is a field extension of K,
    $J \subset \mathcal{Q}$ is the zero-dimensional ideal generated by F,
    $\mathsf{Z} := \{\alpha_1, \ldots, \alpha_s\} = \mathcal{Z}(J) \cap L^r$.
Compute the reduced lexicographical Gröbner basis G of $(f_1, \ldots, f_u)$.
Let $p(Z_1)$ be the unique element in $G \cap K[Z_1]$.
$\mathsf{Z}_1 := \{a \in L : p(a) = 0\}$.
For $i = 2, \ldots, r$ **do**
    $\mathsf{Z}_i := \emptyset$.
    For each $(a_1, \ldots, a_{i-1}) \in \mathsf{Z}_{i-1}$ **do**
        $H := \{g(a_1, \ldots, a_{i-1}, Z_i) : g \in G_i \setminus G_{i-1}\}$.
        $p := \gcd(H)$,
        $\mathsf{Z} := \{a \in L : p(a) = 0\}$,
        $\mathsf{Z}_i := \mathsf{Z}_i \cup \{(a_1, \ldots, a_{i-1}, a) : a \in \mathsf{Z}\}$.
$\mathsf{Z} := \mathsf{Z}_r$.

Figure 39.1. Trinks' Algorithm

Then

$$(a_1, \ldots, a_{i-1}, b) \in \mathcal{Z}(J_i) \implies f(b) = h(a_1, \ldots, a_{i-1}, b) = 0.$$

Conversely, for any $b \in \mathsf{K} : f(b) = 0$ and for any $g(Z_1, \ldots, Z_i) \in J_i$, it holds that $\Phi_\alpha(g) \in \Phi_\alpha(J_i)$, so that

$$g(a_1, \ldots, a_{i-1}, b) = \Phi_\alpha(g)(b) = 0 \quad \text{for each } g \in J_i$$

and $(a_1, \ldots, a_{i-1}, b) \in \mathcal{Z}(J_i)$.   ▣

Algorithm 39.2.2 (Trinks).   Trinks' Algorithm (Figure 39.1) for "solving" a zero-dimensional ideal is based on the theorem above and consists in iteratively computing $\mathcal{Z}(J_i)$ by "solving", for each $\alpha \in \mathcal{Z}(J_{i-1})$, the univariate polynomial generating the principal ideal $\Phi_\alpha(J_i)$.   ▣

Example 39.2.3.   To illustrate Trinks' Algorithm let us consider the zero-dimensional ideal $J \subset \mathbb{Q}[Z_1, Z_2, Z_3]$ discussed in Example 33.2.6 whose lex Gröbner basis is $G := \{g_i, 1 \leq i \leq 8\}$, where (Examples 33.5.1 and 33.5.2)[5]

$$g_1 := 1Z_1^3 - 3Z_1^2 + 2Z_1,$$

$$g_2 := (\mathbf{Z_1^2 - Z_1})Z_2,$$

$$g_3 := \mathbf{Z_1}Z_2^2 - Z_1 Z_2,$$

$$g_4 := 1Z_2^3 - 3Z_2^2 + 2Z_2,$$

$$g_5 := (\mathbf{Z_1^2 - 3Z_1 + 2})Z_3 - 3Z_2^2 - 6Z_2 Z_1 + 9Z_2 - Z_1^2 + 3Z_1 - 2,$$

$$g_6 := (\mathbf{Z_2 + Z_1 - 2})Z_3 + 3Z_2^2 + Z_2 Z_1 - 7Z_2 - 2Z_1^2 + 3Z_1 + 2,$$

---

[5] The *leading polynomial* (see below) $\mathrm{Lp}(g_i)$ is indicated in **bold**.

$$g_7 := (\mathbf{Z_1} - 2)Z_3^2 - 4Z_3 Z_1 + 8Z_3 - 15Z_2^2 - 30Z_2 Z_1 + 45Z_2 + 3Z_1 - 6,$$

$$g_8 := \mathbf{1}Z_3^3 - 3Z_3^2 + 3Z_3 Z_1 - 4Z_3 - 3Z_2^2 - 6Z_2 Z_1 + 9Z_2 - 3Z_1 + 6,$$

and whose roots are $\mathcal{Z}(\mathsf{J}) = \{\mathsf{b}_j, 1 \le j \le 9\}$, with

$$
\begin{array}{lllllll}
\mathsf{b}_1 & = & (0,0,1). & \mathsf{b}_2 & = & (0,1,-2), & \mathsf{b}_3 & = & (2,0,2), \\
\mathsf{b}_4 & = & (0,2,-2), & \mathsf{b}_5 & = & (1,0,3), & \mathsf{b}_6 & = & (1,1,3), \\
\mathsf{b}_7 & = & (1,1,1), & \mathsf{b}_8 & = & (2,0,1), & \mathsf{b}_9 & = & (2,0,0).
\end{array}
$$

Then we have:

$p(Z_1) :=$ g_1, $\mathbf{Z}_1 := \{0,1,2\}$;

$\alpha = (0):$ $\Phi_\alpha(g_2) = \Phi_\alpha(g_3) = 0$; $\Phi_\alpha(g_4) = T^3 - 3T^2 + 2T$;
$\mathbf{Z} := \{0,1,2\}$, $\mathbf{Z}_2 := \{(0,0),(0,1),(0,2)\}$;

$\alpha = (1):$ $\Phi_\alpha(g_2) = 0$; $\Phi_\alpha(g_3) = T^2 - T$; $\Phi_\alpha(g_4) = T^3 - 3T^2 + 2T$;
$\gcd(\Phi_\alpha(g_3), \Phi_\alpha(g_4)) = T^2 - T$;
$\mathbf{Z} := \{0,1\}$, $\mathbf{Z}_2 := \{(0,0),(0,1),(0,2),(1,0),(1,1)\}$;

$\alpha = (2):$ $\Phi_\alpha(g_2) = 2T$; $\Phi_\alpha(g_3) = 2T^2 - 2T$; $\Phi_\alpha(g_4) = T^3 - 3T^2 + 2T$;
$\gcd(\Phi_\alpha(g_i), 2 \le i \le 4) = T$;
$\mathbf{Z} := \{0\}$, $\mathbf{Z}_2 := \{(0,0),(0,1),(0,2),(1,0),(1,1),(2,0)\}$;

$\alpha = (0,0):$ $\Phi_\alpha(g_5) = 2T - 2$; $\Phi_\alpha(g_6) = -2T + 2$;
$\Phi_\alpha(g_7) = -2T^2 + 8T - 6$; $\Phi_\alpha(g_8) = T^3 - 3T^2 - 4T + 6$;
$\gcd(\Phi_\alpha(g_i), 5 \le i \le 8) = T - 1$;
$\mathbf{Z} := \{1\}$, $\mathbf{Z}_2 := \{(0,0,1)\}$;

$\alpha = (0,1):$ $\Phi_\alpha(g_5) = 2T + 4$; $\Phi_\alpha(g_6) = -T - 2$;
$\Phi_\alpha(g_7) = -2T^2 + 8T + 24$; $\Phi_\alpha(g_8) = T^3 - 3T^2 - 4T + 12$;
$\gcd(\Phi_\alpha(g_i), 5 \le i \le 8) = T + 2$;
$\mathbf{Z} := \{-2\}$, $\mathbf{Z}_2 := \mathbf{Z}_2 \cup \{(0,1,-2)\}$;

$\alpha = (0,2):$ $\Phi_\alpha(g_5) = 2T + 4$; $\Phi_\alpha(g_6) = 0$;
$\Phi_\alpha(g_7) = -2T^2 + 8T + 24$; $\Phi_\alpha(g_8) = T^3 - 3T^2 - 4T + 12$;
$\gcd(\Phi_\alpha(g_i), 5 \le i \le 8) = T + 2$;
$\mathbf{Z} := \{-2\}$, $\mathbf{Z}_2 := \mathbf{Z}_2 \cup \{(0,2,-2)\}$;

$\alpha = (1,0):$ $\Phi_\alpha(g_5) = 0$; $\Phi_\alpha(g_6) = -T + 3$;
$\Phi_\alpha(g_7) = -T^2 + 4T - 3$; $\Phi_\alpha(g_8) = T^3 - 3T^2 - T + 3$;
$\gcd(\Phi_\alpha(g_i), 6 \le i \le 8) = T - 3$;
$\mathbf{Z} := \{3\}$, $\mathbf{Z}_2 := \mathbf{Z}_2 \cup \{(1,0,3)\}$;

$\alpha = (1,1):$ $\Phi_\alpha(g_5) = 0$; $\Phi_\alpha(g_6) = 0$;
$\Phi_\alpha(g_7) = -T^2 + 4T - 3$; $\Phi_\alpha(g_8) = T^3 - 3T^2 - T + 3$;
$\gcd(\Phi_\alpha(g_i), 7 \le i \le 8) = T^2 - 4T + 3$;
$\mathbf{Z} := \{1,3\}$, $\mathbf{Z}_2 := \mathbf{Z}_2 \cup \{(1,1,3),(1,1,1)\}$;

$\alpha = (2,0):$ $\Phi_\alpha(g_5) = \Phi_\alpha(g_6) = \Phi_\alpha(g_7) = 0$; $\Phi_\alpha(g_8) = T^3 - 3T^2 + 2T$;
$\gcd(\Phi_\alpha(g_i), 7 \le i \le 8) = T^3 - 3T^2 + 2T$;
$\mathbf{Z} := \{0,1,2\}$, $\mathbf{Z}_2 := \mathbf{Z}_2 \cup \{(2,0,0),(2,0,1),(2,0,2)\}$.

39.3 Gianni–Kalkbrener Algorithm

Apart from the FGLM proposal of indirectly producing the needed lexicographical Gröbner basis via linear algebra from the Gröbner bases w.r.t. an easier-to-compute term ordering, the most relevant improvement on Trinks' Algorithm is based on a deeper analysis performed by Gianni and Kalkbrener on the structure of the lexicographical Gröbner basis of a zero-dimensional ideal.

Remarking that each polynomial $f \in K[Z_1, \ldots, Z_i]$ can be uniquely expressed as

$$f = \sum_{j=0}^{D} h_j(Z_1, \ldots, Z_{i-1})Z_i^j, \quad h_D \neq 0,$$

we recall that the degree of f in the variable Z_i is denoted

$$\deg_{Z_i}(f) := \deg_i(f) := D$$

and that $\mathrm{Lp}(f) := h_d$ is named the *leading polynomial* of f; we observe that, for the lexicographical ordering $\prec$, we have $\mathbf{T}(f) = \mathbf{T}(\mathrm{Lp}(f))Z_i^{\deg_i(f)}$.

We also denote, for each i, $1 \leq i \leq r$, $\delta \in \mathbb{N}$,

$$G_{i\delta} := \{g \in G, g \in K[Z_1, \ldots, Z_i], \deg_i(g) \leq \delta\}$$

and remark that each $G_{i\delta}$ is a section of both $G_{i\delta+1}$ and G_i and that the obvious inclusions hold:

$$G_{11} \subseteq G_{12} \subseteq \cdots \subseteq G_1 \subseteq \cdots \subseteq G_{i-1} \subseteq \cdots \subseteq G_{i\delta} \subseteq G_{i\delta+1} \subseteq \cdots \subseteq G_i \subseteq \cdots.$$

For each i, $1 \leq i \leq r$, $\delta \in \mathbb{N}$, and each $F \subset Q$, we also write

$$\mathrm{Lp}_{i\delta}(F) := \{\mathrm{Lp}(g), g \in F \cap K[Z_1, \ldots, Z_i], \deg_i(g) \leq \delta\}.$$

Theorem 39.3.1 (Gianni–Kalkbrener). *Let $\mathsf{J} \subset Q$ be an ideal and $\prec$ be the lexicographical ordering induced by $Z_1 \prec \cdots \prec Z_r$.*

Let $G := \{g_1, \ldots, g_v\}$ be a Gröbner basis of J w.r.t. $\prec$, enumerated in such a way that

$$\mathbf{T}(g_1) \prec \mathbf{T}(g_2) \prec \cdots \prec \mathbf{T}(g_{v-1}) \prec \mathbf{T}(g_v).$$

Then, with the notation above:
- *for each i, $i \leq r$, G_i is a Gröbner basis of J_i;*
- *for each i, $1 \leq i \leq r$, $\delta \in \mathbb{N}$, $\mathrm{Lp}_{i\delta}(G)$ is a Gröbner basis of $\mathrm{Lp}_{i\delta}(\mathsf{J})$;*
- *for each i, $1 \leq i \leq r$, and each $\alpha := (b_1, \ldots, b_{i-1}) \in \mathcal{Z}(\mathsf{J}_{i-1})$, denoting as σ the minimal value such that $\Phi_\alpha(\mathrm{Lp}(g_\sigma)) \neq 0$ and j, δ the values such that*

$$g_\sigma = \mathrm{Lp}(g_\sigma)Z_j^{\delta+1} + \cdots \in K[Z_1, \ldots, Z_j] \setminus K[Z_1, \ldots, Z_{j-1}],$$

it holds that

(a) $j = i$,

(b) *for each* $g \in G_{i-1}$, $\Phi_\alpha(g) = 0$,

(c) *for each* $g \in G_{i\delta}$, $\Phi_\alpha(g) = 0$,

(d) $\Phi_\alpha(g_\sigma) = \gcd(\Phi_\alpha(g) : g \in G_i) \in \mathsf{K}[T]$,

(e) *for each* $b \in \mathsf{K}$,

$$(b_1, \ldots, b_{i-1}, b) \in \mathcal{Z}(\mathsf{J}_i) \iff \Phi_\alpha(g_\sigma)(b) = 0.$$

Proof. See Sections 26.2 and 34.6. ⊙

Algorithm 39.3.2 (Gianni–Kalkbrener). The Gianni–Kalkbrener improvement to Trinks' Algorithm allows us to avoid, for each $\alpha := (a_1, \ldots, a_{i-1}) \in \mathsf{Z}_{i-1}$, both a complete evaluation $\Phi_\alpha(g)$ of all $g \in G_i \setminus G_{i-1}$ and also the computation of their gcd, reducing this step to the evaluation of the leading polynomials of a suitable subset of such elements (Figure 39.2). ⊙

Example 39.3.3. In Example 39.2.3, the Gianni–Kalkbrener Algorithm computes

$$\mathsf{Z}_1 := \{0, 1, 2\};$$

$$\alpha = (0): \quad \Phi_\alpha(\mathrm{Lp}(g_2)) = \Phi_\alpha(\mathrm{Lp}(g_3)) = 0; \quad \Phi_\alpha(\mathrm{Lp}(g_4)) = 1;$$

$$\Phi_\alpha(g_4) = T^3 - 3T^2 + 2T;$$

$$\mathsf{Z} := \{0, 1, 2\}, \quad \mathsf{Z}_2 := \{(0, 0), (0, 1), (0, 2)\};$$

$\mathsf{Z} := \mathbf{Solve}(F, L)$,
where
    $F := \{f_1, \ldots, f_u\} \subset \mathcal{Q} := K[Z_1, \ldots, Z_r]$,
    $L \supset K$ is a field extension of K,
    $J \subset \mathcal{Q}$ is the zero-dimensional ideal generated by F,
    $\mathsf{Z} := \{\alpha_1, \ldots, \alpha_s\} = \mathcal{Z}(\mathsf{J}) \cap L^r$.
Compute the reduced lexicographical Gröbner basis G of $(f_1, \ldots, f_u)$.
Sort $G := \{g_1, \ldots, g_v\}$ by increasing maximal terms.
$\mathsf{Z}_1 := \{a \in L : g_1(a) = 0\}$,
%% g_1 is the unique element in $G \cap K[Z_1]$.
For $i = 2, \ldots, r$ **do**
    $\mathsf{Z}_i := \emptyset$;
    $g := \min(g \in G_i \setminus G_{i-1})$.
    For each $(a_1, \ldots, a_{i-1}) \in \mathsf{Z}_{i-1}$ **do**
        $h := g$.
        While $\mathrm{Lp}(h)(a_1, \ldots, a_{i-1}) = 0$ **do** $h := \mathbf{Next}(h, G)$,
        $p := h(a_1, \ldots, a_{i-1}, Z_i)$,
        %% $h(a_1, \ldots, a_{i-1}, Z_i) \neq 0$,
        %% $p = \gcd(H)$ for $H := \{g(a_1, \ldots, a_{i-1}, Z_i) : g \in G_i \setminus G_{i-1}\}$,
        $\mathsf{Z} := \{a \in L : p(a) = 0\}$,
        $\mathsf{Z}_i := \mathsf{Z}_i \cup \{(a_1, \ldots, a_{i-1}, a) : a \in \mathsf{Z}\}$.
$\mathsf{Z} := \mathsf{Z}_r$.

Figure 39.2. Trinks' Algorithm, Gianni–Kalkbrener Improvement

$\alpha = (1):$ $\Phi_\alpha(\mathrm{Lp}(g_2)) = 0;$ $\Phi_\alpha(\mathrm{Lp}(g_3)) = 1;$ $\Phi_\alpha(g_3) = T^2 - T;$
          $\mathsf{Z} := \{0, 1\},$ $\mathsf{Z}_2 := \{(0, 0), (0, 1), (0, 2), (1, 0), (1, 1)\};$

$\alpha = (2):$ $\Phi_\alpha(\mathrm{Lp}(g_2)) = 2,$ $\Phi_\alpha(g_2) = 2T;$
          $\mathsf{Z} := \{0\},$ $\mathsf{Z}_2 := \{(0, 0), (0, 1), (0, 2), (1, 0), (1, 1), (2, 0)\};$

$\alpha = (0, 0):$ $\Phi_\alpha(\mathrm{Lp}(g_5)) = 2,$ $\Phi_\alpha(g_5) = 2T - 2;$
          $\mathsf{Z}_2 := \{(0, 0, 1)\};$

$\alpha = (0, 1):$ $\Phi_\alpha(\mathrm{Lp}(g_5)) = 2,$ $\Phi_\alpha(g_5) = 2T + 4;$
          $\mathsf{Z} := \{-2\},$ $\mathsf{Z}_2 := \mathsf{Z}_2 \cup \{(0, 1, -2)\};$

$\alpha = (0, 2):$ $\Phi_\alpha(\mathrm{Lp}(g_5)) = 2,$ $\Phi_\alpha(g_5) = 2T + 4;$
          $\mathsf{Z} := \{-2\},$ $\mathsf{Z}_2 := \mathsf{Z}_2 \cup \{(0, 2, -2)\};$

$\alpha = (1, 0):$ $\Phi_\alpha(\mathrm{Lp}(g_5)) = 0;$ $\Phi_\alpha(\mathrm{Lp}(g_6)) = -1,$ $\Phi_\alpha(g_6) = -T + 3;$
          $\mathsf{Z} := \{3\},$ $\mathsf{Z}_2 := \mathsf{Z}_2 \cup \{(1, 0, 3)\};$

$\alpha = (1, 1):$ $\Phi_\alpha(\mathrm{Lp}(g_5)) = \Phi_\alpha(\mathrm{Lp}(g_6)) = 0;$ $\Phi_\alpha(\mathrm{Lp}(g_7)) = -1,$
          $\Phi_\alpha(g_7) = -T^2 + 4T - 3;$
          $\mathsf{Z} := \{1, 3\},$ $\mathsf{Z}_2 := \mathsf{Z}_2 \cup \{(1, 1, 3), (1, 1, 1)\};$

$\alpha = (2, 0):$ $\Phi_\alpha(\mathrm{Lp}(g_5)) = \Phi_\alpha(\mathrm{Lp}(g_6)) = \Phi_\alpha(\mathrm{Lp}(g_7)) = 0;$ $\Phi_\alpha(\mathrm{Lp}(g_8)) = 1;$
          $\Phi_\alpha(g_8) = T^3 - 3T^2 + 2T;$
          $\mathsf{Z} := \{0, 1, 2\},$ $\mathsf{Z}_2 := \mathsf{Z}_2 \cup \{(2, 0, 0), (2, 0, 1), (2, 0, 2)\}.$

⊙

Remark 39.3.4.   The Cerlienco–Mureddu Correspondence and Algorithm (Chapter 33) can give some hint of the structure of the Gianni–Kalkbrener Algorithm. Our informal discussion assumes that J is radical but holds in general. The Gianni–Kalkbrener Algorithm, for any root $\alpha := (a_1, \ldots, a_{i-1}) \in \mathsf{Z}_{i-1}$, considers the first Gröbner basis element,

$$h = \mathrm{Lp}(h)Z_i^{\delta+1} + \cdots \in K[Z_1, \ldots, Z_i] \setminus K[Z_1, \ldots, Z_{i-1}],$$

whose leading polynomial $\mathrm{Lp}(h)$ does not vanish at α.

In the same mood, the Cerlienco–Mureddu approach considers a new point

$$\mathbf{x} := (a_1, \ldots, a_{i-1}, b) \notin \mathcal{Z}(\mathsf{J}_i)$$

and the first Gröbner basis element which does not vanish at it. Clearly both algorithms are choosing the same polynomial: in fact,

$$\delta + 1 = \#\{\mathbf{y} \in \mathcal{Z}(\mathsf{J}_i) : \pi_{i-1}(\mathbf{y}) = \alpha = \pi_{i-1}(\mathbf{x})\} = d.$$

Then

- $\Phi_\alpha(h) \mid \Phi_\alpha(g)$ for each $g \in G_i \setminus G_{i\delta+1}$, because

$$g(\mathbf{y}) = 0 \quad \text{for each } \mathbf{y} \in \mathcal{Z}(\mathsf{J}_i) : \pi_{i-1}(\mathbf{y}) = \alpha,$$

- $\mathrm{Lp}(g)(\alpha) = 0$ for each $g \in G_{i\delta}$, because there is a $\mathbf{y} \in \mathcal{Z}(\mathsf{J}_i)$ such that

$$\pi_{i-1}(\mathbf{y}) = \alpha = \pi_{i-1}(\mathbf{x}),$$

whence $g(\mathbf{y}) = 0$ and $0 = \mathrm{Lp}(g)(\pi_{i-1}(\mathbf{y})) = \mathrm{Lp}(g)(\alpha)$.

In order to conclude our argument, we need to dispose of the elements $g \in G_{i\delta+1} \setminus G_{i\delta}$, that is, of the Gröbner basis element

$$g = \mathrm{Lp}(g)Z_i^{\delta+1} + \cdots ;$$

to do so, we just remark that

- $g(\alpha)$ and $h(\alpha)$ are associate if $\mathbf{T}(g) \succ \mathbf{T}(h)$, because

$$g(\mathbf{y}) = 0 \quad \text{for each } \mathbf{y} \in \mathcal{Z}(\mathbf{J}_i) : \pi_{i-1}(\mathbf{y}) = \alpha,$$

- and $\mathrm{Lp}(g)(\alpha) = 0$ if $\mathbf{T}(g) \prec \mathbf{T}(h)$, because there is a $\mathbf{y} \in \mathcal{Z}(\mathbf{J}_i)$ such that

$$\pi_{i-1}(\mathbf{y}) = \alpha = \pi_{i-1}(\mathbf{x}),$$

whence $g(\mathbf{y}) = 0$ and $0 = \mathrm{Lp}(g)(\pi_{i-1}(\mathbf{y})) = \mathrm{Lp}(g)(\alpha)$.

39.4 An Ecumenical Notion of Solving

As the decomposition algorithms (Chapter 35) reduce the primary decomposition of multivariate ideals to the factorization of univariate polynomials, Trinks' Algorithm (like most other solving algorithms) reduces multivariate zero-dimensional ideal solving to univariate polynomial solving.

Most of these algorithms are "ecumenical", in the sense that they can be applied to any computational model of the field extension L of K. This allows us, in *endlichvielen Schritten* (a finite number of steps) (see Volume I, p.5),

- to computationally perform the four operations in L,
- and, for each univariate polynomial $p(T) \in L[T]$, to "solve" it, that is, to produce the set $\{a \in L : p(a) = 0\}$ of all the roots of p living in L,

and to use these tools, given F, to "solve" the zero-dimensional ideal $\mathbf{J}$ generated by F, i.e. to produce the set $\mathcal{Z}(\mathbf{J}) \cap L^r$ of all the roots of $\mathbf{J}$ with coordinates in L.

Trinks' Algorithm (Figure 39.2) is a perfect instance of such ecumenical algorithms: for instance, setting $L := \mathbb{R}$, it can be applied *verbatim* to a numerical analysis solver[6] or adapted in order to make use of the Sturm representation and the Thom codification of algebraic reals.[7]

In the same way, Trinks' Algorithm can be easily adapted to make use of Kronecker's (and Duval's) model; obviously the resulting algorithm (Figure 39.3) is a *verbatim* reformulation of the zero-dimensional prime decomposition algorithm discussed in Section 35.2. Such a strict relation between "solving" and decomposing,

---

[6] Of course, such a statement must be taken *cum grano salis*: it ignores the ill-conditioning problem, which requires at least some suitable pre-processing before application of the Algorithm.

[7] In Chapter 13 we discussed these representations and the techniques needed to solve the required polynomials

$$p(Z_i) := h(a_1, \ldots, a_{i-1}, Z_i), h \in \mathbb{Q}[Z_1, \ldots, Z_i],$$

where each $a_i \in \mathbb{R}$ is given by such a representation.

$\mathsf{Z} := \mathbf{Solve}(F)$,
where
> $F := \{f_1, \ldots, f_u\} \subset \mathcal{Q} := K[Z_1, \ldots, Z_r]$,
> $\mathsf{J} \subset \mathcal{Q}$ is the zero-dimensional ideal generated by F,
> $\mathsf{Z} := \{(\mathbf{f}_1, \mathfrak{K}_1, \alpha_1), \ldots, (\mathbf{f}_\mathsf{s}, \mathfrak{K}_\mathsf{s}, \alpha_\mathsf{s})\}$,
> **where**
> > $\mathbf{f}_j = (f_{j1}, \ldots, f_{jr})$ is an admissible sequence (Definition 8.2.2) in $K[Z_1, \ldots, Z_r]$,
> > $\mathfrak{K}_j := K[Z_1, \ldots, Z_r]/(f_{j1}, \ldots, f_{jr})$, $K \subset \mathfrak{K}_j \subset \mathsf{K}$ is the finite extension explicitly given by $\mathbf{f}_j$,
> > $\alpha_j \in \mathfrak{K}_j^r$,
> $\mathcal{Z}(\mathsf{J}) := \{\alpha_1, \ldots, \alpha_\mathsf{s}\} \subset \mathsf{K}^r$.

Compute the reduced lexicographical Gröbner basis G of $(f_1, \ldots, f_u)$.
Sort $G := \{g_1, \ldots, g_v\}$ by increasing maximal terms.
Let $g_1 = \prod_{j=1}^{\sigma} q_j^{e_j}$ be the factorization of g_1 over K.
For $j = 1, \ldots, \sigma$ **let**
> $\mathbf{f}_j := (q_j)$,
> $\mathfrak{K}_j := K[Z_1]/q_j$,
> $\pi_j : K[Z_1] \to \mathfrak{K}_j$ be the canonical projection,
> $\alpha_j := \pi_j(Z_1) \in \mathfrak{K}_j$;

$\mathsf{Z}_1 := \{(\mathbf{f}_1, \mathfrak{K}_1, \alpha_1), \ldots, (\mathbf{f}_\sigma, \mathfrak{K}_\sigma, \alpha_\sigma)\}$.
For $i = 2, \ldots, r$ **do**
> $\mathsf{Z}_i := \emptyset$,
> $g := \min(g \in G_i \setminus G_{i-1})$.
> **For each** $(\mathbf{f}, \mathfrak{K}, \alpha) \in \mathsf{Z}_{i-1}, \mathbf{f} = (f_1, \ldots, f_{i-1}), \alpha = (a_1, \ldots, a_{i-1}),$ **do**
> > $h := g$.
> > **While** $\mathrm{Lp}(h)(a_1, \ldots, a_{i-1}) = 0$ **do** $h := \mathbf{Next}(h, G)$,
> > $p(Z_i) := h(a_1, \ldots, a_{i-1}, Z_i)$.
> > **Let** $p = \prod_{j=1}^{\sigma} q_j^{e_j}$ be the factorization of p over $\mathfrak{K}$.
> > **For** $j = 1, \ldots, \sigma$ **let**
> > > $\mathbf{f}_j := (f_1, \ldots, f_{i-1}, q_j)$,
> > > $\mathfrak{K}_j := \mathfrak{K}[Z_i]/q_j \cong K[Z_1, \ldots, Z_i]/(f_1, \ldots, f_{i-1}, q_j)$,
> > > $\pi_j : \mathfrak{K}[Z_i] \to \mathfrak{K}_j$ be the canonical projection,
> > > $\alpha_j := (a_1, \ldots, a_{i-1}, \pi_j(Z_i)) \in \mathfrak{K}_j$,
> > $\mathsf{Z}_i := \mathsf{Z}_i \cup \{(\mathbf{f}_1, \mathfrak{K}_1, \alpha_1), \ldots, (\mathbf{f}_\sigma, \mathfrak{K}_\sigma, \alpha_\sigma)\}$,
$\mathsf{Z} := \mathsf{Z}_r$.

Figure 39.3. Trinks' Algorithm in Kronecker's model

which had already been stressed in Section 34.5, is just a simple consequence of Kronecker–Duval Philosophy.

Throughout this Part we will preserve this ecumenical approach to the notion of "solving", as much as the relevant solvers will allow; naturally, the solver presented here is an integralist version of the Kronecker–Duval Philosophy.

39.5 Delassus–Gunther Solver

Historical Remark 39.5.1. Trinks' paper, dated 1978, is the first published application of Gröbner bases, except for Buchberger's thesis and paper. His result is an efficient adaptation and improvement of the proof of Hilbert's Nullstellensatz given by Gröbner (Section 20.3).

Gröbner's argument, when restricted to the zero-dimensional case, essentially computes iteratively the roots of each elimination ideal J_i by

- producing (via a suitable generic change of coordinates) a polynomial

$$f_i(Z_1, \ldots, Z_i) := cZ_i^{d_i} + \sum_{j=0}^{d_i-1} h_j(Z_1, \ldots, Z_{i-1})Z_i^j \in J_i, \quad c \neq 0,$$

- solving the univariate polynomials

$$f_i(a_1, \ldots, a_{i-1}, Z_i) \in K[Z_i] \quad \text{for each } (a_1, \ldots, a_{i-1}) \in \mathcal{Z}(J_{i-1})$$

- and including in $\mathcal{Z}(J_i)$ those roots $(a_1, \ldots, a_{i-1}, b)$ which annihilate not only f_i but also a given basis of J_i.

Trinks' proposal makes it possible to obtain the required basis for each J_i and allows the production of univariate polynomials that can be solved without performing a change of coordinates.

Gröbner's argument, in turn, was an adaptation of the argument and solver of Kronecker (Section 20.4), which, in the zero-dimensional case,[8] again consists of

- producing (via a suitable generic change of coordinates) polynomials

$$f_i(Z_1, \ldots, Z_i) := cZ_i^{d_i} + \sum_{j=0}^{d_i-1} h_j(Z_1, \ldots, Z_{i-1})Z_i^j \in I_i, \quad c \neq 0,$$

- solving the univariate polynomials

$$f_i(a_1, \ldots, a_{i-1}, Z_i) \in K[Z_i] \quad \text{for each } (a_1, \ldots, a_{i-1}) \in \mathcal{Z}(I_{i-1})$$

- and including in $\mathcal{Z}(I_i)$ those roots $(a_1, \ldots, a_{i-1}, b)$ which annihilate not only f_i but also a given basis of I_i,

where each ideal I_{i-1} is obtained from I_i ($I_n := J$) via a suitable *resultant* computation.

The resultant is the theoretical tool used in proving an interesting solver which anticipates some ideas by Macaulay: the original version[9] was proposed by Delassus in 1897 but was flawed by a wrong assumption, that the generic initial ideal

---

[8] Unlike Gröbner's argument, which was not intended to produce an effictive solver but was converted into such by Trinks, Kronecker's argument gives an effective solver.

    The restriction to the zero-dimensional case was made here to simplify the argument but is *not* required by Kronecker's solver, which in fact applies also to non-unmixed ideals.

    The details of the general case are discussed in Sections 20.3 and 20.4.

[9] Delassus E., Sur les systèmes algébriques et leurs relations avec certains systèmes d'equations aux dérivées partielles, *Ann. Éc. Norm. 3ᵉ série* **14** (1897), 21–44.

(Definition 37.1.5) of a homogeneous ideal w.r.t. the degrevlex ordering $<$ induced by $X_1 < \cdots < X_n$ consists of the first terms w.r.t. $<$, while in fact it is just Borel (Definition 37.2.7, Corollary 37.2.8); the flaw was found by Gunther[10] and Robinson[11] and fixed by Gunther[12] in 1913.

In order to present the solver proposed by Delassus and Gunther we must slightly adapt the notation they used; we assume $\mathsf{J} \subset \mathcal{Q} = K[Z_1, \ldots, Z_r]$ to be homogeneous and denote, for each $d \in \mathbb{N}$,

$$\mathcal{W}_d := \{\tau \in \mathcal{W}, \deg(\tau) = d\} \quad \text{and}^{13} \quad \mathsf{J}_d := \mathsf{J} \cap \operatorname{Span}_K(\mathcal{W}_d).$$

We assume that a generic change of coordinates has been performed and consider the degrevlex ordering $<$ induced by $Z_1 < Z_2 < \cdots < Z_r$. For each (homogeneous) polynomial $f = \sum_{t \in \mathcal{W}} c(f, t) t \in \mathcal{Q}$ we denote

$$\mathbf{L}_<(f) := \min_<(t \in \mathcal{W} : c(f, t) \neq 0)$$

and, for each (homogeneous) set $F \subset \mathcal{Q}$,

$$\mathbf{L}_<\{F\} := \{\mathbf{L}_<(f) : f \in F\} \quad \text{and} \quad \mathbf{L}_<(F) := \{\tau \mathbf{L}_<(f) : f \in F, \tau \in \mathcal{W}\}.$$

Remark 39.5.2.

(1) If $\prec$ denotes the deglex ordering induced by $Z_1 \succ Z_2 \succ \cdots \succ Z_r$, we have $\mathbf{T}_\prec(f) = \mathbf{L}_<(f)$ for each (homogeneous) polynomial $f \in \mathcal{Q}$ and $\mathbf{T}_\prec(F) = \mathbf{L}_<(F)$ for each (homogeneous) set $F \subset \mathcal{Q}$;

(2) for a homogeneous ideal J and each $d \in \mathbb{N}$, we have

$$\mathbf{L}_<(\mathsf{J})_d = \mathbf{L}_<\{\mathsf{J}\}_d = \mathbf{L}_<\{\mathsf{J}_d\};$$

(3) there is a (minimal) value[14] $D \in \mathbb{N}$ which satisfies, for each $d \in \mathbb{N}$,

$$\mathbf{L}(\operatorname{Span}_K\{\omega f : \omega \in \mathcal{W}_d, f \in \mathsf{J}_D\}) = \{\omega \mathbf{L}(f) : \omega \in \mathcal{W}_d, f \in \mathsf{J}_D\};$$

---

[10] Gunther is the French transcription of the Russian name which transliterates as Gjunter. In the first volume I gave it as Gjunter. Since in this chapter I quote papers written in French either by Janet or by Gjunter himself, I systematically use the French transcription *Gunther* except in the Bibliography, where I use *Gjunter*.

[11] Gunther, N., Sur les caractéristiques des systémes d'equations aux dérivées partialles, *C.R. Acad. Sci. Paris* **156** (1913), 1147–1150, and Robinson, L.B. Sur les systémes d'équations aux dérivées partialles, *C.R. Acad. Sci. Paris* **157** (1913), 106–108.

[12] Gunther, N., Sur la forme canonique des systèmes d'équations homogènes (in Russian), *J. l'Institut des Ponts et Chaussées de Russie, Izdanie Inst. Inž. Putej Soobščenija Imp. Al. I.* **84** (1913) and Gunther, N., Sur la forme canonique des équations algébriques, *C.R. Acad. Sci. Paris* **157** (1913), 577–580.

[13] The notation J_d denotes here the set of homogeneous members of J of degree d and must not be confused with the previous notation, where J_i denotes the members of J depending only on the first i variables.

[14] We can set $D := \max\{\deg(g) : g \in G\}$, where G is a Gröbner basis of J w.r.t. $<$, but the existence can be easily derived (as for the finiteness of Gröbner bases) by Hilbert's Nullstellensatz, and this is the approach used by Gunther.

(4) each set $\mathbf{L}_<\{J_d\}$, $d \geq D$, satisfies[15] for each $\ell, \ell', 1 \leq \ell < \ell' \leq r$,

$$Z_1^{a_1} \cdots Z_r^{a_r} \in \mathbf{L}_<\{J_d\} \implies Z_1^{a_1} \cdots Z_\ell^{a_\ell+1} \cdots Z_{\ell'}^{a_{\ell'}-1} \cdots Z_r^{a_r} \in \mathbf{L}_<\{J_d\}.$$

(5) Denoting, for each d, $\mathbf{N}(J_d) := \mathcal{W}_d \setminus \mathbf{L}_<\{J_d\}$, for each $\tau \in \mathbf{L}_<\{J_d\}$ we have $\mathrm{Can}(\tau, J, <) \in \mathrm{Span}_K(\mathbf{N}(J_d))$, and we can set

$$g_\tau := \tau - \mathrm{Can}(\tau, J, <) \in J_d \quad \text{and} \quad G_d := \{g_\tau : \tau \in \mathbf{L}_<\{J_d\}\},$$

with $\tau = \mathbf{L}_<(g_\tau) = \mathbf{T}_\prec(g_\tau)$ and $\mathbf{L}_<\{G_d\} = \mathbf{L}_<\{J_d\}$.

(6) We have $\mathcal{Z}(J) = \{(a_1, \ldots, a_r) \in K^r : g(a_1, \ldots, a_r) = 0 \text{ for each } g \in G_D\}$.

$\boxed{\odot}$

Theorem 39.5.3 (Delassus–Gunther). *With the present notation and assumptions, let $\gamma_1, \ldots, \gamma_r \in \mathbb{N}$, $\sum_{i=1}^r \gamma_i = D$, be the values such that*

$$\Omega := Z_1^{\gamma_1} \cdots Z_{r-1}^{\gamma_{r-1}} Z_r^{\gamma_r} = \max_<(\mathbf{L}_<\{J_D\}) = \max_<(\mathbf{L}_<\{G_D\}) = \min_\prec(\mathbf{T}_\prec\{J_D\}).$$

We have:

(1) *if $\gamma_1 = 0$ then $\gcd(G_D) = 1$;*

(2) *if $\gamma_1 \neq 0$ then*

    (a) *$h := \gcd(G_D) \neq 1$,*

    (b) *$\mathbf{L}_<(h) = Z_1^{\gamma_1}$,*

    (c) *for each $d \in \mathbb{N}$ it holds that*

$$\mathbf{L}(\mathrm{Span}_K\{\omega g : \omega \in \mathcal{W}_d, g \in G_D\}) = \{\omega \mathbf{L}(g) : \omega \in \mathcal{W}_d, g \in G_D\};$$

(3) *$\max_<\{\mathbf{L}_<(g/h) : g \in G_D\} = Z_2^{\gamma_2} \cdots Z_{r-1}^{\gamma_{r-1}} Z_r^{\gamma_r}$;*

(4) *if $\gamma_1 = 0$ then the set $H := G_D \cap K[Z_2, \ldots, Z_r]$ satisfies, for each $d \in \mathbb{N}$,*

$$\mathbf{L}(\mathrm{Span}_K\{\omega f : \omega \in \mathcal{U}_d, f \in H\}) = \{\omega \mathbf{L}(f) : \omega \in \mathcal{U}_d, f \in H\},$$

where $\mathcal{U}_d := \mathcal{W}_d \cap K[Z_2, \ldots, Z_r]$;

(5) *if $\gamma_1 = 0$ and*

$$(a_2, \ldots, a_r) \in \mathcal{Z}(H) := \{(a_2, \ldots, a_r) \in K^{r-1} : g(a_2, \ldots, a_r) = 0, g \in H\}$$

then $1 \neq h(Z_1, a_2, \ldots, a_r) = \gcd(g(Z_1, a_2, \ldots, a_r) : g \in G_D) \in K[Z_1]$;

(6) *moreover, $h(Z_1, a_2, \ldots, a_r) = (Z_1 - a_1)^{\deg(h)}$ for some $a_1 \in K$.*

Proof.

(1) We remark that the first element w.r.t. $<$ in $\mathbf{L}_<\{J_D\}$ is Z_1^D and that

$$Z_1^D - \mathrm{Can}(Z_1^D, J, <) \in G_D,$$

---

[15] This is a direct consequence of Corollary 37.2.8 applied to the set $\mathbf{T}_\prec(f) = \mathbf{L}_<(f)$ and to the deglex ordering $\prec$ induced by $Z_r \prec \cdots \prec Z_2 \prec Z_1$.

    Delassus' mistake was to assume that $\mathbf{L}_<(J_d) = \mathbf{T}_\prec(J_d)$ is the set $L(d)$ consisting of the first $\#\mathbf{L}_<(J_d)$ terms w.r.t. the degrevlex $<$ induced by $Z_1 < Z_2 < \cdots < Z_r$, which is tantamount to the last $\#\mathbf{T}_\prec(J_d)$ terms w.r.t. the deglex ordering $\prec$ induced by $Z_r \prec \cdots \prec Z_2 \prec Z_1$.

    In his lemma (Section 23.3) Macaulay considered the same set $L(d)$, $\#L(d) = \#\mathbf{L}_<(J_d)$, as Delassus and presented it, like Delassus, in terms of the degrevlex ordering $<$ and not in terms of the deglex ordering $\prec$.

so that

$$h := \gcd(G_D) \neq 1 \implies \gcd(G_D) \in K[Z_2, \ldots, Z_r][Z_1] \setminus K[Z_2, \ldots, Z_r].$$

If $\gamma_1 = 0$ then $\Omega \in K[Z_2, \ldots, Z_r]$ and $g_\Omega \in K[Z_2, \ldots, Z_r]$,[16] so that

$$h = \gcd(G_D) \in K[Z_2, \ldots, Z_r].$$

Thus $\gcd(G_D) = 1$.

(2) Assume now that $\gamma_1 \neq 0$.

(a) Let us consider variables $U_\tau, W_\tau, \tau \in \mathbf{L}_<\{J_D\}$, and polynomials

$$\mathsf{f} := \sum_{\tau \in \mathbf{L}_<\{J_D\}} U_\tau g_\tau, \quad g := \sum_{\tau \in \mathbf{L}_<\{J_D\}} W_\tau g_\tau \in K[U_\tau, W_\tau, Z_2, \ldots, Z_r][Z_1].$$

The Sylvester resultant grants (Proposition 6.6.7) the existence of polynomials $p, q \in K[U_\tau, W_\tau, Z_2, \ldots, Z_r][Z_1]$ such that

$$\mathrm{Res}(\mathsf{f}, g) := p\mathsf{f} + qg \in K[U_\tau, W_\tau, Z_2, \ldots, Z_r];$$

moreover $\mathrm{Res}(\mathsf{f}, g)$ is necessarily linear and homogeneous in terms of members of the set

$$F := \{\omega g_\tau, \tau \in \mathbf{L}_<\{J_D\}, \omega \in \mathcal{W}_d\}$$

where $d := \deg(p) = \deg(q)$. By Remarks 39.5.2(3), (5),

$$\Omega Z_r^d := Z_1^{\gamma_1} \cdots Z_{r-1}^{\gamma_{r-1}} Z_r^{\gamma_r + d} = \max_<(\mathbf{L}_< c\{F\}).$$

As a consequence, since $\gamma_1 \neq 0$, there are elements $f' \in F$, for instance $f' = g_\Omega Z_r^d$, for which

$$\mathbf{L}_<(f') \in K[U_\tau, W_\tau, Z_2, \ldots, Z_r][Z_1] \setminus K[U_\tau, W_\tau, Z_2, \ldots, Z_r]$$

thus giving a contradiction unless $\mathrm{Res}(\mathsf{f}, g) = 0$ and f, g have a common factor in $K[U_\tau, W_\tau, Z_2, \ldots, Z_r][Z_1]$.

Such a factor is necessarily a member of $K[Z_2, \ldots, Z_r][Z_1]$, thus proving that $h := \gcd(G_D) \neq 1$.

(b) We necessarily have $\mathbf{L}_<(h) = Z_1^\gamma$ for some $\gamma \in \mathbb{N}$. Also

$$\max_<\{\mathbf{L}_<(g/h) : g \in G_D\} = Z_1^{\gamma_1 - \gamma} Z_2^{\gamma_2} \cdots Z_{r-1}^{\gamma_{r-1}} Z_r^{\gamma_r}$$

and, setting $F' := \{\omega g/h : g \in G_D, \omega \in \mathcal{W}_d\}$, we have (again by Remarks 39.5.2(3), (5))

$$\max_<\{\mathbf{L}_<(f : f \in F') = Z_1^{\gamma_1 - \gamma} Z_2^{\gamma_2} \cdots Z_{r-1}^{\gamma_{r-1}} Z_r^{\gamma_r + d}.$$

Since clearly $\gcd(F') = 1$ we conclude by (1) and (2a) above that $\gamma_1 - \gamma = 0$, that is $\gamma_1 = \gamma$;

(c) part (2c) is trivial.

(3) This is a trivial consequence of (2).

---

[16] As a consequence of the elimination property of the lex ordering $\prec$, which is explicitly stated by Gunther.

(4) If $\gamma_1 = 0$ then $H := G_D \cap K[Z_2, \ldots, Z_r] \neq \emptyset$ and[17]

$$H = \{g_\tau \in G_D : \tau \in \mathcal{W} \cap K[Z_2, \ldots, Z_r]\}.$$

The claim is then a direct application of Remark 39.5.2(3).

(5) Let us again consider the resultant

$$\mathrm{Res}(\mathsf{f}, g) := p\mathsf{f} + qg \in K[U_\tau, W_\tau, Z_2, \ldots, Z_r]$$

of $\mathsf{f} := \sum_{\tau \in \mathbf{L}_<\{\mathsf{J}_D\}} U_\tau g_\tau$ and $g := \sum_{\tau \in \mathbf{L}_<\{\mathsf{J}_D\}} W_\tau g_\tau$; since $\gcd(G_D) = 1$ we have $\gcd(\mathsf{f}, g) = 1$ and $\mathrm{Res}(\mathsf{f}, g) \neq 0$. Denoting by $\mathcal{V}$ the set of the terms in the variables $\{U_\tau, W_\tau, \tau \in \mathbf{L}_<\{\mathsf{J}_D\}\}$ we therefore have $\mathrm{Res}(\mathsf{f}, g) = \sum_{\upsilon \in \mathcal{V}} c_\upsilon \upsilon$. Each c_υ depends linearly on the elements in

$$F := \{\omega g_\tau, \tau \in \mathbf{L}_<\{\mathsf{J}_D\}, \omega \in \mathcal{W}_d\},$$

where $d := \deg(p) = \deg(q)$, and is independent of Z_1: $c_\upsilon \in K[Z_2, \ldots, Z_r]$. Therefore each c_υ depends linearly on the elements in

$$F' := \{\omega g, g \in H, \omega \in \mathcal{W}_d \cap K[Z_2, \ldots, Z_r]\}$$

and we have $c_\upsilon(a_2, \ldots, a_r) = 0$ and $\mathrm{Res}(\mathsf{f}, g)(U_\tau, W_\tau, a_2, \ldots, a_r) = 0$ for each $(a_2, \ldots, a_r) \in \mathcal{Z}(H)$, so that

$$\gcd\big(\mathsf{f}(U_\tau, W_\tau, a_2, \ldots, a_r, Z_1), g(U_\tau, W_\tau, a_2, \ldots, a_r, Z_1)\big) \neq 1,$$

that is, $1 \neq h(Z_1, a_2, \ldots, a_r) := \gcd(g(Z_1, a_2, \ldots, a_r) : g \in G_D) \in \mathsf{K}[Z_1]$.

(6) Since this is essentially[18] Gröbner's *Allgemeine Nulldimensionale Basissatz* (Theorem 34.2.1) I can skip the interesting, but not immediate, proof proposed by Gunther.[19]

---

[17] Again this is a consequence of the elimination property of the lex ordering $\prec$ explicitly stated by Gunther.

[18] Note that we have performed a change of coordinates and that here Z_1 is the last coordinate w.r.t. the lex ordering.

[19] It begins by considering the substitution

$$Z_1 = U_1, \quad Z_2 = AU_1 + U_2, \quad Z_3 = U_3, \ldots, \quad Z_r = U_r,$$

where A is a variable and the corresponding equations in $K[A][U_1, \ldots, U_r]$, which are *des fonctions holomorphes de A dans la voisinage de A* $= 0$ via holomorphic functions of A in the neighborhood of A.

40

Stetter

The crucial improvement by Gianni and Kalkbrener of Trinks' Algorithm is dated 1987; the year after, Auzinger and Stetter proposed an alternative algorithm for solving a radical zero-dimensional ideal, which was later generalized by Stetter and Möller; the original proposal made no reference to Gröbner techniques,[1] being based on numerical analysis techniques: given a zero-dimensional ideal $J \subset \mathcal{Q}$, the Auzinger–Stetter Theorem states that, for each $f \in A := \mathcal{Q}/J$, the linear form $\Phi_f : A \to A$ describing multiplication by f in A has, as its eigenvalues, the evaluation of f at the roots of J, with the proper multiplicity; moreover, if we fix a K-basis $\mathbf{b} = \{b_1, \ldots, b_s\}$ of A and denote by A_f the matrix representing Φ_f w.r.t. such a basis then $(b_1(\alpha), \ldots, b_s(\alpha))^T$ is an eigenvector of $f(\alpha)$ for each $\alpha \in \mathcal{Z}(J)$.

Thus, provided that A_f is non-derogatory, that is, its Jordan form has a single Jordan block associated with each eigenvalue, it is sufficient to choose $\mathbf{b}$ as a basis which includes the linear basis of the subspace of A consisting of all its linear forms, and use linear dependency to express each variable Z_i in terms of such a linear basis.

After setting the proper notation (Section 40.1) we present the Auzinger–Stetter Theorem (Section 40.2) and discuss how to apply it for solving a zero-dimensional radical ideal (Section 40.3). The extension to the general case being based on duality, I preliminarily discuss the relation between duality and the Auzinger–Stetter technique (Section 40.4) before presenting the Möller–Stetter extension of the Auzinger–Stetter Theorem (Section 40.5).

After specializing this result to the univariate case, thus obtaining the expected statement (Section 40.6) and discussing derogatoriness (Sections 40.7 and 40.13), I finally describe how Stetter's Algorithm can be performed using Gröbner-basis techniques (Section 40.8 and 40.9).

---

[1] While the presentation here is based on the notion of a Gröbner representation, one should remark that such a notion was introduced later in order to provide a convenient frame in which to present the Auzinger–Stetter Algorithm.

40.1 Endomorphisms of an Algebra

Set $\mathcal{Q} := K[Z_1, \ldots, Z_r]$, $\mathcal{W}$ its monomial K-basis and $\mathbf{K}$ the algebraic closure of K. In order to simplify the notation let us wlog assume $K = \mathbf{K}$ to be algebraically closed.

Let $\mathsf{J} \subset \mathcal{Q}$ be a zero-dimensional ideal, $\deg(\mathsf{J}) = s$, and $\mathsf{A} := \mathcal{Q}/\mathsf{J}$ the corresponding quotient algebra, which satisfies $\dim_K(\mathsf{A}) = s$.

For any $f \in \mathcal{Q}$, we will denote by $[f] \in \mathsf{A}$ its residue class modulo J and by Φ_f the endomorphism $\Phi_f : \mathsf{A} \to \mathsf{A}$ defined by

$$\Phi_f([g]) = [fg] \text{ for each } [g] \in \mathsf{A}.$$

Clearly $\Phi_f = \Phi_h$ iff $[f] = [h]$.

If we fix any K-basis $\mathbf{b} = \{[b_1], \ldots, [b_s]\}$ of A such that $\mathsf{A} = \mathrm{Span}_K(\mathbf{b})$ then, for each $g \in \mathcal{Q}$, there is a unique (row) vector, the *Gröbner description of* g (Definition 29.3.3),

$$\mathbf{Rep}(g, \mathbf{b}) := (\gamma(g, b_1, \mathbf{b}), \ldots, \gamma(g, b_s, \mathbf{b})) \in K^s$$

which satisfies

$$[g] = \sum_j \gamma(g, b_j, \mathbf{b})[b_j]$$

and the endomorphism Φ_f is naturally represented by the square matrix

$$M([f], \mathbf{b}) = \left(\gamma(fb_i, b_j, \mathbf{b})\right) : \Phi_f(b_i) = [fb_i] = \sum_j \gamma(fb_i, b_j, \mathbf{b})[b_j].$$

Recall that a *Gröbner representation* (Definition 29.3.3) of J (or, better, of the algebra A) is the assignment of

- a K-basis $\mathbf{b} = \{[b_1], \ldots, [b_s]\} \subset \mathsf{A}$ and
- the square matrices $A_h := \left(a_{ij}^{(h)}\right) = M([Z_h], \mathbf{b})$ for each h, $1 \leq h \leq s$,

and that a Gröbner representation is called a *linear representation* (Definition 29.3.3) iff $\mathbf{b} = \{[1], [\tau_2], \ldots, [\tau_s]\} = \mathbf{N}_<(\mathsf{J})$ w.r.t. a term ordering $<$. We remark that, for each $f(Z_1, \ldots, Z_r) \in \mathcal{Q}$, $M([f], \mathbf{b}) = f(A_1, \ldots, A_r)$.

An alternative way of representing a zero-dimensional ideal $\mathsf{J} \subset \mathcal{Q}$ and the related quotient algebra A is via its *dual space* (Section 28.1)

$$\mathfrak{L}(\mathsf{J}) := \left\{\ell \in \mathcal{Q}^* : \ell(g) = 0 \text{ for each } g \in \mathsf{J}\right\} \subset \mathcal{Q}^*,$$

where $\mathcal{Q}^* := \mathrm{Hom}_K(\mathcal{Q}, K)$ is the K-vector space consisting of all K-linear functionals $\ell : \mathcal{Q} \to K$.

Clearly we have $\dim_K(\mathfrak{L}(\mathsf{J})) = s$, and with each K-basis $\mathbb{L} := \{\lambda_1, \ldots, \lambda_s\}$ of $\mathfrak{L}(\mathsf{J})$ is associated a *Lagrange K-basis* $\mathbf{q} = \{[q_1], \ldots, [q_s]\}$ which is *biorthogonal* to $\mathbb{L}$, that is,

$$\lambda_i(q_j) = \delta_{ij} = \begin{cases} 1 & \text{if } i = j \\ 0 & \text{if } i \neq j. \end{cases}$$

In particular, since, for each i, j, h,

$$\lambda_j(Z_h q_i) = \lambda_j\left(\sum_l a_{il}^{(h)} q_l\right) = \sum_l a_{il}^{(h)} \lambda_j(q_l) = a_{ij}^{(h)},$$

to each basis $\mathbb{L} := \{\lambda_1, \ldots, \lambda_s\}$ of $\mathfrak{L}(\mathsf{J})$ is associated the Gröbner representation

- $\mathbf{q} = \{[q_1], \ldots, [q_s]\} \subset \mathsf{A} : \lambda_i(q_j) = \delta_{ij}$ for each i, j,
- $Q_h := \left(\lambda_j(Z_h q_i)\right)_{ij}$.

Example 40.1.1.   Set $Q := \mathbb{C}[Z_1, Z_2]$ and

$$\mathsf{J} := (Z_1^3 - Z_1^2, Z_1 Z_2 - Z_1 - Z_2 + 1, Z_2^3 + Z_1^2 - 1),$$

which is a Gröbner basis w.r.t. the degree-compatible ordering induced by $Z_1 < Z_2$.

As a consequence we can choose as Gröbner representation the linear representation $\mathbf{b} = \{1, Z_1, Z_2, Z_1^2, Z_2^2\}$,

$$A_1 = \begin{pmatrix} 0 & 1 & 0 & 0 & 0 \\ 0 & 0 & 0 & 1 & 0 \\ -1 & 1 & 1 & 0 & 0 \\ 0 & 0 & 0 & 1 & 0 \\ -1 & 1 & 0 & 0 & 1 \end{pmatrix}, \quad A_2 = \begin{pmatrix} 0 & 0 & 1 & 0 & 0 \\ -1 & 1 & 1 & 0 & 0 \\ 0 & 0 & 0 & 0 & 1 \\ -1 & 0 & 1 & 1 & 0 \\ 1 & 0 & 0 & -1 & 0 \end{pmatrix}.$$

It is easy to verify that $\mathfrak{L}(\mathsf{J}) = \operatorname{Span}_K(\mathbb{L})$, $\mathbb{L} := \{\lambda_1, \ldots, \lambda_5\}$, with

$$\lambda_1(p) = p(0, 1), \quad \lambda_2(p) = \frac{\partial p}{\partial Z_1}(0, 1),$$

$$\lambda_3(p) = p(1, 0), \quad \lambda_4(p) = \frac{\partial p}{\partial Z_2}(1, 0), \quad \lambda_5(p) = \frac{\partial^2 p}{2\partial Z_2^2}(1, 0),$$

whose associated Gröbner representation is $\mathbf{q} = \{q_1, q_2, q_3, q_4, q_5\}$ with

$$q_1 = -Z_1^2 + 1, \quad q_2 = -Z_1^2 + Z_1,$$
$$q_3 = Z_1^2, \quad q_4 = Z_1^2 + Z_2 - 1, \quad q_5 = Z_1^2 + Z_2^2 - 1,$$

$$Q_1 = \begin{pmatrix} 0 & 1 & 0 & 0 & 0 \\ 0 & 0 & 0 & 0 & 0 \\ 0 & 0 & 1 & 0 & 0 \\ 0 & 0 & 0 & 1 & 0 \\ 0 & 0 & 0 & 0 & 1 \end{pmatrix}, \quad Q_2 = \begin{pmatrix} 1 & 0 & 0 & 0 & 0 \\ 0 & 1 & 0 & 0 & 0 \\ 0 & 0 & 0 & 1 & 0 \\ 0 & 0 & 0 & 0 & 1 \\ 0 & 0 & 0 & 0 & 0 \end{pmatrix}.$$

Between the two bases $\mathbf{b}$ and $\mathbf{q}$ there are the basis transformations

$$M_{bq} := \left(\gamma(b_i, q_j, \mathbf{q})\right) \quad \text{and} \quad M_{qb} := \left(\gamma(q_i, b_j, \mathbf{b})\right),$$

so that, for each i,

$$[b_i] = \sum_j \gamma(b_i, q_j, \mathbf{q})[q_j] \quad \text{and} \quad [q_i] = \sum_j \gamma(q_i, b_j, \mathbf{b})[b_j];$$

naturally, we have $M_{bq} = M_{qb}^{-1}$ and

$$M([f], \mathbf{b}) = M_{bq} M([f], \mathbf{q}) M_{qb} = M_{bq} M([f], \mathbf{q}) M_{bq}^{-1}$$

so that $M([f], \mathbf{q})$ and $M([f], \mathbf{b})$ are similar and share the same eigenvalues and Jordan normal form.

Example 40.1.2.   Continuing Example 40.1.1, we have

$$M_{bq} = \begin{pmatrix} 1 & 0 & 1 & 0 & 0 \\ 0 & 1 & 1 & 0 & 0 \\ 1 & 0 & 0 & 1 & 0 \\ 0 & 0 & 1 & 0 & 0 \\ 1 & 0 & 0 & 0 & 1 \end{pmatrix}, \quad M_{qb} = \begin{pmatrix} 1 & 0 & 0 & -1 & 0 \\ 0 & 1 & 0 & -1 & 0 \\ 0 & 0 & 0 & 1 & 0 \\ -1 & 0 & 1 & 1 & 0 \\ -1 & 0 & 0 & 1 & 1 \end{pmatrix},$$

and it is easy to check the relations $M_{bq} = M_{qb}^{-1}$, $M_{bq} Q_1 = A_1 M_{bq}$ and $M_{bq} Q_2 = A_2 M_{bq}$. Setting

$$J_1 = \left(\begin{array}{cc|ccc} 0 & 1 & & & \\ & 0 & & & \\ \hline & & 1 & & \\ & & & 1 & \\ & & & & 1 \end{array}\right), \quad J_2 = \left(\begin{array}{cc|ccc} 1 & & & & \\ & 1 & & & \\ \hline & & 0 & 1 & \\ & & & 0 & 1 \\ & & & & 0 \end{array}\right),$$

we have, for $i = 1, 2$, $M_{bq} J_i = A_i M_{bq}$ and hence $J_i = Q_i$.

From $M_{bq} J_i = A_i M_{bq}$, we obtain $M_{qb} A_i = J_i M_{qb}$ and $A_i^T M_{qb}^T = M_{qb}^T J_i^T = M_{qb}^T Q_i^T$, which allow us to deduce easily the eigenvalues and Jordan normal forms for A_i^T also.

In fact we have $\tilde{M}_{qb} \tilde{J}_i = A_i^T \tilde{M}_{qb}$ with

$$\tilde{M}_{qb} = \begin{pmatrix} -1 & -1 & 0 & 0 & 1 \\ 0 & 0 & 0 & 1 & 0 \\ 0 & 1 & 0 & 0 & 0 \\ 1 & 1 & 1 & -1 & -1 \\ 1 & 0 & 0 & 0 & 0 \end{pmatrix},$$

$$\tilde{J}_1 = \left(\begin{array}{ccc|cc} 1 & & & & \\ & 1 & & & \\ & & 1 & & \\ \hline & & & 0 & 1 \\ & & & & 0 \end{array}\right) \quad \text{and} \quad \tilde{J}_2 = \left(\begin{array}{ccc|cc} 0 & 1 & & & \\ & 0 & 1 & & \\ & & 0 & & \\ \hline & & & 1 & \\ & & & & 1 \end{array}\right).$$

Remark 40.1.3.   Denote P the s-square *backward identity matrix*

$$P := (p_{ij}), \qquad p_{ij} = \begin{cases} 1 & \text{if } i + j = s + 1, \\ 0 & \text{if } i + j \neq s + 1, \end{cases}$$

which satisfies $P^{-1} = P$.

From the relation $MJ = AM$, where A is an s-square matrix, M is invertible, $N := M^{-1}$ and J is the Jordan matrix of A, we obtain $JN = NA$ and

$$A^T(N^T P) = N^T J^T P = (N^T P)(P J^T P);$$

therefore $P J^T P$ is the Jordan matrix of A^T – whose eigenvalues thus are the same (with the same multiplicity) as those of A – and $\widetilde{N} := N^T P$ is its eigenspace matrix.

Given a square matrix $N = (n_{ij})$, the matrix $\widetilde{N} = N^T P = (\tilde{n}_{ij})$ can be obtained from N by, equivalently,

- writing from the right to the left the columns of N^T or
- clock-wise $\frac{\pi}{4}$-rotating N or
- setting $\tilde{n}_{ij} = n_{s-j\ i}$ for each $1 \leq i, j \leq s$. $\boxed{\odot}$

40.2 Toward the Auzinger–Stetter Theorem

With the same notation as in the previous section, let us fix

- a Gröbner representation

$$\mathbf{b} = \{[b_1], \dots, [b_s]\} \subset \mathsf{A}, \qquad A_h := \left(a_{ij}^{(h)}\right) = M([Z_h], \mathbf{b}), \quad 1 \leq h \leq r;$$

- a basis $\mathbb{L} := \{\lambda_1, \dots, \lambda_s\}$ of $\mathfrak{L}(\mathsf{J})$;
- the conjugate Gröbner representation

$$\mathbf{q} = \{[q_1], \dots, [q_s]\} \subset \mathsf{A}, \qquad Q_h := \left(\lambda_j(Z_h q_i)\right)_{ij},$$

where $\mathbf{q}$ is the Lagrange basis satisfying $\lambda_i(q_j) = \delta_{ij}$ for each i, j,

and let us denote

- as $M_{bq} := \left(\gamma(b_i, q_j, \mathbf{q})\right)$ and $M_{qb} := \left(\gamma(q_i, b_j, \mathbf{b})\right)$ the basis transformation matrices;
- as $\widetilde{M}_{qb}$ the matrix obtained from M_{qb} according to the construction described in Remark 40.1.3;
- as J_h and $\widetilde{J}_h$, $1 \leq h \leq r$, the Jordan normal form matrices for A_h and A_h^T;
- for each $f \in \mathcal{Q}/\mathsf{J} = \mathsf{A}$,

$$A_f := M([f], \mathbf{b}) = \left(\gamma(f b_i, b_j, \mathbf{b})\right) : \Phi_f(b_i) = [f b_i] = \sum_j \gamma(f b_i, b_j, \mathbf{b})[b_j];$$

- as J_f and $\widetilde{J}_f$ the Jordan normal form matrices for A_f and A_f^T.

Let us also consider the set

$$\mathcal{Z}(\mathsf{J}) := \{\alpha \in K^r : f(\alpha) = 0 \text{ for each } f \in \mathsf{J}\}.$$

Lemma 40.2.1 (Auzinger–Stetter). *With the present notation it holds that*

$$\gamma(b_i, q_j, \mathbf{q}) = \lambda_j(b_i), \qquad 1 \le i, j \le s.$$

Proof. For each $f \in \mathsf{A}$, $\sum_j \gamma(f, q_j, \mathbf{q})[q_j] = f = \sum_j \lambda_j(f)[q_j]$.

The first equality follows from the definition of γ and the second from the property of the Lagrange basis. The claim then follows by the linear independency of $\mathbf{q}$. $\boxed{\odot}$

Corollary 40.2.2. *Each ith row of M_{bq} is the vector $(\lambda_1(b_i), \ldots, \lambda_s(b_i))$ of the evaluation of the basis element b_i at the functional basis $\mathbb{L}$.*

Each jth column of M_{bq} is the vector $\left(\lambda_j(b_1), \ldots, \lambda_j(b_s)\right)^T$ of the evaluation of the basis $\mathbf{b}$ at the functional λ_j. $\boxed{\odot}$

Lemma 40.2.3 (Auzinger–Stetter). *For each $\alpha \in \mathcal{Z}(\mathsf{J})$, the vector*

$$(b_1(\alpha), \ldots, b_s(\alpha))^T$$

is an eigenvector of the matrix A_f for the eigenvalue $f(\alpha)$.

Proof. For each i, $1 \le i \le s$, we have

$$[fb_i] = \Phi_f([b_i]) = \sum_j \gamma(fb_i, b_j, \mathbf{b})[b_j],$$

so that $f(\alpha)b_i(\alpha) = \sum_j \gamma(fb_i, b_j, \mathbf{b})b_j(\alpha)$ for each i. Thus the claim follows trivially. $\boxed{\odot}$

Lemma 40.2.4 (Möller). *The following holds:*

(1) *for any $\lambda \in K$, λ is an eigenvalue for Φ_f iff $\mathsf{J} : (f - \lambda) \ne \mathsf{J}$;*
(2) *the corresponding eigenspace is the set $\{[h] : h \in \mathsf{J} : (f - \lambda)\}$;*
(3) *$[h] = \sum_i \beta_i[b_i] \in \mathsf{J} : (f - \lambda)$ iff $(\beta_1, \ldots, \beta_s)^T$ is an eigenvector of A_f^T for λ.*

Proof. For each $[h] = \sum_i \beta_i[b_i] \in \mathsf{A}$ we have

$$\Phi_f([h]) = \sum_j \beta_j \Phi_f([b_j])$$

$$= \sum_j \beta_j \sum_i \gamma(fb_j, b_i, \mathbf{b})[b_i]$$

$$= \sum_i \left(\sum_j \beta_j \gamma(fb_j, b_i, \mathbf{b}) \right) [b_i]$$

so that, for $v := (\beta_1, \ldots, \beta_s)^T$, we have

$$\lambda[h] = \Phi_f([h]) \iff \lambda\beta_i = \sum_j \beta_j \gamma(fb_j, b_i, \mathbf{b}) \text{ for each } i \iff \lambda v = A^T v.$$

For any $h \notin J$ we have the obvious equivalences

$$\lambda[h] = \Phi_f([h]) \iff \lambda[h] = [fh]$$
$$\iff (\lambda - [f])[h] = 0$$
$$\iff (\lambda - f)h \in J$$
$$\iff h \in J : (f - \lambda)$$

whence the claim. $\boxed{\odot}$

Definition 40.2.5. *A matrix is called* non-derogatory *if the following equivalent conditions,*

(1) *all its eigenspaces have dimension 1;*
(2) *its Jordan form has a single Jordan block associated with each eigenvalue,*
    hold. $\boxed{\odot}$

Theorem 40.2.6 (Auzinger–Stetter). *The set $\{f(\alpha) : \alpha \in \mathcal{Z}(J)\}$ is the set of eigenvalues of A_f. If A_f is non-derogatory, each eigenspace of A_f for $f(\alpha)$ is spanned by $(b_1(\alpha), \ldots, b_s(\alpha))^T$.*

Proof. This is a direct consequence of Lemmata 40.2.3 and 40.2.4. $\boxed{\odot}$

Corollary 40.2.7. *The set $\{f(\alpha) : \alpha \in \mathcal{Z}(J)\}$ is the set of eigenvalues of A_f^T. If A_f is non-derogatory, so also is A_f^T, and, for each i,*

(1) *the eigenspace of A_f^T for $f(\alpha_i)$ is spanned by $(\gamma(q_i, b_1, \mathbf{b}), \ldots, \gamma(q_i, b_s, \mathbf{b})))^T$.*
(2) $J : (f - f(\alpha_i)) = J + (q_i)$. $\boxed{\odot}$

Example 40.2.8. Continuing Example 40.1.1, the eigenspace of A_1 (respectively, A_2) for the eigenvalue 0 is spanned by $(1, 0, 1, 0, 1)^T$ (respectively, $(1, 1, 0, 1, 0)^T$), while $(1, 1, 0, 1, 0)^T$ (respectively, $(1, 0, 1, 0, 1)^T$) is just an eigenvector for the eigenvalue 1, whose eigenspace has dimension 3 (respectively 2). The eigenspaces for the eigenvalue 0 have dimension 1 for both A_1 and A_2 and those for the eigenvalue 1 have dimension respectively 3 and 2; thus neither matrix is non-derogatory.
  We also have:

- $J : Z_1 = \mathbb{I}(Z_1 - Z_1^2) + J$ and $(0, 1, 0, -1, 0)^T$ spans the eigenspace of A_1^T for the eigenvalue 0;
- $J : \mathbb{I}(Z_1 - 1) = (Z_1^2, Z_2 - 1) + J$ and the eigenspace of A_1^T for the eigenvalue 1 is spanned by

$$\{(-1, 0, 0, 1, 1)^T, (-1, 0, 1, 1, 0)^T, (0, 0, 0, 1, 0)^T\}$$
$$= \{(0, 0, -1, 0, 1)^T, (-1, 0, 1, 0, 0)^T, (0, 0, 0, 1, 0)^T\};$$

- $J : Z_2 = \mathbb{I}(Z_2^2 + Z_1^2 - 1) + J$ and the eigenspace of A_2^T for 0 is spanned by $\{(-1, 0, 0, 1, 1)^T\}$.

- $J : (Z_2 - 1) = \mathbb{I}(Z_1 - 1) + J$ and the eigenspace of A_2^T for 1 is spanned by

$$\{(0, 1, 0, -1, 0)^T, (1, 0, 0, -1, 0)^T\} = \{(1, -1, 0, 0, 0)^T, (0, 1, 0, -1, 0)^T\}.$$

$\boxed{\odot}$

The relevant aspect of the Auzinger–Stetter Theorem 40.2.6 is that, while both the eigenvalues and eigenvectors of A_f depend intrinsically on the roots of J, their actual values are precise functions of the choices of the matrix A_f and of the basis $\mathbf{b}$; one can therefore expect that, for an appropriate choice of f and $\mathbf{b}$, an eigenvalue computation can allow us to deduce the roots of J.

40.3 Auzinger–Stetter: The Radical Case

Let us as a preliminary assume that J is radical and see whether the remark at the end of the previous section leads to something.

The radicality assumption implies that J has $s = \deg(J)$ different roots in K^r:

$$\mathcal{Z}(J) = \{\alpha_1, \ldots, \alpha_s\} \subset K^r, \quad \alpha_j = (a_1^{(j)}, \ldots, a_r^{(j)}).$$

Thus we can wlog identify each functional λ_j with the evaluation at the root α_j:

$$\lambda_j : \mathcal{Q} \to K, \qquad p(Z_1, \ldots, Z_r) \mapsto \lambda_j(p) = p(a_1^{(j)}, \ldots, a_r^{(j)})$$

and $\mathbf{q}$ is the corresponding Lagrange basis.

A matrix A_f is non-derogatory if and only if $f(\alpha_i) \neq f(\alpha_j)$ for each $i \neq j$. Clearly, for a generic linear form $Y = \sum_h c_h Z_h$, A_Y is non-derogatory. Thus if we choose a linear form which *separates* $\mathcal{Z}(J)$, that is, it satisfies the condition

(**AS.1**) $Y = \sum_h c_h Z_h$ is such that $\beta_i := \sum_h c_h a_h^{(i)} \neq \sum_h c_h a_h^{(j)} =: \beta_j$ for each $i \neq j$,

then A_Y and A_Y^T are non-derogatory and have the s distinct eigenvalues

$$\beta_j := \sum_h c_h a_h^{(j)}, \quad 1 \leq j \leq s,$$

whose associated eigenspaces are generated respectively by

$$(b_1(\alpha_j), \ldots, b_s(\alpha_j))^T \quad \text{and} \quad (\gamma(q_j, b_1, \mathbf{b}), \ldots, \gamma(q_j, b_s, \mathbf{b}))^T.$$

In order to deduce the α_j from these eigenvectors, the trick consists in a clever choice of the basis $\mathbf{b}$. An efficient choice is the original one proposed by Auzinger and Stetter: let us denote by V the K-vector space

$$V := \text{Span}_K\{[1], [Z_1], \ldots, [Z_r]\}$$

and let $\delta := \dim_K(V) \leq s$; then, up to a re-enumeration of the variables, we can wlog assume that

- $V = \text{Span}_K\{[1], [Z_1], \ldots, [Z_{\delta-1}]\}$
- $\{[1], [Z_1], \ldots, [Z_{\delta-1}]\}$ is a K-basis of V,
- there are $c_{il} \in K, 0 \leq l < \delta \leq i \leq r$, such that $[Z_i] = c_{i0} + \sum_{l=1}^{\delta-1} c_{il}[Z_l]$.

Moreover, knowledge of the matrices A_h allows us to deduce, by easy linear algebra, both δ and the c_{il}. We can therefore choose a basis $\mathbf{b}$ which satisfies the condition

(AS.2) $\mathbf{b} = ([b_1], \dots, [b_s])$ is such that

$$b_1 = 1, \qquad b_i = Z_{i-1}, \qquad 1 < i \leq \delta = \dim_K(V),$$

so that

$$V := \mathrm{Span}_K\{[1], [Z_1], \dots, [Z_r]\} = \mathrm{Span}_K\{[1], [Z_1], \dots, [Z_{\delta-1}]\}$$
$$= \mathrm{Span}_K\{[b_1], \dots, [b_\delta]\}.$$

Thus the eigenvectors corresponding to $\alpha_j = (a_1^{(j)}, \dots, a_r^{(j)})$ are

$$\left(1, a_1^{(j)}, \dots, a_{\delta-1}^{(j)}, b_{\delta+1}(\alpha_j), \dots, b_s(\alpha_j)\right)^T$$

and the other coordinates of α_j can be deduced from $a_i^{(j)} = c_{i0} + \sum_{l=1}^{\delta-1} c_{il} a_l^{(j)}$.

In conclusion, we have

Theorem 40.3.1 (Auzinger–Stetter). *With the present notation and under the assumption that* J *is radical, it holds that:*

(1) *each jth column* $(b_1(\alpha_j), \dots, b_s(\alpha_j))^T$ *of* M_{bq} *is an eigenvector of each* A_f, $f \in Q$, *for the eigenvalue* $f(\alpha_j)$;
(2) *each jth row* $(\gamma(q_j, b_1, \mathbf{b}), \cdots, \gamma(q_j, b_s, \mathbf{b}))^T$ *of* M_{qb} *is an eigenvector of each* A_f^T, $f \in Q$, *for the eigenvalue* $f(\alpha_j)$;
(3) *for each* $f \in Q$,
    (a) *the eigenvalues of* A_f *and* A_f^T *are* $\{f(\alpha_j) : 1 \leq j \leq s\}$;
    (b) *the eigenspace of* A_f *for* $\lambda \in K$ *is*

$$\mathrm{Span}_K\{(b_1(\alpha_j), \dots, b_s(\alpha_j))^T : f(\alpha_j) = \lambda\};$$

    (c) *the eigenspace of* A_f^T *for* $\lambda \in K$ *is*

$$\mathrm{Span}_K\left\{\left(\gamma(q_j, b_1, \mathbf{b}), \dots, \gamma(q_j, b_s, \mathbf{b})\right)^T : f(\alpha_j) = \lambda\right\};$$

    (d) $[q_j]f(\alpha_j) = [fq_j]$ *for each* j;
    (e) *for each* $\lambda \in K$, $\mathsf{J} : (f - \lambda) = \mathsf{J}$ *iff* $\lambda \notin \{f(\alpha_i) : 1 \leq i \leq s\}$;
    (f) *for each* $\lambda \in K$, $\mathsf{J} : (f - \lambda) = \mathsf{J} + \{q_j : j \in J\}$, *where* $J = \{j : 1 \leq j \leq s, f(\alpha_j) = \lambda\}$.

If, moreover, $Y = \sum_h c_h Z_h$ *satisfies condition* **(AS.1)** *then:*

(4) *the jth column* $(b_1(\alpha_j), \dots, b_s(\alpha_j))^T$ *of* M_{bq} *is the eigenvector for* $\beta_j := \sum_h c_h a_h^{(j)}$ *of* A_Y;
(5) *the jth row* $(\gamma(q_i, b_1, \mathbf{b}), \dots, \gamma(q_j, b_s, \mathbf{b}))^T$ *of* M_{qb} *is the eigenvector for* $\beta_j := \sum_h c_h a_h^{(j)}$ *of* A_Y^T;
(6) $\mathsf{J} : (Y - \beta_j) = \mathsf{J} + \{q_j\}$ *for each* j.

If, further, $\mathbf{b} = \{[1], [Z_1], \ldots, [Z_{\delta-1}], [b_{\delta+1}], \ldots, b_s]\}$ *satisfies condition* **(AS.2)** *then:*

(7) *denoting as* $\{(d_{j1}, \ldots, d_{js})^T, 1 \leq j \leq s\}$ *the eigenvectors of* A_Y *and*

$$\alpha_j := \left(d_{j1}^{-1} d_{j2}, \ldots, d_{j1}^{-1} d_{j\delta}, c_{\delta 0} + \sum_{l=1}^{\delta-1} c_{\delta l} d_{j1}^{-1} d_{jl}, \ldots, c_{n0} + \sum_{l=1}^{\delta-1} c_{nl} d_{j1}^{-1} d_{jl} \right)$$

for each j, *then* $\mathcal{Z}(\mathsf{J}) = \{\alpha_j, 1 \leq j \leq s\}$.

Proof.   Parts (1) and (3a, b) are a direct consequence of Lemma 40.2.3 and items (2), (3c, f) are a direct consequence of Lemma 40.2.4.

For the non-derogatory case, (4) is Theorem 40.2.6 and (5) and (6) are Corollary 40.2.7.

Part (7) is a direct reformulation of (1) applied to the basis that satisfies condition **(AS.2)**.   ⊚

Corollary 40.3.2.   *For each* h, $1 \leq h \leq r$ *we have*

$$A_h M_{bq} = M_{bq} \, \mathrm{diag}(a_h^{(1)}, \ldots, a_h^{(s)}).$$ ⊚

Example 40.3.3.   If we consider the ideal $\mathsf{J} \subset \mathbb{C}[Z_1, Z_2, Z_3]$ discussed in Example 39.2.3, since we have complete knowledge of both the roots and the Gröbner structure all we need to do is to verify the Auzinger–Stetter Theorem for it.

The natural choices for the K-bases $\mathbf{b}$ and $\mathbb{L}$ (see Section 40.1) are (compare Examples 33.2.5 and 33.2.6)

$$\mathbf{b} := \{1, Z_1, Z_2, Z_3, Z_1^2, Z_1 Z_2, Z_2^2, Z_1 Z_3, Z_3^2\}$$

and $\lambda_i(p) := p(\mathbf{b}_i)$ for all i; under these choices,

$$M_{bq} = \begin{pmatrix}
1 & 1 & 1 & 1 & 1 & 1 & 1 & 1 & 1 \\
0 & 0 & 2 & 0 & 1 & 1 & 1 & 2 & 2 \\
0 & 1 & 0 & 2 & 0 & 1 & 1 & 0 & 0 \\
1 & -2 & 2 & -2 & 3 & 3 & 1 & 1 & 0 \\
0 & 0 & 4 & 0 & 1 & 1 & 1 & 4 & 4 \\
0 & 0 & 0 & 0 & 0 & 1 & 1 & 0 & 0 \\
0 & 1 & 0 & 4 & 0 & 1 & 1 & 0 & 0 \\
0 & 0 & 4 & 0 & 3 & 3 & 1 & 2 & 0 \\
1 & 4 & 4 & 4 & 9 & 9 & 1 & 1 & 0
\end{pmatrix},$$

and the matrices related to the Gröbner representation are (compare Examples 33.5.1 and 33.5.2)

$$A_1 = \begin{pmatrix} 0 & 1 & 0 & 0 & 0 & 0 & 0 & 0 & 0 \\ 0 & 0 & 0 & 0 & 1 & 0 & 0 & 0 & 0 \\ 0 & 0 & 0 & 0 & 0 & 1 & 0 & 0 & 0 \\ 0 & 0 & 0 & 0 & 0 & 0 & 0 & 1 & 0 \\ 0 & -2 & 0 & 0 & 3 & 0 & 0 & 0 & 0 \\ 0 & 0 & 0 & 0 & 0 & 1 & 0 & 0 & 0 \\ 0 & 0 & 0 & 0 & 0 & 1 & 0 & 0 & 0 \\ 2 & -3 & -9 & -2 & 1 & 6 & 3 & 3 & 0 \\ 6 & -3 & -45 & -8 & 0 & 30 & 15 & 4 & 2 \end{pmatrix},$$

$$A_2 = \begin{pmatrix} 0 & 0 & 1 & 0 & 0 & 0 & 0 & 0 & 0 \\ 0 & 0 & 0 & 0 & 0 & 1 & 0 & 0 & 0 \\ 0 & 0 & 0 & 0 & 0 & 0 & 1 & 0 & 0 \\ -2 & -3 & 7 & 2 & 2 & -1 & -3 & -1 & 0 \\ 0 & 0 & 0 & 0 & 0 & 1 & 0 & 0 & 0 \\ 0 & 0 & 0 & 0 & 0 & 1 & 0 & 0 & 0 \\ 0 & 0 & -2 & 0 & 0 & 0 & 3 & 0 & 0 \\ -2 & -3 & 9 & 2 & 2 & -3 & -3 & -1 & 0 \\ -8 & -12 & 40 & 8 & 8 & -19 & -12 & -4 & 0 \end{pmatrix},$$

$$A_3 = \begin{pmatrix} 0 & 0 & 0 & 1 & 0 & 0 & 0 & 0 & 0 \\ 0 & 0 & 0 & 0 & 0 & 0 & 0 & 1 & 0 \\ -2 & -3 & 7 & 2 & 2 & -1 & -3 & -1 & 0 \\ 0 & 0 & 0 & 0 & 0 & 0 & 0 & 0 & 1 \\ 2 & -3 & -9 & -2 & 1 & 6 & 3 & 3 & 0 \\ -2 & -3 & 9 & 2 & 2 & -3 & -3 & -1 & 0 \\ -2 & -3 & 9 & 2 & 2 & -1 & -5 & -1 & 0 \\ 6 & -3 & -45 & -8 & 0 & 30 & 15 & 4 & 2 \\ -6 & 3 & -9 & 4 & 0 & 6 & 3 & -3 & 3 \end{pmatrix}.$$

They satisfy

$$A_1 M_{bq} = M_{bq} \operatorname{diag}(0, 0, 2, 0, 1, 1, 1, 2, 2),$$
$$A_2 M_{bq} = M_{bq} \operatorname{diag}(0, 1, 0, 2, 0, 1, 1, 0, 0),$$
$$A_3 M_{bq} = M_{bq} \operatorname{diag}(1, -2, 2, -2, 3, 3, 1, 1, 0)$$

and none is non-derogatory. Instead, the matrix

$$A_Y = \begin{pmatrix} 0 & -3 & 1 & 3 & 0 & 0 & 0 & 0 & 0 \\ 0 & 0 & 0 & 0 & -3 & 1 & 0 & 3 & 0 \\ -6 & -9 & 21 & 6 & 6 & -6 & -8 & -3 & 0 \\ -2 & -3 & 7 & 2 & 2 & -1 & -3 & -4 & 3 \\ 6 & -3 & -27 & -6 & -6 & 19 & 9 & 9 & 0 \\ -6 & -9 & 27 & 6 & 6 & -11 & -9 & -3 & 0 \\ -6 & -9 & 25 & 6 & 6 & -6 & -12 & -3 & 0 \\ 10 & -3 & -99 & -16 & -1 & 69 & 33 & 2 & 6 \\ -44 & 6 & 148 & 44 & 8 & -91 & -48 & -25 & 3 \end{pmatrix},$$

related to the linear form $Y = -3Z_1 + Z_2 + 3Z_3$ is non-derogatory and satisfies

$$A_Y M_{bq} = M_{bq} \, \mathrm{diag}(3, -5, 0, -4, 6, 7, 1, -3, -6).$$

Remark 40.3.4.   We remark here that condition (**AS.2**) of Auzinger and Stetter (see the start of this section) suggests an improvement on the versions of Möller's Algorithm which iterate on terms (Algorithm 28.2.7). Once the first $n + 1$ terms $1, X_1, \ldots, X_n$ have been processed, giving the $\dim(V) - 1$ variables $Z_1, \ldots, Z_{\delta-1} \in \mathbf{T}(G)$ such that $\{[1], [Z_1], \ldots, [Z_{\delta-1}]\}$ generate $V = \mathrm{Span}_K\{[1], [Z_1], \ldots, [Z_n]\}$ and the $n - \delta + 1$ linear relations

$$\ell_i := Z_i - c_{i0} - \sum_{l=1}^{\delta-1} c_{il} Z_l \in G \cap V, \qquad \delta \le i \le n,$$

we know that

$$\mathbf{N} \subset \mathcal{T} \cap k[Z_1, \ldots, Z_{\delta-1}] \quad \text{and} \quad \mathbf{T}(G) \setminus \{Z_\delta, \ldots, Z_n\} \subset \mathcal{T} \cap k[Z_1, \ldots, Z_{\delta-1}].$$

Thus the iterative processing of the terms $t \in \mathcal{T}$ can be correspondingly improved (see Section 40.10).

40.4 Möller: Endomorphisms and Dual Space

If J is radical, setting $\mathbf{b} := \mathbf{q}$ and recalling that $\mathbf{q}$ is the Lagrange basis for the functionals λ_j representing the evaluation at α_j, that is,

$$\lambda_i(q_j) = \delta_{ij} = \begin{cases} 1 & \text{if } i = j, \\ 0 & \text{if } i \ne j, \end{cases}$$

we can reformulate Lemma 40.2.3 as

Corollary 40.4.1.   *For each $\lambda_j \in \mathbb{L}$ the vector*

$$(q_1(\alpha_j), \ldots, q_s(\alpha_j))^T = (\delta_{1j}, \ldots, \delta_{sj})^T = (0, \ldots, 0, 1, 0, \ldots, 0)^T$$

is an eigenvector of the matrix Q_f for the eigenvalue $f(\alpha_j)$.

Proof. For each j, $1 \le i \le s$, we have

$$[fq_j] = \Phi_f([q_j]) = \sum_i \lambda_i(fq_j)[q_i]. \tag{40.1}$$

In the particular case of a radical ideal, where each λ_i is an evaluation at the point α_i, we have further

$$\lambda_i(fq_j) = \lambda_i(f)\lambda_i(q_j) = f(\alpha_i)q_j(\alpha_i) = \begin{cases} f(\alpha_j) & \text{if } i = j, \\ 0 & \text{if } i \ne j, \end{cases}$$

so that $[fq_j] = \Phi_f([q_j]) = \sum_i \lambda_i(fq_j)[q_j] = f(\alpha_j)[q_j]$. $\qquad\qquad\boxed{\odot}$

In order to extend the argument above to the general case, one needs to expand Equation (40.1) to a generic dual basis $\mathbb{L}$ and to its corresponding Lagrange basis $\mathbf{q}$.

The natural choice is to take as $\mathbb{L}$ a Macaulay representation.[2] Following the notation of Chapter 31, for each $\tau = Z_1^{e_1} \cdots Z_r^{e_r} \in \mathcal{W}$, we consider the functionals

$$M(\tau) : \mathcal{Q} \to K, \quad f = \sum_{t \in \mathcal{W}} c(f, t)t \mapsto c(f, \tau)$$

and we restrict ourselves to the set $\mathrm{Span}_K(\mathbb{M}) \subset \mathcal{Q}^*$ of the *Noetherian equations*,[3] where $\mathbb{M} = \{M(\tau) : \tau \in \mathcal{W}\}$.

For any term ordering $<$, for each $\ell = \sum_{t \in \mathcal{W}} c(\ell, t)M(t)$ we set

$$\mathbf{T}_<(\ell) := \min_< \{t : c(\ell, t) \ne 0\}, \quad \mathbf{L}_<(\ell) := \max_< \{t : c(\ell, t) \ne 0\}$$

and we write $\mathbf{T}_<(L) = \{\mathbf{T}_<(\ell) : \ell \in L\}$ and $\mathbf{L}_<(L) = \{\mathbf{L}_<(\ell) : \ell \in L\}$ for each subset $L \subset \mathrm{Span}_K(\mathbb{M})$.

We define, for each $\tau \in \mathcal{W}$, the linear map

$$\sigma_\tau : \mathrm{Span}_K(\mathbb{M}) \to \mathrm{Span}_K(\mathbb{M}) : M(\omega) \mapsto \begin{cases} M(\upsilon) & \text{if } \omega = \tau\upsilon, \\ 0 & \text{if } \tau \nmid \omega \end{cases}$$

and, by linearity, for each $f = \sum_{t \in \mathcal{W}} c(f, t)t \in \mathcal{Q}$ the linear map

$$\sigma_f : \mathrm{Span}_K(\mathbb{M}) \to \mathrm{Span}_K(\mathbb{M}), \quad \ell \mapsto \sigma_f(\ell) = \sum_{t \in \mathcal{W}} c(f, t)\sigma_t(\ell).$$

We recall that a subspace $L \subset \mathbb{M}$ is called *stable* (Definition 31.2.2) iff, for each $\ell \in L$ and $f \in \mathcal{Q}$, $\sigma_f(\ell) \in L$.

Recall that, for each primary ideal $\mathfrak{q} \subset \mathcal{Q}$ at the origin, the corresponding dual space $\mathfrak{L}(\mathfrak{q}) \subset \mathcal{Q}^*$ is a stable subset of $\mathrm{Span}_K(\mathbb{M})$ (Corollary 31.3.3, Proposition 31.3.5) and satisfies $\mathbf{T}_<(\mathfrak{L}(\mathfrak{q})) = \mathbf{N}_<(\mathfrak{q})$ (Corollary 32.1.4); both $\mathbf{T}_<(\mathfrak{L}(\mathfrak{q}))$ and $\mathbf{L}_<(\mathfrak{L}(\mathfrak{q}))$ are ordered ideals.

Moreover the *Leibnitz formula* holds.

---

[2] Compare Section 32.1, Corollary 32.3.3, and Definition 33.2.2.
[3] See Section 30.4.

Lemma 40.4.2. *For any $f, g \in \mathcal{Q}$ and any $\ell \in \mathrm{Span}_K(\mathbb{M})$ we have*

$$\ell(fg) = \sum_{\tau \in \mathcal{W}} M(\tau)(f)\sigma_\tau(\ell)(g).$$

Proof. Compare Corollary 31.4.2.   ⊡

An alternative description of the dual space of a primary is in terms of differential functions (Section 31.5): we denote, for each $\tau = Z_1^{e_1} \cdots Z_r^{e_r} \in \mathcal{W}$, by $D(\tau) : \mathcal{Q} \to \mathcal{Q}$ the differential operator

$$D(\tau) := \frac{1}{e_1! \cdots e_r!} \frac{\partial^{e_1 + \cdots + e_r}}{\partial Z_1^{e_1} \cdots \partial Z_r^{e_r}}$$

and consider the subset $\mathrm{Span}_K(\mathbb{D}) \subset \mathrm{Hom}(\mathcal{Q}, \mathcal{Q})$, $\mathbb{D} = \{D(\tau) : \tau \in \mathcal{W}\}$.

There is an obvious identification

$$\mathrm{ev} : \mathrm{Span}_K(\mathbb{D}) \to \mathrm{Span}_K(\mathbb{M}) : \mathbb{D}(\tau) \mapsto \mathbb{M}(\tau),$$

which satisfies, for each $\delta := \sum_{t \in \mathcal{W}} c(\delta, t)D(t) \in \mathrm{Span}_K(\mathbb{D})$,

$$\mathrm{ev}(\delta)(\cdot) = \delta(\cdot)(0, \ldots, 0) = \sum_{t \in \mathcal{W}} c(\delta, t)M(t)(\cdot). \tag{40.2}$$

Under (40.2) we can impose on $\mathbb{D}$ the same term ordering $<$ as that induced on $\mathbb{M}$, so that

$$D(\tau) \le D(\omega) \iff M(\tau) \le M(\omega) \iff \tau \le \omega$$

and we can set $\mathbf{T}_<(\delta) := \mathbf{T}_<(\mathrm{ev}(\delta))$, $\mathbf{L}_<(\delta) := \mathbf{L}_<(\mathrm{ev}(\delta))$ for each $\delta \in \mathrm{Span}_K(\mathbb{D})$ and $\mathbf{T}_<(D) = \mathbf{T}_<(\mathrm{ev}(D))$, $\mathbf{L}_<(D) = \mathbf{L}_<(\mathrm{ev}(D))$ for each subset $D \subset \mathrm{Span}_K(\mathbb{D})$.

Under this identification we can naturally define the *anti-differential operators*

$$\sigma_\tau(D(\omega)) := \begin{cases} D(\upsilon) & \text{if } \omega = \tau\upsilon, \\ 0 & \text{if } \tau \nmid \omega, \end{cases} \quad \text{for each } \tau, \omega \in \mathcal{W}$$

and $\sigma_f(\delta) = \sum_{t \in \mathcal{W}} c(f, t)\sigma_t(\delta)$ for each $f = \sum_{t \in \mathcal{W}} c(f, t)t \in \mathcal{Q}, \delta \in \mathrm{Span}_K(\mathbb{D})$; a subspace $D \subset \mathbb{D}$ is called *stable* (Definition 31.5.3) iff $\sigma_f(\delta) \in D$ for each $\delta \in D, f \in \mathcal{Q}$.

For each $\alpha = (a_1, \ldots, a_r) \in K^r$, denoting as $\mathfrak{m}_\alpha = (Z_1 - a_1, \ldots, Z_r - a_r)$ the maximal ideal at α and defining

$$\lambda_\alpha : \mathcal{Q} \to \mathcal{Q}, \qquad \lambda_\alpha(f) = f(Z_1 + a_1, \ldots, Z_r + a_r),$$

so that $\lambda_\alpha(\mathfrak{m}_\alpha) = \mathfrak{m} = (Z_1, \ldots, Z_r)$ is the maximal ideal at the origin, we then have

$$(\delta(f))(\alpha) = (\delta(f))(a_1, \ldots, a_r) = \delta(\lambda_\alpha(f))(0, \ldots, 0) = \mathrm{ev}(\delta)(\lambda_\alpha(f))$$

for each $\delta \in \mathrm{Span}_K(\mathbb{D})$ and $f \in \mathcal{Q}$. Thus for each $\mathfrak{m}_\alpha$-primary $\mathfrak{q}_\alpha$, if we write

$$\mathfrak{D}_{\mathfrak{m}_\alpha}(\mathfrak{q}_\alpha) := \{\delta \in \mathrm{Span}_K(\mathbb{D}) : \delta(f)(\alpha) = 0 \text{ for each } f \in \mathfrak{q}_\alpha\} \subset \mathrm{Span}_K(\mathbb{D})$$

then we have $\mathrm{ev}(\mathfrak{D}_{\mathfrak{m}_\alpha}(\mathfrak{q}_\alpha)) = \mathfrak{L}(\lambda_\alpha(\mathfrak{q}_\alpha))$, and $\mathfrak{D}_{\mathfrak{m}_\alpha}(\mathfrak{q}_\alpha)$ is stable (under anti-differentiation).

Under this notation the Leibnitz formula becomes

Corollary 40.4.3. *For any $f, g \in \mathcal{Q}$ and any $\delta \in \mathrm{Span}_K(\mathbb{D})$ we have*

$$\delta(fg) = \sum_{\tau \in \mathcal{W}} D(\tau)(f)\sigma_\tau(\delta)(g). \qquad \boxed{\odot}$$

We have now the tools needed to describe the dual basis $\mathbb{L} := \{\lambda_1, \ldots, \lambda_s\}$ of J and the matrices Q_f describing the effect of each endomorphism Φ_f in terms of its corresponding Lagrange basis $\mathbf{q}$. Using a notation similar to that used in Section 33.2 we set:

- $<$ as any term ordering;
- $\mathcal{Z}(\mathsf{J}) := \{\alpha_1, \ldots, \alpha_\mathsf{s}\} \subset K^r, \quad \alpha_i = (a_1^{(i)}, \ldots, a_r^{(i)}), \mathsf{s} \leq s;$
- for each i, $1 \leq i \leq \mathsf{s}$,

  $\mathfrak{q}_i$ as the $\mathfrak{m}_{\alpha_i}$-primary component of J, so that $\mathsf{J} = \cap_{i=1}^\mathsf{s}\mathfrak{q}_i$,
  $s_i := \deg(\mathfrak{q}_i)$, so that $\sum_{i=1}^\mathsf{s} s_i = s$,
  $L_i := \mathfrak{L}(\lambda_{\alpha_i}(\mathfrak{q}_i)) \subset \mathrm{Span}_K(\mathbb{M})$,
  for each $\upsilon \in \mathbf{N}_<(\lambda_{\alpha_i}(\mathfrak{q}_i))$,

$$\ell_{\upsilon\alpha_i} = M(\upsilon) + \sum_{\tau \in \mathcal{W}} c(\tau, \ell_{\upsilon\alpha_i})M(\tau) \in L_i$$

  as the unique element (Definition 32.1.3) for which
  $\circ \tau \in \mathbf{N}_<(\lambda_{\alpha_i}(\mathfrak{q}_i)) = \mathbf{T}_<(L_i) \implies c(\tau, \ell_{\upsilon\alpha_i}) = 0$ and
  $\circ \mathbf{T}_<(\ell_{\upsilon\alpha_i}) = \upsilon$,
  so that
  $\circ$(in particular) $\ell_{1\alpha_i} = M(1)$ and
  $\circ\{\ell_{\upsilon\alpha_i} : \upsilon \in \mathbf{N}_<(\lambda_{\alpha_i}(\mathfrak{q}_i))\}$ is the Macaulay basis of

$$L_i = \mathrm{Span}_K\{\ell_{\upsilon\alpha_i} : \upsilon \in \mathbf{N}_<(\lambda_{\alpha_i}(\mathfrak{q}_i))\};$$

- $\mathbb{L} := \{\lambda_1, \ldots, \lambda_s\} := \{\ell_{\upsilon\alpha_i}\lambda_{\alpha_i} : \upsilon \in \mathbf{N}_<(\lambda_{\alpha_i}(\mathfrak{q}_i)), 1 \leq i \leq \mathsf{s}\}$, ordered so that,

$$\text{for } \lambda_x := \ell_{\upsilon_x\alpha_{i_x}}\lambda_{\alpha_{i_x}}, \lambda_y := \ell_{\upsilon_y\alpha_{i_y}}\lambda_{\alpha_{i_y}}, \text{ it holds that } x < y$$

$$\iff \begin{cases} i_x < i_y & \text{or} \\ i_x = i_y, & \text{and } \upsilon_x < \upsilon_y; \end{cases}$$

- $n_1 := 1, n_{i+1} := 1 + \sum_{l=1}^i s_l, 1 < i < \mathsf{s}, n_{r+1} = 1 + s;$
- $N_i := \{h : \lambda_h = \ell_{\tau\alpha_i}\lambda_{\alpha_i}\} = \{n_i, \ldots, n_{i+1} - 1\}$ for each i, $1 \leq i \leq \mathsf{s};$
- $\mathbf{q} = \{q_1, \ldots, q_s\}$, the set biorthogonal to $\mathbb{L}$, so that $\lambda_i(q_j) = \delta_{ij};$
- for each h, the element $\delta_h \in \mathbb{D}$ such that $\lambda_h = \mathrm{ev}(\delta_h)\lambda_{\alpha_i}, h \in N_i.$

We recall that, under these assumptions:

(1) $[p] = \sum_i \lambda_i(p)[q_i]$ for each $p \in Q$;
(2) $J_\sigma = \{f \in Q : \lambda_i(f) = 0, 1 \le i \le \sigma\}$ is an ideal for each $\sigma \le s$;
(3) $J_1 \supset J_2 \supset \cdots \supset J_s$ (cf. Corollary 32.3.3);
(4) by definition, for each i and each $f \in Q$,

$$\ell_{1\alpha_i}\lambda_{\alpha_i}(f) = M(1)\lambda_{\alpha_i}(f) = \lambda_{\alpha_i}(f)(0,\ldots,0) = f(\alpha_i).$$

Lemma 40.4.4 (Möller). *For each $\lambda_j = \ell_{\upsilon\alpha_i}\lambda_{\alpha_i} \in \mathbb{L}$, writing*

$$T_j := \{\tau \in \mathbf{N}_<(\lambda_{\alpha_i}(q_i)) : \tau < \upsilon\} \setminus \{1\},$$

the following hold:

(1) for each $f, g \in Q$,

$$\lambda_j(fg) = \delta_j(fg)(\alpha_i) = f(\alpha_i)\delta_j(g)(\alpha_i) + \sum_{\tau\in T_j} D(\tau)(f)\sigma_\tau(\delta_j)(g)(\alpha_i)$$

$$= f(\alpha_i)\lambda_j(g) + \sum_{\tau\in T_j} M(\tau)(f)\sigma_\tau(\ell_{\upsilon\alpha_i})\lambda_{\alpha_i}(g)$$

$$= f(\alpha_i)\lambda_j(g) + \sum_{x=n_i}^{j-1} c_x\lambda_x(g)$$

$$= f(\alpha_i)\delta_j(g)(\alpha_i) + \sum_{x=n_i}^{j-1} c_x\delta_x(g)(\alpha_i)$$

for suitable $c_x \in K$;
(2) if $f = Z_\iota$ then, for each $g \in Q$,

$$\lambda_j(Z_\iota g) = \delta_j(Z_\iota g)(\alpha_i) = a_\iota^{(i)}\delta_j(g)(\alpha_i) + \sigma_{Z_\iota}(\delta_j)(g)(\alpha_i)$$

$$= a_\iota^{(i)}\lambda_j(g) + \sigma_{Z_\iota}(\ell_{\upsilon\alpha_i})\lambda_{\alpha_i}(g);$$

(3) if $\lambda \in L_i$ is such that $\lambda(Z_\iota g) = a_\iota^{(i)}\lambda(g)$ for each $g \in Q$ and each ι, $1 \le \iota \le r$, we have $\lambda = \lambda_{\alpha_i}$.

Proof.

(1) This is a direct consequence of the Leibnitz formula (Lemma 40.4.2 and Corollary 40.4.3) and of the remark that

$$\sigma_\tau(\ell_{\upsilon\alpha_i}) \in \mathrm{Span}_K\{\ell_{\varsigma\alpha_i} : \deg(\varsigma) < \deg(\upsilon)\} \quad \text{for each } \tau, \upsilon.$$

(2) Consider $D(\tau)(Z_\iota) = M(\tau)(Z_\iota) = \begin{cases} 1 & \text{if } \tau = Z_\iota, \\ 0 & \text{if } \tau \neq Z_\iota. \end{cases}$

(3) For $\lambda = \ell\lambda_{\alpha_i}$, $\ell = \sum_{\tau\in\mathcal{W}} c(\tau, \ell)M(\tau) \in \mathrm{Span}_K(\mathbb{M})$, the assumption is equivalent to $\sigma_{Z_\iota}(\ell) = 0$ for each ι and, in turn, to

$$c(\tau, \ell) \neq 0 \implies Z_\iota \nmid \tau \text{ for each } \iota;$$

that is $\ell = M(1) = \ell_{1\alpha_i}$.   ◎

Corollary 40.4.5 (Möller). *For $A_f = M([f], \mathbf{b})$, we have the following.*

(1) *For each $l, 1 \leq l \leq s, i, 1 \leq i \leq \mathsf{s}$, and $j \in N_i, \lambda_j := \ell_{v\alpha_i}\lambda_{\alpha_i}$ satisfies on the one hand*

$$A_f\lambda_j(b_l) = A_f\delta_j(b_l)(\alpha_i) = f(\alpha_i)\delta_j(b_l)(\alpha_i) + \sum_{x=n_i}^{j-1} c_x\delta_x(b_l)(\alpha_i)$$

$$= f(\alpha_i)\lambda_j(b_l) + \sum_{x=n_i}^{j-1} c_x\lambda_x(b_l)$$

for suitable $c_x \in K$.

(2) *In particular, for each i,*

$$A_f\lambda_{n_i}(b_l) = A_f\delta_{n_i}(b_l)(\alpha_i) = f(\alpha_i)\delta_{n_i}(b_l)(\alpha_i) = f(\alpha_i)\lambda_{n_i}(b_l), \quad 1 \leq l \leq s.$$

(3) *On the other hand, for each $i, 1 \leq i \leq \mathsf{s}$, and each $j \in N_i, j \neq n_i$, there are at least one $l, 1 \leq l \leq s$, and an $\iota, 1 \leq \iota \leq r$, for which*

$$A_\iota\lambda_j(b_l) \neq a_\iota^{(i)}\lambda_j(b_l).$$

Proof.

(1) This follows by applying λ_j to each equation $\sum_i \gamma(fb_l, b_i, \mathbf{b})[b_i] = A_f[b_l] = [fb_l]$ and expanding $\lambda_j(fb_l)$ via Lemma 40.4.4(1).
(2) This is the special case $j = n_i$, for which the summation is empty.
(3) This follows directly from Lemma 40.4.4(3). $\qquad\qquad\qquad\qquad\qquad\boxdot$

Now denote, for each $\rho \in \mathbb{N}$,

$$\nabla_\rho := \operatorname{Span}_K (M(\tau), \deg(\tau) \leq \rho) \quad \text{and} \quad \Delta_\rho := \operatorname{Span}_K (D(\tau), \deg(\tau) \leq \rho)$$

and set $\nabla := \nabla_1 \setminus \nabla_0 = \operatorname{Span}_K (M(Z_h), 1 \leq h \leq r)$ and

$$\Delta := \operatorname{Span}_K (D(Z_h), 1 \leq h \leq r) = \operatorname{Span}_K \left(\frac{\partial}{\partial Z_h}, 1 \leq h \leq r\right).$$

Proposition 40.4.6 (Möller–Stetter). *The following hold:*

(1) *for $\delta := \sum_{h=1}^r a_h \frac{\partial}{\partial Z_h} \in \Delta$ and each linear form $g = \sum_{h=1}^r b_h Z_h \in \mathcal{Q}$,*

$$\delta(g) = \operatorname{ev}(\delta)(g) = \sum_{h=1}^r a_h b_h \quad \text{and} \quad \operatorname{ev}(\delta) = \sum_{h=1}^r a_h M(Z_i) \in \nabla;$$

(2) *for each $i, 1 \leq i \leq \mathsf{s}, \ell \in L_i \cap \nabla, \delta = \operatorname{ev}(\ell)$ and $g \in \mathcal{Q}$,*

$$A_f\ell\lambda_{\alpha_i}(g) = A_f\delta(g)(\alpha_i) = f(\alpha_i)\ell\lambda_{\alpha_i}(g) + \ell\lambda_{\alpha_i}(f)g(\alpha_i)$$

$$= f(\alpha_i)\delta(g)(\alpha_i) + \delta(f)(\alpha_i)g(\alpha_i);$$

(3) *for each $i, 1 \leq i \leq \mathsf{s}$, if $\dim(L_i \cap \nabla) > 1$ then there are $\ell \in L_i$ and $\delta = \operatorname{ev}(\ell) \in \Delta$ which satisfy, for each $g \in \mathcal{Q}$,*

$$A_f\ell\lambda_{\alpha_i}(g) = A_f\delta(g)(\alpha_i) = f(\alpha_i)\ell\lambda_{\alpha_i}(g) = f(\alpha_i)\delta(g)(\alpha_i);$$

(4) *for each $i, 1 \leq i \leq$ s, if $\dim(L_i \cap V) = 1$ then $\delta_{n_i+j} \in \Delta_j$ for each $j, 1 \leq j < s_i$;*

(5) *for each $i, 1 \leq i \leq$ s, and each $j, 1 \leq j < s_i$, if $\dim(L_i \cap V) = 1$ and $\deg(\tau) = x$ then $\sigma_\tau(\delta_{n_i+j}) = \delta_{n_i+j-x}$.*

Proof.

(1) This is trivial.

(2) This is proved by a direct application of Lemma 40.4.4(2).

(3) Let us consider two linearly independent elements $\ell_1, \ell_2 \in L_i$ and set $b_\iota := \ell_\iota \lambda_{\alpha_i}(f), \iota \in \{1, 2\}$. If $b_\iota = 0$ then ℓ_ι satisfies the required formula. If $b_1 \neq 0 \neq b_2$ then $\ell := b_2 \ell_1 - b_1 \ell_2$ satisfies

$$\ell \lambda_{\alpha_i}(f) = b_2 \ell_1 \lambda_{\alpha_i}(f) - b_1 \ell_2 \lambda_{\alpha_i}(f) = 0;$$

hence $A_f \ell \lambda_{\alpha_i}(g) = f(\alpha_i)\ell \lambda_{\alpha_i}(g)$ for each $g \in \mathcal{Q}$.

(4) Let $\prec$ be any degree-compatible term ordering. Then, for each $\ell \in \mathrm{Span}_K(\mathbb{M})$,

$$\ell \in \nabla_\rho \setminus \nabla_{\rho-1} \iff \deg(\mathbf{L}_\prec(\ell)) = \rho.$$

Since $\mathbf{L}_\prec(L_i)$ is an ordered ideal, if, for some $\rho \in \mathbb{N}$, $\dim(L_i \cap \nabla_\rho) > 1$ then $\dim(L_i \cap \nabla) > 1$.

(5) is a direct consequence of (4). $\qquad\boxed{\odot}$

Corollary 40.4.7 (Möller–Stetter). *For each $i, 1 \leq i \leq$ s, $j, 1 \leq j < s_i$, and each $f, g \in \mathcal{Q}$, if $\dim(L_i \cap V) = 1$ then it holds that*

$$\lambda_{n_i+j}(fg) = \delta_{n_i+j}(fg)(\alpha_i)$$

$$= f(\alpha_i)\lambda_{n_i+j}(g) + \sum_{x=0}^{j-1} \lambda_{n_i+j-x}(f)\lambda_{n_i+x}(g)$$

$$= f(\alpha_i)\delta_{n_i+j}(g)(\alpha_i) + \sum_{x=0}^{j-1} \delta_{n_i+j-x}(f)(\alpha_i)\delta_{n_i+x}(g)(\alpha_i).$$

Proof. This follows from a reformulation of Lemma 40.4.4(1) via Proposition 40.4.6(5). $\qquad\boxed{\odot}$

40.5 Möller–Stetter: The General Case

Let us now consider the general case in which J is not radical and some roots are not simple.

With the notation of Sections 40.2 and 40.4 let us now also set:

- $\mathcal{Z}(\mathsf{J}) := \{\alpha_1, \dots, \alpha_\mathsf{s}\} \subset K^r, \ \alpha_i = (a_1^{(i)}, \dots, a_r^{(i)}), \mathsf{s} \leq s$;
- for each $i, 1 \leq i \leq r$,

- o $\lambda_{\alpha_i} : \mathcal{Q} \to \mathcal{Q}$, the translation $\lambda_{\alpha_i}(Z_j) = Z_j + a_j^{(i)}$, for each j,
  - o $\mathfrak{m}_{\alpha_i} = (Z_1 - a_1^{(i)}, \dots, Z_r - a_r^{(i)})$, the maximal ideal at α_i,
  - o $\mathfrak{q}_i$, the $\mathfrak{m}_{\alpha_i}$-primary component of J, so that $\mathsf{J} = \cap_{i=1}^s \mathfrak{q}_i$,
  - o $L_i := \mathcal{L}(\lambda_{\alpha_i}(\mathfrak{q}_i)) \subset \mathrm{Span}_K(\mathbb{M})$,
  - o $s_i := \mathrm{mult}(\alpha_i, \mathsf{J}) = \mathrm{deg}(\mathfrak{q}_i) = \dim_K(L_i)$, the multiplicity in J of α_i, so that $s = \sum_{i=1}^s s_i$,
  - o $n_{i+1} := 1 + \sum_{l=1}^i s_l$, so that $n_1 = 1$ and $n_{r+1} = s + 1$,
  - o $N_i := \{n_i, \dots, n_{i+1} - 1\}$;
- $\mathbb{L} := \{\lambda_1, \dots, \lambda_s\}$, the basis biorthogonal to $\mathbf{q}$ defined in Section 40.4, so that in particular, for each i, $L_i = \mathrm{Span}_K\{\lambda_j : j \in N_i\}$, λ_{n_i} is the evaluation at α_i and $\lambda_i(q_j) = \delta_{ij}$ for each i, j;
- for each i and each $h \in N_i$, the element $\delta_h \in \mathbb{D}$ such that $\lambda_h(\cdot) = \delta_h(\cdot)(\alpha_i)$;
- $v_{lj}(\mathbf{b}) := \lambda_j(b_l) = \delta_j(b_l)(\alpha_i)$, $1 \le l, j \le s$, $j \in N_i$;
- $v_j(\mathbf{b}) = (\lambda_j(b_1), \dots, \lambda_j(b_s))^T = (\delta_j(b_1)(\alpha_i), \dots, \delta_j(b_l)(\alpha_i))^T$, $1 \le j \le s$, $j \in N_i$;
- $U(\mathbf{b})$, the $s \times s$ matrix $U := (v_{lj}(\mathbf{b}))$.

Proposition 40.5.1 (Möller–Stetter). *Under the above notation it holds that:*

(1) $U(\mathbf{b}) = M_{bq}$;

(2) *each matrix* $A_f = M([f], \mathbf{b})$ *satisfies the relation* $A_f U(\mathbf{b}) = U(\mathbf{b}) Q_f$;

(3) *each matrix* $Q_f = M([f], \mathbf{q}) = (q_{lj})$ *is a block diagonal matrix where the* i*th diagonal block,* $1 \le i \le s$*, is an upper-triangular* $s_i \times s_i$ *matrix* U_i *whose diagonal entries are* $f(\alpha_i)$ *and which covers the rows and columns indexed by* $N_i := \{n_i, \dots, n_{i+1} - 1\}$ *for each* i*,* $1 \le i \le s$;

(4) *in particular, under the assumption* $\dim(L_i \cap \nabla) = 1$*, it holds that*

$$
q_{\iota\kappa} = \begin{cases} \delta_{\iota-\kappa}(f)(\alpha_i) & \text{if } n_i \le \iota \le \kappa < n_{i+1}, \\ 0 & \text{otherwise;} \end{cases}
$$

(5) *for each* i*,* $1 \le i \le s$*, and each* $f \in \mathcal{Q}$*,*

$$
v_{n_i}(\mathbf{q}) = (\delta_{1n_i}, \dots, \delta_{sn_i})^T = (0, \dots, 0, 1, 0, \dots, 0)^T
$$

is an eigenvector for $f(\alpha_i)$ *of* Q_f;

(6) *for each* i*,* $1 \le i \le s$*, and each* $f \in \mathcal{Q}$*,*

$$
v_{n_i}(\mathbf{b}) = (b_1(\alpha_i), \dots, b_s(\alpha_s))^T
$$

is an eigenvector for $f(\alpha_i)$ *of* A_f;

(7) *for* $\mathbf{b} = \{[1], [Z_1], \dots [Z_{\delta-1}], [b_{\delta+1}], \dots, [b_s]\}$ *satisfying condition* (**AS.2**) *and the values* $c_{\iota l} \in K, 0 \le l < \delta \le \iota \le r$*, such that* $[Z_\iota] = c_{\iota 0} + \sum_{l=1}^{\delta-1} c_{\iota l}[Z_l]$,

it holds, for each i, $1 \leq i \leq s$, that if $(d_{i1}, \ldots, d_{is})^T$ is an eigenvector for $f(\alpha_i)$ of a non-derogatory matrix A_f, $f \in Q$, then

$$\alpha_i := \left(d_{i1}^{-1} d_{i2}, \ \ldots, \ d_{i1}^{-1} d_{i\delta}, \ c_{\delta 0} + \sum_{l=1}^{\delta-1} c_{\delta l} d_{i1}^{-1} d_{il}, \ \ldots, \ c_{n0} + \sum_{l=1}^{\delta-1} c_{nl} d_{i1}^{-1} d_{il} \right);$$

(8) $[q_{n_i}] f(\alpha_i) = [f q_{n_i}]$ *for each i, $1 \leq i \leq s$;*

(9) *for each $f \in Q$ and $\lambda \in K$, $J : (f - \lambda) = J$ iff $\lambda \notin \{f(\alpha_i) : 1 \leq i \leq s\}$;*

(10) *for each $f \in Q$, if A_f is non-derogatory then $J : (f - f(\alpha_i)) = J + \{q_{n_i}\}$ for each i, $1 \leq i \leq s$.*

Proof. (1) and (2) are trivial; (3) is a direct consequence of Corollary 40.4.5 and (4) of Corollary 40.4.7; (5) and (6) are trivial consequences of (3); (7) is a direct reformulation of (6) applied to the basis satisfying condition (**AS.2**); (8)–(10) amount to Lemma 40.2.4.   ◉

Corollary 40.5.2. *It holds that:*

- *the trace of A_f is $\mathrm{Tr}(A_f) := \sum_{i=1}^{s} s_i f(\alpha_i)$;*
- *the determinant of A_f is $\det(A_f) := \prod_{i=1}^{s} f(\alpha_i)^{s_i}$;*
- *the characteristic polynomial of A_f is $\chi_f(T) := \prod_{i=1}^{s} (T - f(\alpha_i))^{s_i}$;*
- *the minimal polynomial of A_f is $m_f(T) := \prod_{i=1}^{s} (T - f(\alpha_i))^{\rho_i}$, where ρ_i denotes the characteristic number (Definition 27.2.13) of q_i.*

Proof. All the claims are trivial except the last, which is a consequence of the facts that $q_i \supseteq m_{\alpha_i}^{\rho} \iff \mathcal{L}(\lambda_{\alpha_i}(q_i)) \subseteq \Delta_\rho$, that the multiplicity μ of the eigenvalue $f(\alpha_i)$ in the minimal polynomial is the minimal value for which $U_i = \ker\left(\Phi_f - f(\alpha_i)\right)^\mu = 0$ and that Lemma 40.4.4(1) holds.   ◉

Remark 40.5.3 (Monico). For simplicity we have assumed that $K = \mathsf{K}$ throughout this chapter, but the construction and the stated results hold also if $K \subsetneq \mathsf{K}$. Clearly, in this case, conjugate roots have the same multiplicity both in χ_f and in m_f.

Moreover, as a direct application of the Chinese Remainder Theorem, if $\chi_f := \prod_{i=1}^{r} p_{if}(T)^{r_i}$ is an irreducible factorization in $K[T]$ then

$$J = \bigcup_{i=1}^{r} \left(J + (p_{if}(f)^{r_i}) \right)$$

is an irreducible primary decomposition in $K[Z_1, \ldots, Z_r]$.

Of course, if f is an *allgemeine* coordinate then the corresponding primary decomposition algorithm is that proposed by Alonso and Raimondo and reported in Section 35.5.1.   ◉

40.6 The Univariate Case

As a short *intermezzo* before discussing derogatoriness, let us briefly show how the Auzinger–Stetter procedure reformulates the classical elementary linear algebra results for a univariate polynomial.

For a polynomial

$$f = X^s + \sum_{i=0}^{s-1} a_i X^i = \sum_{i=1}^{s} (X - \xi_i)^{s_i}, \quad s = \sum_{i=1}^{s} s_i,$$

the linear representation of $J = (f)$ comprises an assignment of the normal basis $N(J) := \{1, X, \ldots, X^{s-1}\}$ and of the Frobenius companion matrix

$$A_1 = \begin{pmatrix} 0 & 1 & 0 & \cdots & 0 \\ 0 & 0 & 1 & \cdots & 0 \\ \vdots & \vdots & \vdots & \ddots & \vdots \\ 0 & 0 & 0 & \cdots & 1 \\ -a_0 & -a_1 & -a_2 & \cdots & -a_{s-1} \end{pmatrix},$$

whose characteristic polynomial is of course f, so that the eigenvalues of A_1 coincide (up to multiplicity) with the roots of f.

Recalling that ξ_i is a root of f of multiplicity s_i iff $f^{(j)}(\xi_i) = 0$ for $0 \le j < s_i$, the natural choice for the dual space $\mathbb{L} = \{\lambda_1, \ldots, \lambda_s\}$ is

$$\lambda_{n_i}(p) := p(\xi_i), \quad \lambda_{n_i+j}(p) := \frac{p^{(j)}(\xi_i)}{j!}, \quad 1 \le j < s_i, 1 \le i \le s,$$

where we have set $n_{i+1} := 1 + \sum_{l=1}^{i} s_l, 0 \le i < s$, and the biorthogonal dual basis $\mathbf{q}$ is the associated Lagrange basis.

If f is squarefree then M_{qb} is the Vandermonde matrix,

$$M_{qb} = \begin{pmatrix} 1 & 1 & \cdots & 1 \\ \xi_1 & \xi_2 & \cdots & \xi_s \\ \vdots & \vdots & \ddots & \vdots \\ \xi_1^{s-1} & \xi_2^{s-1} & \cdots & \xi_s^{s-1} \end{pmatrix};$$

in general, each block of the so called *generalized Vandermonde matrix* has the shape

$$\begin{pmatrix} 1 & 0 & 0 & \cdots & 0 & \cdots & 0 \\ \xi_i & 1 & 0 & \cdots & 0 & \cdots & 0 \\ \xi_i^2 & 2\xi_i & 1 & \cdots & 0 & \cdots & 0 \\ \xi_i^3 & 3\xi_i^2 & 3\xi_i & \cdots & 0 & \cdots & 0 \\ \vdots & \vdots & \vdots & \ddots & \vdots & \ddots & \vdots \\ \xi_i^j & j\xi_i^{j-1} & \binom{j}{2}\xi_i^{j-2} & \cdots & 1 & \cdots & 0 \\ \vdots & \vdots & \vdots & \ddots & \vdots & \ddots & \vdots \\ \xi_i^{s-1} & (s-1)\xi_i^{s-2} & \binom{s-1}{2}\xi_i^{s-3} & \cdots & \binom{s-1}{j-1}\xi_i^{s-j} & \cdots & \binom{s-1}{s_i-1}\xi_i^{s-s_i} \end{pmatrix}$$

and is related to the ith diagonal block of Q_1, which is a classical Jordan block

$$\begin{pmatrix} \xi_i & 1 & & & & 0 \\ & \xi_i & 1 & & & \\ & & \ddots & \ddots & & \\ & & & \ddots & 1 \\ 0 & & & & \xi_i \end{pmatrix}.$$

Example 40.6.1. For

$$f = X^8 - X^7 - X^6 + 3X^5 + 9X^4 - 3X^3 - 7X^2 + X + 2$$
$$= (X - 2)(X + 1)^4(X - 1)^3,$$

we have

$$A_1 = \begin{pmatrix} 0 & 1 & 0 & 0 & 0 & 0 & 0 & 0 \\ 0 & 0 & 1 & 0 & 0 & 0 & 0 & 0 \\ 0 & 0 & 0 & 1 & 0 & 0 & 0 & 0 \\ 0 & 0 & 0 & 0 & 1 & 0 & 0 & 0 \\ 0 & 0 & 0 & 0 & 0 & 1 & 0 & 0 \\ 0 & 0 & 0 & 0 & 0 & 0 & 1 & 0 \\ 0 & 0 & 0 & 0 & 0 & 0 & 0 & 1 \\ -2 & -1 & 7 & 3 & -9 & -3 & 5 & 1 \end{pmatrix}$$

whose eigenvalues are 2 (simple), -1 (with mulitiplicity 4) and 1 (with mulitiplicity 3). The generalized Vandermonde matrix is

$$M_{qb} = \left(\begin{array}{c|cccc|ccc} 1 & 1 & 0 & 0 & 0 & 1 & 0 & 0 \\ 2 & -1 & 1 & 0 & 0 & 1 & 1 & 0 \\ 4 & 1 & -2 & 1 & 0 & 1 & 2 & 1 \\ 8 & -1 & 3 & -3 & 1 & 1 & 3 & 3 \\ 16 & 1 & -4 & 6 & -4 & 1 & 4 & 6 \\ 32 & -1 & 5 & -10 & 10 & 1 & 5 & 10 \\ 64 & 1 & -6 & 15 & -20 & 1 & 6 & 15 \\ 128 & -1 & 7 & -21 & 35 & 1 & 7 & 21 \end{array} \right)$$

and is related to the Jordan matrix

$$J_1 = \left(\begin{array}{c|cccc|ccc} 2 & & & & & & & \\ \hline & -1 & 1 & & & & & \\ & & -1 & 1 & & & & \\ & & & -1 & 1 & & & \\ & & & & -1 & & & \\ \hline & & & & & 1 & 1 & \\ & & & & & & 1 & 1 \\ & & & & & & & 1 \end{array} \right).$$

Moreover, we have

$$\mathbf{q} = (q_1, \ldots, q_s)^T = M_{qb}^{-1}(1, X, \ldots, X^{s-1})^T$$

allowing us to deduce the Lagrange basis by inverting M_{qb}.

We note that $J_1 = Q_1 = M([X], \mathbf{q})$ is the multiplication matrix of $K[X]/J$ w.r.t. the Lagrange basis $\mathbf{q}$.

Example 40.6.2.　The example above is too hard to deal with by hand, so let us restrict ourselvs to the easier case $f = X^5 - X^3 = X^3(X + 1)(X - 1)$, where we have

$$A_1 = \begin{pmatrix} 0 & 1 & 0 & 0 & 0 \\ 0 & 0 & 1 & 0 & 0 \\ 0 & 0 & 0 & 1 & 0 \\ 0 & 0 & 0 & 0 & 1 \\ 0 & 0 & 0 & 1 & 0 \end{pmatrix}, \qquad M_{qb} = \left(\begin{array}{ccc|c|c} 1 & 0 & 0 & 1 & 1 \\ 0 & 1 & 0 & -1 & 1 \\ 0 & 0 & 1 & 1 & 1 \\ 0 & 0 & 0 & -1 & 1 \\ 0 & 0 & 0 & 1 & 1 \end{array} \right)$$

$$J_1 = \left(\begin{array}{cc|c|c} 0 & 1 & & \\ & 0 & 1 & \\ & & 0 & \\ \hline & & -1 & \\ \hline & & & 1 \end{array} \right), \qquad M_{qb}^{-1} = \begin{pmatrix} 1 & 0 & 0 & 0 & -1 \\ 0 & 1 & 0 & -1 & 0 \\ 0 & 0 & 1 & 0 & -1 \\ 0 & 0 & 0 & -\frac{1}{2} & \frac{1}{2} \\ 0 & 0 & 0 & \frac{1}{2} & \frac{1}{2} \end{pmatrix}.$$

In the matrix M_{qb}^{-1} one reads, along the rows, the Lagrange basis

$$\{1 - X^4,\ X - X^3,\ X^2 - X^4,\ -\tfrac{1}{2}(X^3 - X^4),\ \tfrac{1}{2}(X^3 + X^4)\}.$$

Finally, for $q \in K[X]$, the ith diagonal block of Q_q related to the root ξ_i is

$$\begin{pmatrix} q(\xi_i) & q^{(1)}(\xi_j) & \frac{q^{(2)}(\xi_i)}{2} & \cdots & & \frac{q^{(s_i-1)}(\xi_i)}{(s_i-1)!} \\ & q(\xi_i) & q^{(1)}(\xi_j) & \frac{q^{(2)}(\xi_i)}{2} & & \vdots \\ & & \ddots & \ddots & \ddots & \vdots \\ & & & \ddots & \ddots & \frac{q^{(2)}(\xi_i)}{2} \\ & & & & \ddots & q^{(1)}(\xi_i) \\ 0 & & & & & q(\xi_i) \end{pmatrix}.$$

Example 40.6.3.　In Example 40.6.2, for $q := 1 + X^2$ we have $A_q M_{qb} = M_{qb} Q_q$ with

$$A_q = \begin{pmatrix} 1 & 0 & 1 & 0 & 0 \\ 0 & 1 & 0 & 1 & 0 \\ 0 & 0 & 1 & 0 & 1 \\ 0 & 0 & 0 & 2 & 0 \\ 0 & 0 & 0 & 0 & 2 \end{pmatrix}, \qquad Q_q = \left(\begin{array}{ccc|c|c} 1 & 0 & 1 & & \\ & 1 & 0 & & \\ & & 1 & & \\ \hline & & & 2 & \\ \hline & & & & 2 \end{array} \right).$$

40.7 Derogatoriness

The crucial assumption in Theorem 40.3.1 and Proposition 40.5.1, that Φ_f is non-derogatory, is easily met in the radical case by a generic linear form but the general case is more involved.

Example 40.7.1. Continuing Example 40.1.1 we remark that, while being *allgemeine* coordinates, neither Z_1 nor Z_2 has a derogatory matrix.

However, $f = Z_1 - Z_2$ is non-derogatory; in fact, with

$$A_f = \begin{pmatrix} 0 & 1 & -1 & 0 & 0 \\ 1 & -1 & -1 & 1 & 0 \\ -1 & 1 & 1 & 0 & -1 \\ 1 & 0 & -1 & 0 & 0 \\ -2 & 1 & 0 & 1 & 1 \end{pmatrix} \quad \text{and} \quad J := \left(\begin{array}{cc|cc} -1 & 1 & & \\ & -1 & & \\ \hline & & 1 & -1 \\ & & & 1 & -1 \\ & & & & 1 \end{array} \right),$$

we have $M_{bq} J = A_f M_{bq}$.

The reason why f is a good choice will be explained in Corollary 40.13.6 below.

In general the commuting family $\{A_f : f \in \mathcal{Q}\}$ does not possess any non-derogatory matrix, as can be seen in the following, trivial, examples:

Example 40.7.2. Let us consider the ideal

$$J = (Z_1, Z_2)^2 = (Z_1^2, Z_1 Z_2, Z_2^2) \in K[Z_1, Z_2].$$

For a generic $[f] := [a + bZ_1 + cZ_2]$, the matrix Φ_f is represented, via the basis $\{1, Z_1, Z_2\}$, as

$$\Phi_f = \begin{pmatrix} a & b & c \\ 0 & a & 0 \\ 0 & 0 & a \end{pmatrix};$$

it has the single eigenvalue a with multiplicity 3 and the eigenspace $\mathrm{Span}\{(1, 0, 0)^T, (0, c, -b)^T\}$, except in the trivial case $b = c = 0$.

Example 40.7.3. In order to dispel the impression that the bad behaviour of the example above could be overcome by the reducibility of J, we repeat the same argument for the irreducible primary ideal

$$J = (Z_1^2, Z_2^2) \in K[Z_1, Z_2].$$

For a generic $[f] := [a + bZ_1 + cZ_2 + dZ_1 Z_2]$, the matrix Φ_f is represented, via the basis $\{1, Z_1, Z_2, Z_1 Z_2\}$, as

$$\Phi_f = \begin{pmatrix} a & b & c & d \\ 0 & a & 0 & c \\ 0 & 0 & a & b \\ 0 & 0 & 0 & a \end{pmatrix}.$$

The single eigenvalue is a, with multiplicity 4, and

- if $b^2 + c^2 \neq 0$, the eigenspace is $\mathrm{Span}\{(1, 0, 0, 0)^T, (0, c, -b, 0)^T\}$,
- if $b = 0 = c, d \neq 0$, the eigenspace is $\mathrm{Span}\{(1, 0, 0, 0)^T, (0, 1, 0, 0)^T, (0, 0, 1, 0)^T\}$. ◎

We remark that, in each of the above examples, the eigenspaces share an eigenvector: $(1, 0, 0)^T$ and $(1, 0, 0, 0)^T$ respectively.

Example 40.7.4.   The same can be said for Example 40.1.1, where $(1, 0, 1, 0, 1)^T$ is an eigenvector of A_1 for the eigenvalue 0 and of A_2 for the eigenvalue 1, while $(1, 1, 0, 1, 0)^T$ is an eigenvector of A_1 for 1 and of A_2 for 0; in general it is easy to verify that, for each $f \in \mathcal{Q}$, $(1, 0, 1, 0, 1)^T$ is an eigenvector of A_f for $f(0, 1)$ and $(1, 1, 0, 1, 0)^T$ is an eigenvector of A_f for $f(1, 0)$. ◎

The fact that a whole family shares *at least* the eigenvectors

$$\{v_{n_i}(\mathbf{b}), 1 \leq i \leq r\}$$

is already granted by Proposition 40.5.1(5). But there is something more: the set $\{A_i, 1 \leq i \leq n\}$ and so, *a fortiori*, $\{A_f, f \in \mathcal{Q}\}$ cannot share any further eigenvector, as a consequence of Corollary 40.4.5(3).

In other words, in all these examples, each non-trivial common eigenspace for the whole family has dimension 1.

Definition 40.7.5.   *A commuting family of matrices $\mathfrak{A}$ is called* non-derogatory *if each joint eigenspace has dimension at most 1:*

$$\dim_K\{v \in \mathbf{K}^s : Av = \lambda v, Bv = \mu v\} \leq 1 \quad \textit{for each } \lambda, \mu \in K, A, B \in \mathfrak{A}.$$

A zero-dimensional ideal $\mathbf{J} \subset \mathcal{Q}$ is called a non-derogatory *ideal if there is an endomorphism $\Phi_f : \mathbf{A} \to \mathbf{A}$ for which the matrix A_f is non-derogatory.* ◎

Corollary 40.7.6.   *The family $\{\Phi_\iota : 1 \leq \iota \leq n\}$ is non-derogatory and*

$$v_{n_i}(\mathbf{b}) := (b_1(\alpha_i), \ldots, b_s(\alpha_i))^T, \quad i = 1, \ldots, r$$

are joint eigenvectors of all the matrices $M([f], \mathbf{b})$, with associated eigenvalue $f(\alpha_i)$.

Proof.   This is a direct consequence of Corollary 40.4.5(3). ◎

As a consequence, once the eigenspaces of a matrix A_Y, $Y = \sum_{j=1}^r c_j Z_j$, are obtained, if some eigenspaces have dimension greater than 1^4 then one performs the

---

[4] The possible reasons are two, as follows.

(1) Either Y is not sufficiently generic and does not satisfy condition (**AS.1**); in the next steps the variables Y_i will separate the roots via the eigenspace intersection method.

(2) The ideal is not radical and, even if Y satisfies condition (**AS.1**), the eigenspace corresponding to $\sum_{j=1}^r c_j a_j^{(i)}$ for A_Y has dimension greater then 1.

The eigenspace intersection method covers this case also, thanks to Corollary 40.4.5. Alternatively, the multiplicity of the roots can be decreased by a proper application of Gianni's Proposition (Proposition 35.6.1), e.g. enlarging the ideal $\mathbf{J}$ to $\mathbf{J} + (\sqrt{g(Y)})$ where $g(A_Y)$ is the characteristic polynomial of A_Y (see Remark 40.8.1 below).

same computation for different matrices A_{Y_i}, $Y_i = \sum_{j=1}^r c_{ij} Z_j$, $i = 2, \ldots, n$, since they are linearly independent forms. Next, one repeatedly applies the *eigenspace intersection method*, based on a direct application of Lemma 40.7.7 below, to repeatedly compute eigenspace intersections until each eigenspace has dimension 1.

Lemma 40.7.7. *Let M, N be two s-square matrices, λ an eigenvalue of M with associated eigenspace U and $\{u_1, \ldots, u_l\}$ an orthonormal basis of U.*

Define, for each i, j, $1 \le i, j \le l$, $a_{ij} := u_i^T N u_j$ and set $A := (a_{ij})$.

If $(d_1, \ldots, d_l)^T$ is any eigenvector of A for μ then $u := \sum_j d_j u_j$ is a simultaneous eigenvector of M for λ and of N for μ. $\quad\boxed{\odot}$

Proof.   We have that
$$u \text{ is an eigenvector to } \mu \text{ for } N$$
$$\Longleftrightarrow \mu \sum_j d_j u_j = \mu u = Nu = N \sum_j d_j u_j = \sum_j d_j N u_j$$
$$\Longleftrightarrow \mu d_i = u_i^T \mu \sum_j d_j u_j = \sum_j d_j u_i^T N u_j = \sum_j a_{ij} d_j \text{ for each } i$$
$$\Longleftrightarrow (d_1, \ldots, d_l)^T \text{ is an eigenvector of } A \text{ for } \mu. \qquad\boxed{\odot}$$

40.8 Stetter Algorithm via Grobnerian Technology

We can assume that the zero-dimensional ideal J is given by means of the Gröbner basis[5] w.r.t. an ordering $<$; thus we obtain also the linear representation

$$\mathbf{N}_<(\mathsf{J}) = \{\tau_1, \ldots, \tau_s\}, M([Z_h], \mathbf{N}_<(\mathsf{J})),$$

allowing us to compute, with good complexity, the corresponding Gröbner description (compare Definition 29.3.3),

$$\mathbf{Rep}(g, \mathbf{N}_<(\mathsf{J})) := (\gamma(g, \tau_1, \mathbf{N}_<(\mathsf{J})), \ldots, \gamma(g, \tau_s, \mathbf{N}_<(\mathsf{J}))) \in K^s,$$

$$[g] = \sum_j \gamma(g, \tau_j, \mathbf{N}_<(\mathsf{J}))[\tau_j],$$

for each $g \in \mathcal{Q}$.

Remark 40.8.1.   Since the Stetter Algorithm is improved if J is radical and the matrix A_Y is given w.r.t. a linear form Y satisfying condition (**AS.1**), these results can be found efficiently, $\mathcal{O}(n^2 s^3)$, from an FGLM-like linear algebra version of Gianni's proposition 35.6.1 obtained by merging the algorithms of Alonso and Raimondo (Algorithm 35.7.1) and Traverso (Algorithm 29.3.8) as follows:[6]

---

[5] An alternative, Gröbner-free, approach with good complexity for affine complete intersection ideals and which gives both a Gröbner representation of J and the corresponding Gröbner description of g will be discussed in Section 41.15.

    The discussion in this section does not require that the obtained representation is *linear*: Algorithm 35.7.1 applies equally to the data of Section 41.15.

[6] An important improvement of this algorithm will be discussed in Section 42.9.

(1) set $\ell := \sum_i a_i Z_i$;

(2) by linear algebra on the Gröbner descriptions of

$$[1], [\ell], [\ell^2], \ldots, [\ell^s]$$

compute the minimal polynomial $g_0[Y] \in K[Y]$ such that

$$g_0(Y) \in \mathsf{J}^+ := \mathsf{J} + \left(Y - \sum_i a_i Z_i\right);$$

(3) if g_0 is not squarefree,

    (a) set $g_0 := \sqrt{g_0}$, $d := \deg(g_0)$ and $\mathsf{J} := \mathsf{J} + (g_0(\ell))$;

    (b) compute, by linear algebra using Traverso's Algorithm the Gröbner basis of J and deduce the corresponding linear representation and Gröbner descriptions and the value $\deg \mathsf{J}$;

(4) if $d := \deg(g_0) < \deg(\mathsf{J})$, which is equivalent to the existence of a value i, $1 < i \le r$ for which $[Z_i], [1], [\ell], [\ell^2], \ldots, [\ell^{d-1}]$ are linearly independent and $Z_i \in \mathbf{N}_<(\mathsf{J})$, then

    (a) read in $\mathbf{N}_<(\mathsf{J})$, the minimal such value i,

    (b) set $\ell := \ell + cZ_i$, $a_i := a_i + c$ and go to (2);

(5) if g_0 is squarefree and $\deg(g_0) = \deg(\mathsf{J})$ then

    • J is radical,

    • $\ell := \sum_i a_i Z_i$ is a separating linear form, thus satisfying condition (**AS.1**).

Note that, for each i, $1 \le i \le r$, linear algebra on the natural descriptions of

$$[Z_i], [1], [\ell], [\ell^2], \ldots, [\ell^{d-1}]$$

gives the relation $Z_i - h_i(Y) \in \mathsf{J}^+$, $\deg(h_i) < d$, so that

$$[Z_i] = [h_i(\ell)] \text{ for } i = 1, \ldots, r. \qquad \boxed{\odot}$$

We can therefore assume that we have a linear form $Y = \sum_i a_i Z_i$ satisfying condition (**AS.1**) and a *radical* zero-dimensional ideal J,[7] which is given by means of the Gröbner basis w.r.t. $<$ and via the linear representation

$$\mathbf{N}_<(\mathsf{J}) = \{\tau_1, \ldots, \tau_s\}, \quad M([Z_h], \mathbf{N}_<(\mathsf{J})),$$

thus allowing us to compute the Gröbner description $\mathbf{Rep}(g, \mathbf{N}_<(\mathsf{J}))$ for each $g \in \mathcal{Q}$. Thus, by linear algebra on the Gröbner representations of

$$[1], [Z_1], \ldots, [Z_r],$$

one can obtain with complexity $\mathcal{O}(ns^2)$ both the K-basis $\{[1], [Z_1], \ldots, [Z_{\delta-1}]\}$ of V and the linear representations $[Z_i] = c_{i0} + \sum_{l=1}^{\delta-1} c_{il}[Z_l]$, $i \ge \delta$; further, linear algebra on the Gröbner representation extends this set to a basis

---

[7] Notwithstanding these assumptions, A_Y is not necessarily non-derogatory; as is shown by Examples 40.7.2 and 40.7.3 and explained in Corollary 40.13.6, this requires that each primary $\mathfrak{q}_i$ of J has a good shape.

$$\mathbf{b} = \{1, Z_1, \ldots, Z_{\delta-1}, b_{\delta+1}, \ldots, b_s\} = \{b_1, \ldots, b_s\}$$

that satisfies condition (**AS.2**).

If now we denote by $\mathbb{L} := \{\ell_1, \ldots, \ell_s\}$ the set of the functionals $\ell_i(\cdot) := \gamma(\cdot, b_i, \mathbf{b})$ such that

$$[g] = \ell_1(g) + \sum_{i=2}^{\delta} \ell_i(g)[Z_{i-1}] + \sum_{i=\delta+1}^{s} \ell_i(g)[b_i], \quad \forall g \in \mathcal{Q},$$

then $\mathbb{L}$ is biorthogonal to $\mathbf{b}$; therefore it is sufficient simply to adapt the Enhanced Möller Algorithm (Figure 29.4) to obtain, among the other information provided by that algorithm, the matrices $M([Z_i], \mathbf{b})$, $1 \leq i \leq n$. This is all one needs to obtain the matrix $M([Y], \mathbf{b}) = \sum_i a_i M([Z_i], \mathbf{b})$.

Once $M([Y], \mathbf{b})$ is obtained, the eigenvalue and eigenspace computation is dealt with by numerical analysis and the joint eigenvectors are obtained, if needed, via the eigenspace intersection method (Lemma 40.7.7).

40.9 Stetter Algorithm

The numerical analysis aspects of an efficient solution of the eigenproblem are beyond my competence,[8] so I limit myself to noting that efficiency is measured by the *condition number*

$$\kappa(M_{bq}) = \|M_{bq}\| \|M_{bq}^{-1}\| = \|M_{bq}\| \|M_{qb}\|$$

and is therefore influenced by the choice of the basis $\mathbf{b}$; in general, κ becomes large if the column vectors $v_j(\mathbf{b}) = (\lambda_j(b_1), \ldots, \lambda_j(b_s))^T$ are nearly linearly dependent (the near linear dependency of the rows naturally has the same effect).

It is interesting to remark that, for a choice $\mathbf{b} := \mathbf{N}_{\prec}(\mathsf{J})$ w.r.t. a suitable term ordering $\prec$, not surprisingly the choice of a degree-compatible ordering gives a better condition number than a lexicographical ordering.

40.10 Lundqvist: Analysis and Improvements
of Möller's Algorithm

On the basis of Remark 40.3.4, a better analysis of the complexity of Möller's Algorithm (Figures 28.2 and 29.4) can be performed.

Remark 40.10.1 (Lundqvist).   In particular, the analysis performed in Remark 29.2.5, which correctly stated that we have to perform $\mathcal{O}(s^2)$ operations and s evaluations of functionals for at most ns terms,[9] should be refined by stating that we have to deal with at most $\#\mathbf{N} + \#G = s + \#G$ terms and by giving a more precise value for $\#G$, using the fact that, in the notation of Remark 40.3.4,

---

[8] For that, see Stetter H., *Numerical Polynomial Algebra*, SIAM (2004).
[9] Here $s = \#\mathsf{J}$ and n is the number of variables.

$$G = \{\ell_\delta, \ldots, \ell_n\} \sqcup G', \quad \mathbf{T}(\ell_i) = Z_i, \mathbf{T}(G')$$
$$= \mathbf{T}(G) \cap k[Z_1, \ldots, Z_{\delta-1}] \subset \mathcal{T} \cap k[Z_1, \ldots, Z_{\delta-1}] =: \mathcal{U}. \tag{40.3}$$

Since #$\mathbf{N}$ contains $\delta - 1 \leq s - 1$ variables, we have the relations

$$\delta - 1 < \min(s - 1, n) \quad \text{and} \quad n - \min(s - 1, n) \leq n - \delta + 1 \leq n.$$

Moreover, in each step in which we add a new term $t \in \mathbf{N}$ and enlarge $\mathbf{B}$ instead of performing the merging $\mathbf{B} := \mathbf{B} \cup \{X_h t : 1 \leq h \leq n\}$, we perform the shorter merging $\mathbf{B} := \mathbf{B} \cup \{Z_h t : 1 \leq h \leq \delta - 1\}$.

Since the algorithm adds s elements to $\mathbf{N}$, the evaluation of the terms in $\mathbf{T}(G')$ which are not variables naturally involves at most $(\delta - 1)s$ terms. Thus #$\mathbf{N}$ + #G = $(\delta - 1)s + n - \delta + 1 < s \min(s - 1, n) + n$.

As a consequence, the complexity of the part of Algorithm 28.2.7 which evaluates and Gauss-reduces the vectors can be evaluated as

$$\mathcal{O}\left(ns^2 + \min(s - 1, n)s^3\right).$$

The advantage over $\mathcal{O}(ns^3)$ becomes relevant for applications in which $s \ll n$.

$$\boxed{\odot}$$

This suggests that we should reconsider the part of the algorithm which enlarges and reorders $\mathbf{B}$ each time a new element t is added to $\mathbf{N}$.

The analysis presented in Remark 29.2.5 gave the evaluation $\mathcal{O}(n^2 s^2)$ as the result of s = #$\mathbf{N}$ times merging in an orderly way the ordered list $\mathbf{B}$ of at most $n(s - 1)$ elements with the ordered list $\{X_h t : 1 \leq h \leq n\}$, assuming that each of the $n(s - 1) + n$ comparisons of elements requires n values to be compared. Such an analysis, which is correct for lexigraphical ordering, was wlog extended to a generic term ordering since the requirement was to compare the values of the weight functions which characterize the term orderings, according to Erdős' characterization (Corollary 24.9.5).

The computational analysis, as throughout the book, required merely the counting of the arithmetical complexity; a better analysis would also take into account the size of the elements considered. Thus, associating with each integer $a \in \mathbb{Z}$ the value of its bits as

$$\text{numbits}(a) = \begin{cases} \lfloor \log_2(|a|) \rfloor + 2 & \text{if } a \neq 0, \\ 2 & \text{if } a = 0, \end{cases}$$

we have

Lemma 40.10.2 (Lundqvist). *For $\alpha = (a_1, \ldots, a_n) \in \mathbb{Z}^n$ with $m = \sum_i |a_i|$ it holds that*

$$\text{numbits}(\alpha) := \sum_i \text{numbits}(a_i) \leq \begin{cases} n \log_2\left(\frac{m}{n}\right) + 2n & \text{if } m > 2n, \\ 3n & \text{if } m \leq 2n. \end{cases}$$

Proof. Let k be the number of non-zero entries and wlog assume that they are the first entries; so, since $\prod_{i=1}^{k} a_i \le (m/k)^k$, we have

$$\text{numbits}(\alpha) = \sum_{i=1}^{k} \text{numbits}(a_i) + 2(n - k)$$

$$= \sum_{i=1}^{k} \left\lfloor \log_2\left(|a_i|\right) \right\rfloor + 2n$$

$$\le \log_2 \left(\left| \prod_{i=1}^{k} a_i \right| \right) + 2n$$

$$\le \log_2 \left(\left(\frac{m}{k}\right)^k \right) + 2n$$

$$\le k \log_2 \left(\frac{m}{k}\right) + 2n.$$

Moreover, since $x := m/2$ is the maximum of the fuction $f(x) := x \log_2(m/x) + 2n$ and f is a monotone increasing function on $[1, m/2]$ we have

- if $m > 2n$ then $\text{numbits}(\alpha) \le f(n) = n \log_2(m/n) + 2n$, while
- if $m \le 2n$ then $\text{numbits}(\alpha) \le f(m/2) = m/2 + 2n \le 3n$.

$\qquad\qquad\qquad\qquad\qquad\qquad\qquad\qquad\qquad\qquad\qquad\qquad\qquad\qquad$ ◉

Corollary 40.10.3 (Lundqvist). *Let $\alpha = (a_1, \ldots, a_n)$, $\beta = (b_1, \ldots, b_n)$ be vectors of integers and set*

$$m := \max\left(\sum_i |a_i|, \sum_i |b_i| \right).$$

With time complexity $\mathcal{O}(n \max(\log(m/n)), 1)$ we can determine the first index, if any, at which α and β differ.

Proof. We need to iteratively compare a_i and b_i, by increasing the value of i until either $a_i \ne b_i$ or $\alpha = \beta$. This requires us to compare at most all entries, that is,

$$\text{numbits}(\alpha) + \text{numbits}(\beta) \le 2 \max\left(\sum_i \text{numbits}(a_i), \sum_i \text{numbits}(b_i) \right)$$

$$\le 2n \left(\log_2\left(\frac{m}{n}\right) + 2 \right).$$

Thus the time complexity is proportional to $\mathcal{O}(n \max(\log(m/n)), 1)$. $\qquad$ ◉

Given two n-tuples of elements in an ordered set Σ, $\alpha = (a_1, \ldots, a_n) \in \Sigma^n$ and $\beta = (b_1, \ldots, b_n) \in \Sigma^n$, denote as $\Delta(\alpha, \beta)$ the first index where α and β differ, setting $\Delta(\alpha, \beta) = n + 1$ when $\alpha = \beta$.

Lemma 40.10.4 (Lundqvist). *For $\alpha = (a_1, \ldots, a_n), \beta = (b_1, \ldots, b_n), \gamma = (c_1, \ldots, c_n)$ it holds that:*

(1) $\alpha < \beta, \alpha < \gamma, \Delta(\alpha, \gamma) < \Delta(\alpha, \beta) \implies \alpha < \beta < \gamma$ *and* $\Delta(\beta, \gamma) = \Delta(\alpha, \gamma)$;

(2) $\alpha < \beta, \alpha < \gamma \implies \Delta(\beta, \gamma) \geq \min(\Delta(\alpha, \beta), \Delta(\alpha, \gamma))$.

Proof.

(1) We have $b_i = a_i = c_i$ for $i < j := \Delta(\alpha, \gamma)$ and $b_j = a_j \neq c_j$.

(2) The claim is a trivial consequence of (1) when $\Delta(\alpha, \beta) \neq \Delta(\alpha, \gamma)$. Otherwise $\Delta(\alpha, \beta) = \Delta(\alpha, \gamma) =: j$ would imply that β and γ agree at least in the first j positions.

$\qquad\qquad\qquad\qquad\qquad\qquad\qquad\qquad\qquad\qquad\qquad\qquad\qquad$ $\boxed{\odot}$

Proposition 40.10.5 (Lundqvist).   *Let* $(\alpha_1, \ldots, \alpha_t)$ *be an ordered list of n-tuples in* Σ^n, *lexicographically sorted in increasing order, and assume that we are given* $\Delta(\alpha_i, \alpha_{i+1})$ *for each* i, $1 \leq i < t$.

For any $\beta = (b_1, \ldots, b_n) \in \Sigma^n$ *it is possible, using* $\mathcal{O}(t + n)$ *comparisons in* Σ *and a time proportional to* $t\log(n)$, *to find an index* i, $0 \leq i \leq t$, *such that*

$$\alpha_1 \leq \cdots \leq \alpha_i < \beta \leq \alpha_{i+1} \leq \cdots \leq \alpha_t$$

and to find $\Delta(\alpha_i, \beta)$ *(when* $i \neq 0$*) and* $\Delta(\beta, \alpha_{i+1})$ *(when* $i \neq t$*).*
   When

Σ *is the set of non-negative integers,*
$\sum_{j=1}^n b_j \leq m$ *and*
$\sum_{j=1}^n a_j^{(i)} \leq m$ *for each* $\alpha_i = (a_1^{(i)}, \ldots, a_n^{(i)})$, $1 \leq i \leq t$,

an upper bound for the time complexity is

$$\mathcal{O}\left(n \max\left(\log\left(\frac{m}{n}\right), 1\right) + t\log(\max(n, m)) \right).$$

Proof.   We begin by computing $\Delta(\alpha_1, \beta)$; if $\alpha_1 < \beta$ we iterate but if $\beta \leq \alpha_1$ then we stop, returning $i = 0$ and $\Delta(\beta, \alpha_1) := \Delta(\alpha_1, \beta)$.

   Iteratively we can assume that $\alpha_i < \beta$ and that $\Delta(\alpha_i, \beta)$ is computed, and we can use the results of Lemma 40.10.4, as follows.

- If $\Delta(\alpha_i, \alpha_{i+1}) = n + 1$, so that $\alpha_i = \alpha_{i+1}$, we set $\Delta(\alpha_{i+1}, \beta) := \Delta(\alpha_i, \beta)$ and iterate.

- If $\Delta(\alpha_i, \beta) < \Delta(\alpha_i, \alpha_{i+1}) \leq n$ then $\alpha_i < \alpha_{i+1} < \beta$ and $\Delta(\alpha_{i+1}, \beta) = \Delta(\alpha_i, \beta)$; thus, we set $\Delta(\alpha_{i+1}, \beta) = \Delta(\alpha_i, \beta)$ and iterate.

- If $\Delta(\alpha_i, \beta) > \Delta(\alpha_i, \alpha_{i+1})$ then $\alpha_i < \beta < \alpha_{i+1}$ and $\Delta(\beta, \alpha_{i+1}) = \Delta(\alpha_i, \alpha_{i+1})$; thus we stop, returning i, $\Delta(\alpha_i, \beta)$ and $\Delta(\beta, \alpha_{i+1}) = \Delta(\alpha_i, \alpha_{i+1})$.

- If $\Delta(\alpha_i, \beta) = \Delta(\alpha_i, \alpha_{i+1}) =: j$ we iteratively compare b_ι and a_ι^{i+1}, $\iota = j, j + 1, \ldots$, and so on until we conclude that

  o $b_\iota < a_\iota^{i+1}$, in which case $\alpha_i < \beta < \alpha_{i+1}$ and we stop, returning i, $\Delta(\alpha_i, \beta)$ and $\Delta(\beta, \alpha_{i+1}) := \iota$, or

○ $b_\iota > a_\iota^{i+1}$, in which case $\beta > \alpha_{i+1}$ and we continue, setting $\Delta(\beta, \alpha_{i+1}) := \iota$, or

○ $\beta = \alpha_{i+1}$, in which case we stop, returning i, $\Delta(\alpha_i, \beta)$ and $\Delta(\alpha_{i+1}, \beta) := n+1$.

At termination ($i = t$) we have $\alpha_t < \beta$ and terminate, returning t and $\Delta(\alpha_t, \beta)$.

There are two key indices that we update during the algorithm. The first, i, refers to a position in the list $(\alpha_1, \ldots, \alpha_t)$; the second, ι, refers to a position in the vector β. Since after each comparison either i or ι increases, the number of comparisons of the elements in Σ is at most $n + t$. Since the number of comparisons between the values $\Delta(\cdot, \cdot)$ is one for each i, that is, at most t, and since every such comparison consists of comparing integers bounded by n, we can conclude that the time needed for the integer comparisons is proportional to $t \log(n)$.

Now let Σ be the set of non-negative integers. Every time we increase i we make a comparison of an integer bounded by m, so this gives a time proportional to $t \log(m)$. However, when increasing ι, we could be in a situation where $b_\iota = a_\iota$; hence, the total time used for increasing b_ι and a_ι^{i+1} is proportional to $n \max (\log (m/n), 1)$ (by Corollary 40.10.3). Only once during the algorithm will we compare b_ι and a_ι and conclude that they differ; this cost is $\log(m)$. It follows that an upper bound for the algorithm is proportional to

$$t \log(n) + t \log(m) + n \max \left(\log \left(\frac{m}{n} \right), 1 \right) + \log(m)$$

$$= t \log(\max(n, m)) + n \max \left(\log \left(\frac{m}{n} \right), 1 \right).$$

⊙

Theorem 40.10.6 (Lundqvist). *Let $(\alpha_1, \ldots, \alpha_t)$ and $(\beta_1, \ldots, \beta_s)$ be two ordered lists of n-tuples in Σ^n, each lexicographically sorted in increasing order and assume that we are given $\Delta(\alpha_i, \alpha_{i+1})$ for each i, $1 \leq i < t$, and $\Delta(\beta_i, \beta_{i+1})$ for each i, $1 \leq i < s$. We can merge the two lists into a new ordered list $(\gamma_1, \ldots, \gamma_{s+t})$ and compute $\Delta(\gamma_i, \gamma_{i+1})$ for each i, $1 \leq i < t + s$, using $\mathcal{O}(sn + t)$ comparisons in Σ and a time proportional to $t \log(n)$. When*

> *Σ is the set of non-negative integers,*
> *$\sum_{j=1}^{n} b_j^{(i)} \leq m$ for each $\beta_i = (b_1^{(i)}, \ldots, b_n^{(i)})$, $1 \leq i \leq s$, and*
> *$\sum_{j=1}^{n} a_j^{(i)} \leq m$ for each $\alpha_i = (a_1^{(i)}, \ldots, a_n^{(i)})$, $1 \leq i \leq t$,*

an upper bound for the time complexity is

$$\mathcal{O} \left(\min(s, t) n \max \left(\log \left(\frac{m}{n} \right), 1 \right) + \min(s, t) \log(n + m) \right).$$

Proof. Suppose wlog that $s < t$ and that the algorithm described in Proposition 40.10.5 returns an index l_1, an ordered list $(\alpha_1, \ldots, \alpha_{l_1}, \beta_1)$, with $\alpha_{l_1} < \beta_1 \leq \alpha_{l_1+1}$, and the required Δs, in particular $\Delta(\beta_1, \alpha_{l_1+1})$.

Iteratively, we can apply the algorithm described in Proposition 40.10.5 to the list $(\alpha_{l_{j-1}+1}, \ldots, \alpha_t)$ and β_j, returning the index l_j, the list

$$\alpha_1 \leq \cdots \leq \alpha_{l_1} < \beta_1 \leq \alpha_{l_1+1} \leq \cdots \leq \alpha_{l_2} < \beta_2 \leq \alpha_{l_2+1} \leq \cdots$$

$$\leq \alpha_{l_j} < \beta_j \leq \alpha_{l_j+1} \leq \cdots \leq \alpha_t$$

and the required Δs up to $\Delta(\beta_j, \alpha_{l_j+1})$.

We need to check carefully what happens when $l_j = l_{j+1}$; in this case $b_j < b_{j+1}$ and the value $\Delta(\beta_j, \beta_{j+1})$ is available.

Note that when the first step of the algorithm described in Proposition 40.10.5 is applied to $(\alpha_{l_{j-1}+1}, \ldots, \alpha_t)$ and β_j, we are required to compute $\Delta(\alpha_{l_{j-1}+1}, \beta_j)$; this step can be slightly improved (although the complexity is not improved) by remarking that (Lemma 40.10.4(2))

$$\Delta(\alpha_{l_{j-1}+1}, \beta_j) \geq \max\left(\Delta(\alpha_{l_{j-1}+1}, \beta_{j-1}), \Delta(\beta_{j-1}, \beta_j)\right).$$

We make s calls to the algorithm described in Proposition 40.10.5, performing

$$(l_1 + 1) + n + \sum_{j=2}^{s}(l_j + 1) - (l_{j-1} + 1) + n = (l_s + 1) + sn < t + sn$$

comparisons of elements in Σ; the time complexity is proportional to

$$(l_1 + 1)\log(n) + \sum_{j=2}^{s}\left((l_j + 1) - (l_{j-1} + 1)\right)\log(n) = (l_s + 1)\log(n) < t\log(n).$$

If Σ is the set of non-negative integers, the time complexity of the algorithm becomes

$$\mathcal{O}\left(sn \max\left(\log\left(\frac{m}{n}\right), 1\right) + t\log(n + m)\right).$$

$$\boxed{\odot}$$

Remark 40.10.7 (Lundqvist).

(1) Since each term ordering, according to Erdős' characterization (Corollary 24.9.5), is given through a $\rho \times n$ matrix M ($\rho \leq n$) whose rows are the ρ vectors w_i defining the weight functions characterizing the term ordering, the vectors α, β which we were comparing in the results above are, in the context of Möller's Algorithm, the evaluations

$$v(\omega) := \left(v_{\mathsf{w}_1}(\omega), \ldots, v_{\mathsf{w}_\rho}(\omega)\right)$$

of the terms ω of the weight functions v_{w_i} and are thus elements in Σ^ρ, $\rho \leq n$.

(2) Given a term ω, the vector $v(\omega)$ and a variable Z_i, $i \leq \delta - 1$, the cost of computing the vector $v(Z_i\omega)$ is $\mathcal{O}(\rho\log(mc))$, where $c = \max\left(|v_{\mathsf{w}_j}(X_i)| \mid i \leq \delta_1, j \leq \rho\right)$ and $m = \sum_i v_{\mathsf{w}_i}(\omega) \leq c\deg(\omega)$, and, in the context of Möller's Algorithm, $\omega \in \mathbf{N}$, $\deg(\omega) \leq s$ (see Remark 40.10.1).

(3) Thus, assuming $s < n$, the cost of computing all the vectors

$$v(Z_i\omega), \qquad \omega \in \mathbb{N},\ 1 \le i \le \delta - 1,$$

can be evaluated as $\mathcal{O}(s\delta n \log(sc))$.

(4) In an application of the result above to produce $\mathbf{N}$ and $\mathbf{B}$, the set $\{Z_h t : 1 \le h \le \delta - 1\}$ is assumed to be properly sorted. This is trivial for orderings (such as lex, deg, deglex or degrevlex) where the variable order is part of the definition of term ordering, but in general sorting all the vectors $v(Z_i)$, $1 \le i < \delta$, is required, each comparison costing $\mathcal{O}\left(\rho \log\left(\rho c / \rho\right) + \rho\right) = \mathcal{O}(\rho \log(c))$; since sorting requires $\mathcal{O}(n \log(n))$, we have an upper bound $\mathcal{O}(n^2 \log(c) \log(n))$.

(5) Again for the case in which $\delta \le s < n$, (40.3) holds and all terms considered belong to $\mathcal{U} := \mathcal{T} \cap k[Z_1, \ldots, Z_{\delta-1}]$, the $\rho \times n$ matrix M ($\rho \le n$) representing the term ordering can be trivially restricted to a matrix with $\delta - 1$ instead of n columns, by extracting from the matrix M the columns corresponding to the elements Z_i, $i \le \delta - 1$, giving a $\rho \times \delta - 1$ matrix ($\rho \le n$).

(6) Each of the $s - 1$ mergings, assuming $s < n$, of the, at most, δs elements of $\mathbf{B}$ with $\{Z_h t : 1 \le h \le \delta - 1\}$ thus costs

$$\mathcal{O}\left(n\delta \log(sc) + \delta s \log(\delta + s)\right) = \mathcal{O}(ns \log(sc)).$$

(7) In order to detect the $\delta - 1$ variables which, together with 1, form a basis of $V = \mathrm{Span}_K\{[1], [Z_1], \ldots, [Z_n]\}$, one can linearly reduce the n vectors of the evaluation of each variable to s functionals with complexity $\mathcal{O}(fns^2)$, where f denotes the cost of evaluating a functional at a polynomial.[10]

(8) Thus, assuming $s < n$, the cost of a monomial manipulation which includes the vector evaluation (3), the variable reordering (4) and the mergings (6) can be evaluated as

$$\mathcal{O}\left(s\delta n \log(sc) + n^2 \log(c) \log(n) + ns^2 \log(sc)\right)$$
$$= \mathcal{O}\left(ns^2 \log(sc) + n^2 \log(c) \log(n)\right).$$

$\boxed{\odot}$

Remark 40.10.8 (Lundqvist).   A better model still, when $s < n$, can be obtained through a preprocessing, by row Gauss-reducing the ρ vectors $\mathbf{w}_i$. The complexity is $\mathcal{O}(ns^2)$ and there are $\bar\rho$ linearly independent weight functions $\bar{\mathbf{w}}_j$, $j \le \bar\rho$, on $\mathcal{U}$ and a $\bar\rho \times \delta - 1$ matrix ($\bar\rho \le \delta - 1 < s$). In this model the evaluation of all vectors (3),

$$\bar v(Z_i\omega) := (v_{\bar{\mathbf{w}}_1}(Z_i\omega), \ldots, v_{\bar{\mathbf{w}}_{\bar\rho}}(Z_i\omega)), \qquad \omega \in \mathbb{N},\ 1 \le i \le \delta - 1,$$

costs $\mathcal{O}(s\delta^2 \log(sc))$. Each of the $s - 1$ mergings (6) of the, at most, δs elements of $\mathbf{B}$ with $\{Z_h t : 1 \le h \le \delta - 1\}$ thus costs

$$\mathcal{O}\left(\delta^2 \log(sc) + \delta s \log(\delta + s)\right) = \mathcal{O}(s^2 \log(sc)).$$

---

[10] An evaluation for more important classes of functionals is given in Algorithm 29.4.2.

Thus, in this model, the cost of the monomial manipulation is

$$\mathcal{O}\left(s\delta^2 \log(sc) + n^2 \log(c) \log(n) + s^3 \log(sc)\right) = \mathcal{O}\left(s^3 \log(sc) + n^2 \log(c) \log(n)\right).$$

Remark 40.10.9 (Lundqvist).   When $n \leq s$, the two representations of the term ordering coincide and the evaluation of all vectors (3),

$$v(Z_i\omega) := (v_{\mathsf{w}_1}(Z_i\omega), \ldots, v_{\mathsf{w}_{\bar\rho}}(Z_i\omega)), \quad \omega \in \mathbb{N}, 1 \leq i \leq n,$$

costs $\mathcal{O}(sn^2 \log(sc))$. Each of the $s - 1$ mergings (6) of the, at most, ns elements of B with $\{Z_h t : 1 \leq h \leq n\}$ costs

$$\mathcal{O}\left(n^2 \log(sc) + ns \log(n + s)\right).$$

Thus the cost of the monomial manipulation is

$$\mathcal{O}\left(sn^2 \log(sc) + ns^2 \log(sc) + n^2 \log(c) \log(n)\right).$$

Now adapting the analysis given for Algorithm 29.4.2, the complexity of the manipulation can be evaluated as follows.

Theorem 40.10.10 (Lundqvist).   *Writing $s = \#\mathbf{N}$, n as the number of variables, f as the cost of evaluating a functional to a polynomial, and $\delta = \dim(V)$, $V = \mathrm{Span}_K \{[1], [Z_1], \ldots, [Z_n]\}$, the arithmetical complexity of Möller's Algorithm (Figures 28.2 and 29.4) is*

$$\mathcal{O}\left(\min(s, n)s^3 + ns^2 + fsn + fs^2 \min(s, n)\right),$$

where

- *the linear algebra operations cost $\mathcal{O}\left(ns^2 + \min(s - 1, n)s^3\right)$,*
- *the evaluation of the s functionals at the $n + \min(s, n)s + s$ polynomials costs $\mathcal{O}\left(fsn + fs^2 \min(s, n) + fs^2\right)$ when $n > s$.*

When the functionals are evaluated at points $f = 1$ and the arithmetical complexity is

$$\mathcal{O}\left(\min(s, n)s^3 + ns^2\right),$$

the complexity of the monomial manipulation can be evaluated as

- *$\mathcal{O}\left(ns^2 \log(sc) + n^2 \log(c) \log(n)\right)$ using a $\rho \times n$ matrix for characterizing the term ordering on $\mathcal{U}$ and*
- *$\mathcal{O}\left(s^3 \log(sc) + n^2 \log(c) \log(n)\right)$ using a $\rho \times \delta - 1$ matrix for characterizing the term ordering on $\mathcal{U}$, to which we must add $\mathcal{O}(ns^2)$ operations to extract the required matrix,*

when $n > s$; when $n \leq s$ the two matrices agree and the complexity is

- *$\mathcal{O}\left(sn^2 \log(sc) + ns^2 \log(sc) + n^2 \log(c) \log(n)\right)$.*

Corollary 40.10.11 (Lundqvist). *The arithmetical complexity of Möller's Algorithm when the functionals are evaluated at points is*

$$\mathcal{O}\left(\min(s, n)s^3 + ns^2\right)$$

and, when $s < n$, can be expressed as

$$\mathcal{O}\left(s^4 + ns^2 + s^3 \log(sc)\right). \qquad \boxed{\odot}$$

40.11 Lex Game: Analysis and Improvements on Cerlienco–Mureddu Correspondence

Remark 40.11.1 (Felszerghy *et al.*).   Apparently, the cost of the procedure suggested by Cerlienco and Mureddu for computing their correspondence (Section 33.2) for a given finite ordered set of s points in n coordinates is $\mathcal{O}(s^2 n^2)$, a result of its being a recursive algorithm which deals with all the coordinates of each point.

In order to improve this complexity, Felszerghy *et al.* gave an alternative approach:

(1) they organize the relevant information on the finite set of points

$$V := \{\mathsf{a}_1, \ldots, \mathsf{a}_s\} \subset k^n, \quad \mathsf{a}_j := (a_{j1}, \ldots, a_{jn}),$$

by building a *trie* (the *point trie*),[11]

(2) reading this, they produce an alternative trie (the *lex trie*), and a proper reordering of the points,

(3) where one can easily read the Cerlienco–Mureddu correspondence.

Remember that, given a finite ordered set $V := \{\mathsf{a}_1, \ldots, \mathsf{a}_s\} \in k^n$ of points, writing

$$\mathsf{I} := \mathcal{I}(V) := \{p \in k[X_1, \ldots, X_n] : p(\mathsf{a}_j), 1 \leq j \leq s\}$$

and denoting as $<$ the lexicographical ordering definied by $X_1 < \cdots < X_n$,[12] the Cerlienco–Mureddu Correspondence produces the *escalier* $\mathsf{N}_<(\mathsf{I})$ and a correspondence $\Phi : V \to \mathsf{N}_<(\mathsf{I})$. Furthermore, the order ideal $\mathsf{N}_<(\mathsf{I})$ depends only on the set V while Φ depends on its ordering.

Thus the effect of step (2) is to give a smoother ordering which avoids the recursive step of the procedure given by Cerlienco and Mureddu. $\qquad \boxed{\odot}$

As in Section 33.2 we will denote by π_m the projection $\pi_m : k^n \to k^m$, $\pi_m(x_1, \ldots, x_n) = (x_1, \ldots, x_m)$.

The point trie is the trie associating a leaf with each $\mathsf{a}_j := (a_{j1}, \ldots, a_{jn}) \in V$ in such a way that the unique path from the root to the leaf is the one whose edge labels are $a_{j1}, \ldots, a_{jn}$.

---

[11] That is, a rooted tree in which there is a symbol, from a fixed alphabet, written on every edge.

[12] More precisely, the procedure applies to each lexicographical ordering; in order to have the solution for the lex ordering induced by $X_{i_1} < \cdots < X_{i_n}$ one needs to suitably reorder the coordinates and use a_{ji_k} rather than a_{jk}.

With each vertex is naturally associated the set of all points associated with a leaf linked to this vertex. In particular, at the ith level the vertices are associated with the equivalence classes of the relation $\mathbf{a}_j \sim \mathbf{a}_l \iff \pi_i(\mathbf{a}_j) = \pi_i(\mathbf{a}_l)$; clearly each equivalence class of $\pi_i(V)$ partitions into equivalence classes of $\pi_{i+1}(V)$.

With each pair $1 \leq j < l \leq n$ we can associate the unique value $c_{jl} := i$, $1 \leq c_{jl} \leq n$, for which $\pi_{i-1}(\mathbf{a}_j) = \pi_{i-1}(\mathbf{a}_l)$ and $\pi_i(\mathbf{a}_j) \neq \pi_i(\mathbf{a}_l)$, setting $c_{jl} := 0$ if $\mathbf{a}_j = \mathbf{a}_l$.

Definition 40.11.2 (Lundqvist). *The* witness matrix *is the upper triangular matrix of the values c_{jl}.*

The witness list W *comprises the non-zero entries of the witness matrix.*   ◉

The construction of the lex trie is based on a rereading of the point trie. For $i = 1$ to $n-1$ we construct the vertices at the ith level by partitioning each set Σ associated with a vertex at the $(i + 1)$th level as $\Sigma = \sqcup_{j=0}^s \Sigma_j$; in order to do so we read each of the $(n - i)$th equivalence classes S of $\pi_{n-i}(V)$ and a point $\mathbf{a}_\ell \in S \cap \Sigma$ is inserted into Σ_j, where $j = \#\{\mathbf{a}_l \in S \cap \Sigma, l < \ell\}$. The vertex associated with each Σ_j is linked to the vertex associated with Σ and is labeled X_{n-i}^j. Reading the path from the leaf to the root returns the corresponding monomial.

Example 40.11.3.   Let us consider the points

$$\mathbf{a}_1 = (1,1,2,3), \quad \mathbf{a}_2 = (1,1,2,4), \quad \mathbf{a}_3 = (1,1,2,5), \quad \mathbf{a}_4 = (1,2,1,1), \quad \mathbf{a}_5 = (1,2,1,2),$$
$$\mathbf{a}_6 = (1,2,2,1), \quad \mathbf{a}_7 = (1,2,2,2), \quad \mathbf{a}_8 = (3,1,1,2), \quad \mathbf{a}_9 = (3,1,2,2), \quad \mathbf{a}_{10} = (3,1,2,3),$$
$$\mathbf{a}_{11} = (3,3,1,1), \quad \mathbf{a}_{12} = (3,4,1,1), \quad \mathbf{a}_{13} = (3,4,1,2).$$

The related point trie is

1	2	3	4	5	6	7	8	9	10	11	12	13	{1,2,3,4,5,6,7,8,9,10,11,12,13}
1	1	1	1	1	1	1	3	3	3	3	3	3	{1,2,3,4,5,6,7},{8,9,10,11,12,13}
1	1	1	2	2	2	2	1	1	1	3	4	4	{1,2,3},{4,5,6,7},{8,9,10},{11},{12,13}
2	2	2	1	1	2	2	1	2	2	1	1	1	{1,2,3},{4,5},{6,7},{8},{9,10},{11},{12,13}
3	4	5	1	2	1	2	2	2	3	1	1	2	{1},{2},{3},{4},{5},{6},{7},{8},{9},{10},{11},{12},{13}

This gives the related lex trie

{1},{2},{3},{4},{5},{6},{7},{8},{9},{10},{11},{12},{13}	{1,2,3,4,5,6,7,8,9,10,11,12,13}
{1,2,3},{4,5},{6,7},{8},{9,10},{11},{12,13}	{1,4,6,8,9,11,12},{2,5,7,10,13},{3}
{1,2,3},{4,5,6,7},{8,9,10},{11},{12,13}	{1,4,8,11,12},{6,9},{2,5,10,13},{7},{3}
{1,2,3,4,5,6,7},{8,9,10,11,12,13}	{1,8},{4,11},{12},{6,9},{2,10},{5,13},{7},{3}
{1,2,3,4,5,6,7,8,9,10,11,12,13}	{1},{8},{4},{11},{12},{6},{9},{2},{10},{5},{13},{7},{3}

from which we read

{1,2,3,4,5,6,7,8,9,10,11,12,13}	1	8	4	11	12	6	9	2	10	5	13	7	3
{1,4,6,8,9,11,12},{2,5,7,10,13},{3}	1	1	1	1	1	1	1	X_4	X_4	X_4	X_4	X_4	X_4^2
{1,4,8,11,12},{6,9},{2,5,10,13},{7},{3}	1	1	1	1	1	X_3	X_3	1	1	1	1	X_3	1
{1,8},{4,11},{12},{6,9},{2,10},{5,13},{7},{3}	1	1	X_2	X_2	X_2^2	1	1	1	1	X_2	X_2	1	1
{1},{8},{4},{11},{12},{6},{9},{2},{10},{5},{13},{7},{3}	1	X_1	1	X_1	1	1	X_1	1	X_1	1	X_1	1	1

◉

Remark 40.11.4 (Felszerghy *et al.*).   The correctness of the procedure is based on the following remarks:

- for each leaf of the point trie, the vertices in the lex trie which contain the leaf form a path from the root to the leaf;
- if two leaves of the point trie are associated with the same vertex on the ith-level in the lex trie then their ancestors on the $(n-i)$th-level of the point trie are different;
- the monomials read in the lex trie form an order ideal.

$$\boxed{\odot}$$

Denoting by r the maximal number of edges from a vertex of the point trie, the inductive step which adds a point to the current point trie requires in the worst case r comparisons for each level and an $\mathcal{O}(nr)$ time for each point; the time can be reduced to $\mathcal{O}(n\log(r))$ if the edges are ordered according to an ordering on k. Since the procedure which reads the point trie and constructs the lex trie costs $\mathcal{O}(ns)$, the complexity is either $\mathcal{O}(nsr)$ or $\mathcal{O}(ns\log(r))$.

Remark 40.11.5 (Lundqvist).

(1) In connection with the results of the previous section it is worthwhile to improve the complexity by substituting $\min(s, n)$ for s as much as possible.

   This can be done via an alternative procedure for producing the point tree. At the ith level we begin by copying all the equivalence classes of $\pi_{i-1}(V)$[13] and marking them as "undone". Then, until all classes are marked as "done", we pick an undone class, with, say, first element a_j, and compare a_{ji} and a_{li}, for all a_l in the class and which thus satisfy $\pi_{i-1}(\mathsf{a}_j) = \pi_{i-1}(\mathsf{a}_l)$, thereby constructing two lists:

   - a list containing all the values l (including j) such that $\pi_i(\mathsf{a}_j) = \pi_i(\mathsf{a}_l)$, which is marked as "done"; and
   - a list containing all the values l such that $\pi_i(\mathsf{a}_j) \neq \pi_i(\mathsf{a}_l)$, which is marked as "undone" unless it contains only a single element; at the same time we update the witness matrix, setting $c_{jl} = i$ for all l in this second list.

   Since at the terminations of the procedure we have leaves (and not lists of leaves), we can split a set into two non-empty subsets at most $s-1$ times, each requiring at most s comparisons. Moreover, each time there are at most s comparisons returning no splitting; moreover, no splitting of a set can happen at most n times. This gives $\mathcal{O}(ns + s^2)$ in total.

(2) A more careful analysis shows that both procedures for producing the point tree perform the same comparisons. In fact, according to both procedures, each element a_{ji} is compared with a_{li} exactly when either

   - $l < j$ and $a_{li} \neq a_{hi}$ for each $h < l$, or
   - $l > j$ and $a_{li} \neq a_{hi}$ for each $h < j$.

---

[13] Setting $\pi_0(V) = V$.

We therefore conclude that the construction of the point tree can be computed using at most

$$\mathcal{O}(ns + s\min(s, rn))$$

comparisons.

(3) Note that for each value i, $1 \le i \le n$, which does not appear in the witness list $W := \{h_1, \ldots, h_v\}$, $v = \#W$, there is no splitting; therefore, in Möller's Algorithm, $X_i \in \mathbf{T}(\mathsf{I})$ for each $i \notin W$ and in the Cerlienco–Mureddu Correspondence we have

$$\mathbf{N}_<(\mathsf{I}) \subset \mathcal{T} \cap k[X_{h_1}, \ldots, X_{h_v}].$$

(4) Within Stetter's algorithm, an *allgemeine* coordinate $Y = \sum_{i=1}^{n} c_i X_i$ is a solving tool and can be produced only probabilistically. Within the Cerlienco–Mureddu setting, where the solved points are available, it can be restricted to the relevant variables X_{h_i}, $h_i \in W$, as $Y = \sum_{i=1}^{v} c_i X_{h_i}$ and can be easily determined iteratively by consulting the point tree.

Let $\iota \in W$ and assume that we have inductively obtained $Y = \sum_{\substack{i=1 \\ h_i < \ell}}^{v} c_i X_{h_i}$ such that, writing $b_j := \sum_{\substack{i=1 \\ h_i < \iota}}^{v} c_i a_{jh_i}$, it holds that $\pi_{\iota-1}(a_j) = \pi_{\iota-1}(a_l) \iff b_j = b_l$. Our aim is to fix a value c_ι such that $\pi_\iota(a_j) = \pi_\iota(a_l) \iff b_j + ca_{j\iota} = b_l + ca_{l\iota}$; thus we need to discard 0 and the values $c = (b_j - b_l)/(a_{l\iota} - a_{j\iota})$ for each pair (l, j), $1 \le l \le j \le s$, for which both $\pi_\iota(a_j) \neq \pi_\iota(a_l)$ and $b_j \neq b_l$.

Constructing the two ordered lists $L_1 = \{(l, j) : \pi_\iota(a_j) \neq \pi_\iota(a_l)\}$ and $L_2 = \{(l, j) : b_j \neq b_l\}$ and producing $L_1 \cap L_2$ while merging them costs $\mathcal{O}(s^2 \log(s^2)) = \mathcal{O}(s^2 \log(s))$. Computing the values c costs $\mathcal{O}(s^2)$; if k (see item (3) above) is ordered then we further order the lists, with $\mathcal{O}(s^2 \log(s))$ complexity.

The potential candidates can be extracted from an (ordered) list of at least $\binom{s}{2}$ elements in k.

If k is ordered then we have to perform $\mathcal{O}(s^2)$ check to produce the first elements which potentially satisfy the test. Each checks costs $\mathcal{O}(\log(s))$. If k is not ordered then we have to perform $\mathcal{O}(s^4)$ checks.

Thus this procedure returns an *allgemeine* coordinate with $\mathcal{O}(\min(s, n)s^2 \log(s))$ operations if k is ordered and $\mathcal{O}(\min(s, n)s^4)$ otherwise, to which we must add $\mathcal{O}(ns)$ for the lex trie construction.   ⊙

Remark 40.11.6.   While the construction of a point trie, which costs $\mathcal{O}(s\min(s, rn))$, is iterative, the construction of the lex trie is not iterative and costs $\mathcal{O}(ns)$.

This is in contrast with the Cerlienco–Mureddu procedure, which is fully iterative. However, it is clear that an iterative version of the Lex Game would cost at most $\mathcal{O}(ns^2)$, which is still an improvement on the Cerlienco–Mureddu procedure.   ⊙

40.12 Lundqvist: A Gröbner-free Solver

The techniques introduced by Auzinger and Stetter allow us to solve effectively zero-dimensional ideals but they also give a completely different perspective to duality and to the tools that were discussed in Part four (in particular Chapters 28, 29 and 33).

Using the standard notation, let $J \subset \mathcal{Q}$ be a zero-dimensional ideal, $\deg(J) = s$, and set $A := \mathcal{Q}/J$, given via a K-basis $\mathbb{L} := \{\lambda_1, \ldots, \lambda_s\}$ of its dual basis $\mathfrak{L}(J)$; denote as $\{q_1, \ldots, q_s\}$ the Lagrange basis biorthogonal to $\mathbb{L}$.

If we consider any K-basis $\mathbf{b} = \{[b_1], \ldots, [b_s]\} \subset A$ of A then the matrix $B := M_{bq} := (\lambda_i(b_l))$ (Lemma 40.2.1) allows us:

- to compute each matrix $A_h := M([Z_h], \mathbf{b})$ via Proposition 40.5.1 and, in the radical case, by the easier Corollary 40.3.2;
- thus giving the Gröbner representation $\mathbf{b}, A_h$;
- and, moreover, allowing us to deduce explicitly the biorthogonal Lagrange basis $\{q_1, \ldots, q_s\}$.

In fact (compare the proof of Theorem 28.1.18, (2) $\Longrightarrow$ (1)) it holds that

Lemma 40.12.1.   *Setting*

$$(q_1, \ldots, q_s) := (b_1, \ldots, b_s)B^{-1},$$

the set $\{q_1, \ldots, q_s\}$ is biorthogonal to $\mathbb{L}$.

Proof.   Let $B^{-1} = (a_{lj})$, so that $\delta_{ij} = \sum_l \lambda_i(b_l)a_{lj}$ and $q_j = \sum_l b_l a_{lj}$; thus $\lambda_i(q_j) = \sum_l \lambda_i(b_l)a_{lj} = \delta_{ij}$.   ◉

A more subtle application of the matrix B was proposed in Lundqvist:[14]

The main tool used to compute vanishing ideals of points is [Möller's Algorithm], which gives a Gröbner basis for the ideal, and, as a side effect, a vector space basis of the quotient [A]. However, in many applications it turns out that it is the vector space basis, rather than the Gröbner basis of the ideal, which is of interest. For instance, it is preferable to compute normal forms using vector space methods rather than Gröbner basis methods.

A [...] bound for the arithmetic complexity of [Möller's Algorithm] is given in [Corollary 40.10.11] as $\left[\mathcal{O}\left(s^4 + ns^2\right)\right]$.[15] We discuss [for] constructions of vector space bases [...] all of which perform better than [Möller's Algorithm].

Proposition 40.12.2 (Lundqvist).   *It holds that* $\mathbf{Rep}(g, \mathbf{b}) = B^{-1}(\lambda_1(g), \ldots, \lambda_s(g))$.

Proof.   By Lagrange, $g = (q_1, \ldots, q_s)(\lambda_1(g), \ldots, \lambda_s(g))^T = (b_1, \ldots, b_s)B^{-1}(\lambda_1(g), \ldots, \lambda_s(g))^T$.   ◉

Given $\mathbb{L}$, once an efficient procedure has returned a K-basis $\mathbf{b}$ of A and appropriate preprocessing has produced both $B := M_{bq} := (\lambda_i(b_l))$ and $B^{-1} = (a_{lj})$, the

---

[14] Lundqvist S., Vector space bases associated to vanishing ideals of points, *J. Pure Appl. Algebra* **214** (2010), 309–321.

[15] Note that we are speaking here of lex ordering and thus $c = 1$.

computation of the normal form $\mathbf{Rep}(g, \mathbf{b})$ of g w.r.t. $\mathbf{b}$ costs $\mathcal{O}(sf + s^2)$, where f denotes the cost of evaluating f at each of the s functionals λ_i.

Lundqvist considers only evaluation at points for which we have $f = \mathcal{O}\left(\#\operatorname{supp}(g) \cdot \deg(g)\right)$ and compares Möller's Algorithm as a tool for returning $\mathbf{b}$ with four applications of the Lex Game variations of the Cerlienco–Mureddu correspondence, with the aim of producing a solution whose complexity is better than the (Corollary 40.10.11)

$$\mathcal{O}\left(\min(s, n)s^3 + ns^2\right)$$

of Möller's Algorithm.

In his analysis, Lundqvist assumes that the point trie had already been produced and processed; thus he also has the lex trie and the witness matrix and list.

(1) A first proposal is to deduce a family of separators from the witness matrix. In fact the polynomials

$$q_i = \prod_{j \neq i} \frac{X_{c_{ij}} - a_{jc_{ij}}}{a_{ic_{ij}} - a_{jc_{ij}}}, \quad 1 \leq j \leq s,$$

form exactly the Lagrange basis biorthogonal to the point evaluations.

Note that the witness matrix can be produced with $\mathcal{O}(ns + s\min(s, rn))$ comparisons and that to express each q_i we need to compute the s^2 denomerators, giving $\mathcal{O}(ns + s^2)$.

(2) A natural basis is $\{1, Y, \ldots, Y^{s-1}\}$, where Y is the *allgemeine* coordinate $Y = \sum_{i=1}^{n} c_i X_i$ produced with $\mathcal{O}(\min(s, n)s^2 \log(s) + ns)$ operations if k is ordered and $\mathcal{O}(\min(s, n)s^4 + ns)$ operations otherwise.

(3) The monomial basis $\mathbf{N}(\mathbf{I})$ can be deduced with $\mathcal{O}(\min(s, n)s + ns)$ operations by the Lex Game procedure.

(4) Finally, using Remark 40.11.5(3), once the point trie is produced and processed to give the witness list, with complexity $\mathcal{O}(\min(s, nr)s + ns)$, the Möller Algorithm can be applied to $k[X_{h_1}, \ldots, X_{h_v}]$ with respect to any term ordering on $\mathcal{T} \cap k[X_{h_1}, \ldots, X_{h_v}]$.

Its cost is $\mathcal{O}(\min(s, v)s^3 + vs^2)$ but since $v \leq \min(s, n)$ this becomes $\mathcal{O}(\min(s, n)s^3)$; thus the final cost is $\mathcal{O}(\min(s, n)s^3 + ns)$.

For a lexicographical ordering it is better to reapply the Lex Game procedure, with complexity $\mathcal{O}(\min(s, nr)s + ns)$.

40.13 Derogatoriness and Allgemeine Coordinates

With the same notation as in Section 40.5, we recall that a linear form

$$Y := \sum_{h=1}^{r} c_h Z_h$$

is said to be an *allgemeine* coordinate (Definition 34.4.7) for the zero-dimensional ideal $J = \cap_{i=1}^{s} q_i$ iff

(a) there are polynomials $g_i \in K[Y], 0 \le i \le n$, g_0 monic, $\deg(g_i) < \deg(g_0)$, such that

$$G := (g_0(Y), Z_1 - g_1(Y), Z_2 - g_2(Y), \ldots, Z_r - g_r(Y))$$

is the reduced Gröbner basis of the ideal

$$J^+ := J + \left(Y - \sum_h c_h Z_h \right) \subset K[Y, Z_1, \ldots, Z_r]$$

w.r.t. the lex ordering induced by $Y < Z_1 < \cdots < Z_r$

and that condition (a) implies, among other things, that (Corollary 34.4.6):

(b) $Q/J \cong K[Y]/g_0(Y)$;
(c) for each $i, 1 \le i \le s$, $\beta_i := \sum_{h=1}^{r} c_h a_h^{(i)}$ is a root of g_0 with multiplicity s_i and
(d) $a_j^{(i)} = g_j(\beta_i)$ for each $i, 1 \le i \le s$, and each $j, 1 \le j \le r$;
(e) $g_0(Y) = \prod_{i=1}^{r}(Y - \beta_i)^{s_i}$;
(f) $f \in J \iff \mathbf{Rem}(f(g_1(Y), \ldots, g_r(Y)), g_0(Y)) = 0$.

Moreover, there is a Zarisky open set $U \subset K^n$ such that $Y := \sum_{h=1}^{r} c_h Z_h$ is an *allgemeine* coordinate for J iff $(c_1, \ldots, c_r) \in U$. For each $f = \sum_{t \in W} c(f, t)t$: $f(0, \ldots, 0) = 0$ set

$$\text{lin}(f) := \sum_{h=1}^{r} c(f, Z_i)Z_i$$

and, for each primary ideal q at the origin, let

$$\text{lin}\{q\} := \{\text{lin}(f) : f \in q\} \quad \text{and} \quad \Lambda_1(q) := \{\ell \in \nabla_1 : \ell(g) = 0, g \in \text{lin}\{q\}\},$$

where $\nabla_1 := \text{Span}_K (M(Z_i), 1 \le h \le r)$.

Proposition 40.13.1. *For each primary ideal $q \subset Q$, $\deg(q) = s$, at the origin, the following conditions are equivalent:*

(1) *there is an* allgemeine *coordinate for q;*
(2) $\dim_K (\text{lin}\{q\}) = r - 1$;
(3) $\dim_K (\Lambda_1(q)) = 1$;
(4) $\dim_K (\text{lin}\{q\}) = r - 1$ *and, for each linear form*

$$Y_1 := \sum_{h=1}^{r} c_h Z_h \notin \text{lin}\{q\}$$

and each basis $(Y_2, \ldots, Y_r)$ of $\text{lin}\{q\}$, there are polynomials $g_\kappa \in K[Y_1], 2 \le \kappa \le r$, such that

$$q = \left(Y_1^s, Y_2 - g_2, \ldots, Y_r - g_r\right) \subset K[Y_1, \ldots, Y_r] = K[Z_1, \ldots, Z_r];$$

(5) $\dim_K(\Lambda_1(q)) = 1$ *and for each linear form* $Y_1 := \sum_{h=1}^{r} c_h Z_h \notin \mathrm{lin}\{q\}$ *and each basis* $(Y_2, \ldots, Y_r)$ *of* $\mathrm{lin}\{q\}$ *there are polynomials* $g_\kappa \in K[Y_1], 2 \le \kappa \le r$, *such that*

$$\mathfrak{L}(q) = \{\delta_l, 1 \le l \le s\}, \qquad \delta_l(f) := \frac{\bar{f}^{(l-1)}(0)}{(l-1)!}, \qquad 1 \le l \le s$$

where $\bar{\ } : \mathcal{Q} \to \mathrm{Span}_K\{1, Y_1, \ldots, Y_1^{s-1}\}$ *denotes the projection*

$$f(Y_1, Y_2, \ldots, Y_r) \mapsto \bar{f} := \mathbf{Rem}(f(Y_1, g_2(Y_1), \ldots, g_r(Y_1)), Y_1^s);$$

(6) $\dim_K(\mathrm{lin}\{q\}) = r - 1$ *and, for each linear form* $Y := \sum_{h=1}^{r} c_h Z_h \notin \mathrm{lin}\{q\}$, *there are polynomials* $g_\kappa = \sum_{l=1}^{s-1} c_{\kappa l} Y^l \in K[Y], 1 \le \kappa \le r$, *such that* $q = (Y^s, Z_1 - g_1, \ldots, Z_r - g_r);$

(7) $\dim_K(\Lambda_1(q)) = 1$ *and, for each linear form* $Y := \sum_{h=1}^{r} c_h Z_h \notin \mathrm{lin}\{q\}$, *there are polynomials* $g_\kappa \in K[Y], 1 \le \kappa \le r$, *such that, writing*

$$\check{\ } : \mathcal{Q} \mapsto K[Y]/(Y^s) : f(Z_1, \ldots, Z_r) \to \check{f} := \mathbf{Rem}(g_1(Y), \ldots, g_r(Y)), Y^s),$$

we have $\mathfrak{L}(q) = \{\delta_l, 1 \le l \le s\}, \delta_{l+1}(f) := \frac{\check{f}^{(l)}(0)}{l!}, 0 \le l < s.$

Proof. The scheme of the proof is

$$\begin{array}{ccccc} & & (1) & & \\ & \swarrow & & \searrow & \\ (2) & \leftrightarrow & (4) & \to & (6) \\ \updownarrow & & \updownarrow & & \updownarrow \\ (3) & & (5) & & (7) \end{array}$$

The implications (2) $\iff$ (3), (4) $\iff$ (5), (6) $\iff$ (7), (4) $\implies$ (2) and (6) $\implies$ (1) hold trivially. If Y is an *allgemeine* coordinate for q and $g_\kappa = \sum_{l=1}^{s-1} a_{lh} Y^l, 1 \le h \le r$, are such that

$$G := (g_0(Y), Z_1 - g_1(Y), Z_2 - g_2(Y), \ldots, Z_r - g_r(Y))$$

is the reduced Gröbner basis of the ideal

$$q^+ := q + \left(Y - \sum_{h=1}^{r} c_h Z_h)\right) \subset K[Y, Z_1, \ldots, Z_r]$$

w.r.t. the lex ordering induced by $Y < Z_1 < \ldots < Z_r$ then

$$Y \notin \mathrm{lin}\{q^+\} = \mathrm{Span}_K\left(\left\{Y - \sum_{h=1}^{r} c_h Z_h\right\} \cup \{Z_\kappa - a_{1\kappa h} Y, 1 \le \kappa \le r\}\right)$$

so that $\dim_K(\mathrm{lin}\{q\}) = \dim_K(\mathrm{lin}\{q^+\}) = r - 1$ and $Y = \sum_{h=1}^{r} c_h Z_h \notin \mathrm{lin}\{q\}$. Thus we have (1) $\implies$ (2).

Moreover, for $Y_1, Y_2, \ldots, Y_r$ as in (4), the Gröbner basis w.r.t. the lex ordering induced by $Y_1 < Y_2 < \cdots < Y_r$ of $q' := qK[Y_1, \ldots, Y_r]$ necessarily satisfies $\deg(q') = \deg(q) = s$ and $\mathbf{T}(q') = (Y_1^s, Y_2, \ldots, Y_r)$, thus giving also (2) $\implies$ (4).

We are therefore left with proving that (4) $\implies$ (6). Let us consider $\mathrm{lin}\{q\} \subset K[Z_1, \ldots, Z_r]$. There are two cases: either

(i) $\mathrm{lin}\{q\} = \mathrm{Span}_K\{Z_2 - d_2 Z_1, \ldots, Z_r - d_r Z_1\}$ or

(ii) there is a Z_j, wlog we can say Z_1, such that $\mathrm{lin}\{q\} = \mathrm{Span}_K\{Z_2, \ldots, Z_r\}$.

For case (i) we set $Y_\kappa := Z_\kappa - d_\kappa Z_1, 2 \leq \kappa \leq r$. By (4) we know that there are polynomials $g_\kappa = \sum_{l=1}^{s-1} c_{\kappa l} Y^l \in K[Y], 2 \leq \kappa \leq r$ such that

$$Y_\kappa - g_\kappa(Y) = Z_\kappa - d_\kappa Z_1 - c_{\kappa 1} Y + \sum_{l=2}^{s-1} c_{\kappa l} Y^l \in q$$

and $Z_\kappa - d_\kappa Z_1 - c_{\kappa 1} Y \in \mathrm{lin}\{q\}$. Since

$$\left(c_1 + \sum_{\kappa=2}^{r} c_\kappa d_\kappa \right) Z_1 + \sum_{\kappa=2}^{r} c_\kappa Y_\kappa = \sum_{h=1}^{r} c_h Z_h = Y \notin \mathrm{lin}\{q\},$$

then $c := c_1 + \sum_{\kappa=2}^{r} c_\kappa d_\kappa \neq 0$ and, setting $g_1 := c^{-1} Y - \sum_{\kappa=2}^{r} c^{-1} c_\kappa g_\kappa$, we have

$$c(Z_1 - g_1) \equiv c Z_1 - Y + \sum_{\kappa=2}^{r} c_\kappa Y_\kappa = 0 \pmod{q} \quad \text{and} \quad Z_1 - g_1 \in q.$$

For case (ii) we set $Y_\kappa := Z_\kappa, 2 \leq \kappa \leq r$, and, by (4), we know that there are polynomials $g_\kappa = \sum_{l=1}^{s-1} c_{\kappa l} Y^l \in K[Y], 2 \leq \kappa \leq r$, such that

$$Y_\kappa - g_\kappa(Y) = Z_\kappa - c_{\kappa 1} Y + \sum_{l=2}^{s-1} c_{\kappa l} Y^l \in q \quad \text{and} \quad Z_\kappa - c_{\kappa 1} Y \in \mathrm{lin}\{q\}.$$

Since

$$Y = c_1 Z_1 + \sum_{\kappa=2}^{r} c_\kappa Z_\kappa \notin \mathrm{lin}\{q\},$$

we have $c_1 \neq 0$ and $c_{\kappa 1} = 0$ for each $\kappa > 1$; setting $g_1 := c_1^{-1} Y - \sum_{\kappa=2}^{r} c_1^{-1} c_\kappa g_\kappa$ we obtain

$$c_1(Z_1 - g_1) \equiv c_1 Z_1 - Y + \sum_{\kappa=2}^{r} c_\kappa Y_\kappa = 0 \pmod{q} \quad \text{and} \quad Z_1 - g_1 \in q.$$

$\boxed{\odot}$

Example 40.13.2. Let $q := (Z_2^5, Z_1 - Z_2^3) \subset K[Z_1, Z_2]$ be a Gröbner basis for the lex ordering induced by $Z_2 < Z_1$. Thus

$$\deg(q) = 5, \qquad \Lambda_1\{q\} = \mathrm{Span}_K\left\{ \frac{\partial}{\partial Z_2} \right\} \quad \text{and} \quad \mathrm{lin}\{q\} = Z_1.$$

For $Y_1 := a Z_1 + Z_2, Y_2 = Z_1$ we have $q' = (Y_1^5, Y_2 - Y_1^3)$. Indeed, Z_1 is not an *allgemeine* coordinate since, for the lex ordering induced by $Z_1 < Z_2$, we have $q = (Z_1^2, Z_1 Z_2^2, Z_2^3 - Z_1)$.

$\boxed{\odot}$

Remark 40.13.3.   As a direct application of Corollary 40.4.7, we have that the functionals $\delta_{l+1}(f) := \frac{\check{f}^{(l)}(0)}{l!}, 0 \le l < s$, satisfy

$$\delta_{l+1}(fg)(\alpha_i) = f(\alpha_i)\delta_{l+1}(g)(\alpha_i) + \sum_{x=0}^{l-1} \delta_{1+l-x}(f)(\alpha_i)\delta_{1+x}(g)(\alpha_i).$$

Thus we have the related matrix

$$Q_f = \begin{pmatrix} f(\alpha_i) & \delta_2(f)(\alpha_i) & \delta_3(f)(\alpha_i) & \cdots & & \delta_{l+1}(f)(\alpha_i) \\ & f(\alpha_i) & \delta_2(f)(\alpha_i) & \delta_3(f)(\alpha_i) & & \vdots \\ & & \ddots & \ddots & \ddots & \vdots \\ & & & \ddots & \ddots & \delta_3(f)(\alpha_i) \\ & & & & \ddots & \delta_2(f)(\alpha_i) \\ 0 & & & & & f(\alpha_i) \end{pmatrix}.$$

$$\tag{40.4}$$

$$\boxed{\odot}$$

Definition 40.13.4.   *A primary ideal* $\mathfrak{q} \subset Q$ *at the origin which satisfies the equivalent conditions of Proposition 40.13.1 is called a* curvilinear primary with derivative $\delta_2(f) := \check{f}'(0)$.

Now, for $\alpha = (a_1, \ldots a_r) \in K^r$, write

$$\lambda_\alpha : Q \to Q, \quad \lambda_\alpha(f) = f(Z_1 + a_1, \ldots, Z_r + a_r),$$

and $\mathfrak{m}_\alpha = (Z_1 - a_1, \ldots, Z_r - a_r)$, the maximal ideal at α.

Theorem 40.13.5.   *Let* J *be a zero-dimensional ideal.*

Write $\mathcal{Z}(\mathsf{J}) := \{\alpha_1, \ldots, \alpha_s\} \subset K^r$ *and, for each* i, $\alpha_i = (a_1^{(i)}, \ldots, a_r^{(i)})$, $\mathfrak{q}_i$ *the* $\mathfrak{m}_{\alpha_i}$*-primary component of* J *and* $s_i := \mathrm{mult}(\alpha_i, \mathsf{J}) = \deg(\mathfrak{q}_i)$, *so that* $\mathsf{J} = \cap_{i=1}^s \mathfrak{q}_i$ *and* $\deg(\mathsf{J}) = \sum_{i=1}^s s_i := s$.

A linear form $Y := \sum_{h=1}^r c_h Z_h$ *is an* allegemeine *coordinate for* J *iff, denoting* $\beta_i := \sum_{h=1}^r c_h a_h^{(i)}$, *the following conditions hold:*

(1) *for each primary component* $\mathfrak{q}_i$ *of* J,
   (a) *either* $\mathfrak{q}_i$ *is simple, so that* $\mathfrak{q}_i = \mathfrak{m}_{\alpha_i}$ *and* $s_i = 1$,
   (b) *or the primary ideal* $\mathfrak{q} := \lambda_{\alpha_i}(\mathfrak{q}_i)$ *at the origin is curvilinear and* $Y \notin \mathrm{lin}\{\mathfrak{q}_i\}$;
(2) $\beta_i \ne \beta_j$ *if* $i \ne j$.

Proof.   Condition (1) implies that, for each primary component $\mathfrak{q}_i$ of J, the ideal

$$\mathfrak{q}_i^+ := \mathfrak{q}_i + \left(Y - \sum_{h=1}^r c_h Z_h\right) \subset K[Y, Z_1, \ldots, Z_r]$$

has as its basis

(a) either $(Y - \beta_i, Z_1 - a_1^{(i)}, \ldots Z_r - a_r^{(i)})$ if q_i is maximal, in which case we set
$f_{ih}(Z_1) := 0$ for each h, $1 \leq h \leq r$,

(b) or $\left((Y - \beta_i)^{s_i}, Z_1 - a_1^{(i)} - f_{i1}(Y), \ldots, Z_r - a_r^{(i)} - f_{ir}(Y)\right)$, if q_i is multiple,
with $f_{ih}(Y) := g_h(Y)$ where $g_h(Y)$ are the polynomials whose existence is
implied in Proposition 40.13.1(6).

Since, by assumption (2), the β_i are all different, $J \cap K[Y]$ is generated by $g_0(Y) = \prod_j (Y - \beta_i)^{s_i}$. By the Chinese Remainder Theorem there is then for each $h \geq 1$ a
unique polynomial $g_h(Y)$, $\deg(g_h) < s = \deg(g_0)$, such that

$$g_h \equiv a_h^{(i)} + f_{ih} \mod (Y - \beta_i)^{s_i} \text{ for each } i.$$

Then $(g_0(Y), Z_1 - g_1(Y), \ldots, Z_r - g_r(Y))$ is the required Gröbner basis of $J^+ = J + \left(Y - \sum_h c_h Z_h\right)$, implying that Y is an *allgemeine* coordinate for J.

Conversely, if J^+ has a basis $(g_0(Y), Z_1 - g_1, \ldots, Z_r - g_r)$ then, for each primary
component q_i of J, q_i^+ has a basis $(f(Y), Z_1 - f_2, \ldots, Z_r - f_r)$ where f runs over
the irreducible-power factors of g_0 and $f_h = \mathbf{Rem}(g_h, f)$ for each h. Thus each
component which is not simple has Y as an *allgemeine* coordinate. $\odot$

A better description of the derogatoriness of Φ_f is the following:

Corollary 40.13.6 (Möller–Stetter).   *Let J be a zero-dimensional ideal and let $f \in Q$. With the notation of Theorem 40.13.5, Φ_f is non-derogatory iff the following
conditions hold:*

(1) $f(\alpha_i) \neq f(\alpha_j)$ *if* $i \neq j$;
(2) *for each primary component* q_i *of* J
    (a) *either* q_i *is simple, so that* $q_i = m_{\alpha_i}$,
    (b) *or the primary ideal* $\lambda_{\alpha_i}(q_i)$ *at the origin is curvilinear with derivative* ℓ_i;
(3) *for each multiple component* q_i, $\ell_i(f) \neq 0$.

Proof.   Assume that Φ_f is non-derogatory. Then:

(1) For $\alpha_i \neq \alpha_j$ we have $v_{n_i}(\mathbf{b}) \neq v_{n_j}(\mathbf{b})$; thus necessarily $f(\alpha_i) \neq f(\alpha_j)$
otherwise the two vectors are independent eigenvectors for $f(\alpha_i) = f(\alpha_j)$.
(2) For a multiple component q_i, Proposition 40.4.6(3) implies that if $L_i \cap V = \Lambda_1(\lambda_{\alpha_i}(q_i))$ has dimension > 1 then there is an eigenvector for $f(\alpha_i)$ that is
linearly independent of $v_{n_i}(\mathbf{b})$; thus q_i must be curvilinear.
(3) If, instead, $L_i \cap V = \Lambda_1(\lambda_{\alpha_i}(q_i))$ has dimension 1, Corollary 40.4.7 shows that
if $\ell_i(f) = 0$ then $v_{n_i+1}(\mathbf{b})$ is a further eigenvector for $f(\alpha_i)$.

Conversely, by (2) the matrix Q_f has $\mathbf{s}$ blocks; the ith block has Equation (40.4)
as its form; thus, since by (3) $\ell_i(f) = \delta_2(f)(\alpha_i) \neq 0$, each block contributes a single
eigenvector. Finally, (1) grants that different eigenvectors correspond to different
eigenvalues. $\odot$

Remark 40.13.7 (Möller–Stetter). If $\delta_2(f)(\alpha_i) = \cdots = \delta_j(f)(\alpha_i) = 0 \neq \delta_{j+1}(f)(\alpha_i)$ then the ith block in (40.4) contributes j linearly independent eigenvectors. ⊙

Corollary 40.13.8. *A zero-dimensional ideal is a non-derogatory ideal if each multiple primary component is curvilear.* ⊙

Proof. Each linear form $Y = \sum_h c_h Z_h$ is non-derogatory provided that it satisfies $Y(\alpha_i) \neq Y(\alpha_j)$ for each $i \neq j$ and $\delta_2(Y) \neq 0$.[16]

Both conditions define a Zarisky open set. ⊙

---

[16] Since both Y and δ_2 are linear, $\delta_2(Y)$ is a constant.

41

Macaulay IV

> The device that follows [...] finally eliminates from algebraic geometry the last traces of Elimination Theory.
>
> *A. Weil*

> Eliminate, eliminate, eliminate,
> Eliminate the eliminators of elimination theory.
>
> *S.S. Abhyankar*

After having discussed the two "standard" recent solving algorithms, this chapter is devoted to the old-fashoned tools of *resultants* and *resolvants*.

After recalling the notion and the main properties of resultants (Section 41.1) we focus on Macaulay's approach[1] for computing resultants. Macaulay defines the term resultant as the gcd of all determinants of a particular matrix, *Macaulay's matrix* (Section 41.3), obtained by expanding a proper generating set which can be deduced by a result of Bézout (Section 41.2), and proves that the resultant is obtained by dividing out from any such determinant an *extraneous factor* (Section 41.4), which he is able to characterize precisely; finally, we are able to prove that Macaulay's definition truly gives the resultant (Section 41.5).

The knowledge of the resultant of a set of forms (homogeneous polynomials) allows one to compute the roots of the ideal

$$\mathsf{J} := \mathbb{I}(f_1, \ldots, f_r) \subset K[Z_1, \ldots, Z_r]$$

by computing in $K[U_1, \ldots, U_r][Z]$ the resultant (*u-resultant*) of the polynomials $f_1, \ldots, f_r, f_u$, where $f_u := Z - \sum_{i=1}^{r} U_i Z_i$, and factorizing it into linear factors $Z - \sum_{i=1}^{r} U_i \alpha_i$; each such linear factor returns a root $(\alpha_1, \ldots, \alpha_r) \in \mathcal{Z}(\mathsf{J})$ (Section 41.6).

---

[1] I mainly follow the original result

    Macaulay F.S., Some formulae in elimination, *Proc. London Math. Soc.* (1) **35** (1903), 3–27,

supported by his book

    Macaulay F.S., *The Algebraic Theory of Modular Systems*, Cambridge University Press (1916).

I next discuss[2] another nineteenth century solver, Kronecker's resolvent (Section 41.7). If, given an ideal

$$\mathsf{I} := \mathbb{I}(f_1, \ldots, f_s) \subset k[X_1, \ldots, X_n],$$

one computes the resolvent (*u-resolvant*) in $k[\Lambda_1, \ldots, \Lambda_n][X_1, \ldots, X_{n-1}][X]$ of the polynomials

$$\Lambda_n^{\deg(f_l)} f_l \left(X_1, \ldots, X_{n-1}, \Lambda_n^{-1} \left(X - \sum_{i=1}^{n-1} \Lambda_i X_i \right) \right), \quad 1 \le l \le s,$$

and factorizes it into linear factors as

$$X - \Lambda_1 X_1 - \cdots - \Lambda_{\nu-1} X_{\nu-1} - \Lambda_\nu \xi_\nu - \cdots - \Lambda_n \xi_n,$$

each factor for which each ξ_i is independent of the Λs corresponds to an $\Omega(k)$-prime component of dimension $\nu - 1$ (Section 41.8).

Notwithstanding the fact that Macaulay strongly criticizes Kronecker's resolvent for its being doubly exponential rather than singly exponential, like the resultant, Kronecker's approach provided (in the nineteenth century!) a parametrization[3]

$$\begin{cases} q(X_1, \ldots, X_{\nu-1}, U) & = & 0, \\[2mm] X_\nu & = & \dfrac{w_\nu(X_1, \ldots, X_{\nu-1}, U)}{\dfrac{\partial q}{\partial U}(X_1, \ldots, X_{\nu-1}, U)}, \\ & \vdots & \\ X_n & = & \dfrac{w_n(X_1, \ldots, X_{\nu-1}, U)}{\dfrac{\partial q}{\partial U}(X_1, \ldots, X_{\nu-1}, U)}, \end{cases} \tag{41.1}$$

of a radical equidimensional ideal (Section 41.9) which is at the core of the most efficient of today's solvers: Rouillier's Rational Univariate Representation (Section 42.9) and the TERA Kronecker Package (Chapter 44).

First, I give an *intermezzo*, where I cover the history of resultants from Bézout to the English algebra school (Section 41.10) and, in particular, report Cayley's interpretation of the resultant of two polynomials

$$U(X), V(X) \in k[X], \quad \deg(U) = \deg(V) = n,$$

as the determinant of the matrix $(\alpha_{\rho\sigma})$ defined by the relation

$$\frac{U(X)V(Y) - U(Y)V(X)}{X - Y} = \sum_{\rho=0}^{n-1} \sum_{\sigma=0}^{n-1} \alpha_{\rho\sigma} X^{n-\rho-1} Y^{n-\sigma-1}.$$

Then I briefly discuss a different representation of the resultant, due to Dixon, which extended Cayley's interpretation to more than two polynomials (Section 41.11).

---

[2] Still following the guideline provided by Macaulay's book.
[3] See F. S. Macaulay, *The Algebraic Theory*, *op. cit.*, pp. 27–28.

Dixon's resultant was recently revised and reproposed by Kapur *et al.*; at the same time, Cardinal considered Cayley's formula and Dixon's matrix (Section 41.12) and proposed an algorithm (Section 41.13) which performed a series of transformations on Dixon matrices defined by the polynomials

$$f_1, \ldots, f_n, X_i, \quad 1 \le i \le n, \quad f_l \in k[X_1, \ldots, X_n].$$

Cardinal conjectured that if $f_1, \ldots, f_n$ is a complete intersection then the output of his algorithm is a Gröbner representation of the ideal

$$\mathsf{J} := \mathbb{I}(f_1, \ldots, f_n) \subset k[X_1, \ldots, X_n].$$

Recently Mourrain proposed an improved version of Cardinal's algorithm and gave a complete proof of Cardinal's conjecture for this abridged version of the algorithm (Section 41.14).

The result not only returns, with good complexity, a Gröbner representation of the ideal J but also, with the same complexity, allows one to test ideal membership and, for a polynomial $f \in \mathsf{J}$, returns a representation $f = \sum_{i=1}^{n} g_i f_i$ (Section 41.15).

41.1 The Resultant of r Forms in r Variables

Set $\mathcal{Q} := K[Z_1, \ldots, Z_r]$, $\mathcal{W} := \{Z_1^{a_1} \cdots Z_r^{a_r} : (a_1, \ldots, a_r) \in \mathbb{N}^r\}$ as its monomial K-basis and K as the algebraic closure of K.

For each $d \in \mathbb{N}$ we also set

$$\mathcal{W}_d := \{t \in \mathcal{W} : \deg(t) = d\} \qquad \text{and} \qquad \mathcal{W}(d) := \{t \in \mathcal{W} : \deg(t) \le d\}.$$

Let us also denote r integers as $d_1, \ldots, d_r$.

Our aim being to considering r "generic" forms (homogeneous polynomials) $f_1, \ldots, f_r \in \mathcal{Q}$, $\deg(f_i) = d_i$, we apply the same notation and approach as that of the English algebra school (see Sections 6.4–6.7).

Since each homogeneous polynomial $f \in \mathcal{Q}$, $\deg(f) = d$, can be uniquely expressed as $f = \sum_{\tau \in \mathcal{W}_d} c(f, \tau)\tau$, we introduce indeterminate coefficients $a_{i,\tau}$, $1 \le i \le r$, $\tau \in \mathcal{W}_{d_i}$, and consider

> the domain $\mathbb{D} := \mathbb{Z}[a_{i,\tau}, 1 \le i \le r, \tau \in \mathcal{W}_{d_i}]$,
> its quotient field $\mathbb{K} := \mathbb{Q}(a_{i,\tau}, 1 \le i \le r, \tau \in \mathcal{W}_{d_i})$ and
> the "generic" forms $F_i = \sum_{\tau \in \mathcal{W}_{d_i}} a_{i,\tau}\tau, 1 \le i \le r.$

For any set $\mathbf{f} := \{f_1, \ldots, f_r\}$ of r concrete homogeneous forms

$$f_i = \sum_{\tau \in \mathcal{W}_{d_i}} c(f_i, \tau)\tau$$

of degree d_i, we denote by

$$\Xi_{\mathbf{f}} : \mathbb{D}[Z_1, \ldots, Z_r] \to K[Z_1, \ldots, Z_r]$$

the *ansatz*

$$\Xi_{\mathbf{f}}(a_{i,\tau}) = c(f_i, \tau) \qquad \text{for each } 1 \leq i \leq r, \tau \in \mathcal{W}_{d_i},$$

so that $\Xi_{\mathbf{f}}(F_i) = f_i$ for each i.

Definition 41.1.1. *A polynomial*

$$\mathrm{Res} := \mathrm{Res}(d_1, \ldots, d_r) := \mathrm{Res}(F_1, \ldots, F_r) \in \mathbb{D}$$

is called the resultant *of $F_1, \ldots, F_r$ if, for any set $\mathbf{f} := \{f_1, \ldots, f_r\}$ of r homogeneous forms f_i of degree d_i,*

$$\Xi_{\mathbf{f}}(\mathrm{Res}) = 0 \iff \text{there exists } \alpha \in \mathbb{P}^{r-1}(\mathbf{K}) : f_1(\alpha) = \cdots = f_r(\alpha) = 0.$$

$\boxed{\odot}$

Fix a variable, say Z_r, and define a *weight* on the variables $a_{i,\tau}$ by setting $\mathrm{wt}(a_{i,\tau}) = \deg_r(\tau)$, that is, $\mathrm{wt}(a_{i,\tau}) = a_r$ for each $\tau = Z_1^{a_1} \cdots Z_r^{a_r}$.
Set $D := \prod_{i=1}^r d_i$, $D_i := D/d_i$ for each i and

$$d := 1 - r + \sum_{i=1}^r d_i = 1 + \sum_{i=1}^r (d_i - 1).$$

Fact 41.1.2. *With the present notation the following holds:*

(1) For each $r - 1$ homogeneous polynomials $f_1, \ldots, f_{r-1} \in K[Z_1, \ldots, Z_r]$ which generate a homogeneous ideal $\mathbf{J} := (f_1, \ldots, f_{r-1})$ having only a finite number of (projective) zeros, we have $\#\mathcal{Z}(\mathbf{J}) = D_r = \prod_{i=1}^{r-1} d_i$.
(2) The resultant $\mathrm{Res}(F_1, \ldots, F_r)$ is
    (a) homogeneous of degree D_i in the variables $a_{i,\tau}$, for each i, and
    (b) isobaric[4] of weight D.

Proof.

(1) This is trivial for $r = 2$ and is obtained for $r > 2$ by considering the polynomials f_i as elements in $K[Z_1][Z_2, Z_3, \ldots, Z_r]$; thus (2) implies that $\mathrm{Res}(f_1, \ldots, f_{r-1}) \in K[Z_1]$ is a univariate polynomial of degree D_r.
(2) For $r = 2$, it is a homogeneous reformulation of the description of the structure of the Sylvester resultant. For

$$F_1(Z_1, Z_2) = a_0 \prod_{i=1}^{d_1}(Z_1 - \alpha_i Z_2) = a_0 Z_1^{d_1} + a_1 Z_1^{d_1-1} Z_2 + \cdots + a_{d_1} Z_2^{d_1},$$

$$F_2(Z_1, Z_2) = b_0 \prod_{j=1}^{d_2}(Z_1 - \beta_j Z_2) = b_0 Z_1^{d_2} + b_1 Z_1^{d_2-1} Z_2 + \cdots + b_{d_2} Z_2^{d_2},$$

---

[4] A function $f \in \mathbb{D}$ is called *isobaric* iff all of its terms are of the same weight.

if we consider the F_i as elements of $K[Z_2][Z_1]$, part (a) comes from Proposition 6.6.8 and part (b) from Proposition 6.7.1. This returns

$$\mathrm{Res}(F_1, F_2) = a_0^{d_2} b_0^{d_1} \prod_{i=1}^{d_1} \prod_{j=1}^{d_2} ((\alpha_i - \beta_j) Z_2) = a_0^{d_2} b_0^{d_1} Z_2^{d_1 d_2} \prod_{i=1}^{d_1} \prod_{j=1}^{d_2} (\alpha_i - \beta_j)$$

$$= a_0^{d_2} \prod_{i=1}^{d_1} F_2(\alpha_i Z_2) = a_0^{d_2} Z_2^{d_1} \prod_{i=1}^{d_1} g(\alpha_i)$$

$$= (-1)^{d_1 d_2} b_0^{d_1} \prod_{j=1}^{d_2} f(\beta_j Z_2) = (-1)^{d_1 d_2} b_0^{d_1} Z_2^{d_2} \prod_{j=1}^{d_2} f(\beta_j);$$

thus, $\mathrm{Res}(F_1, F_2) = Z_2^{d_1 d_2} \mathrm{Res}(f, g)$ is obtained by substituting each instance of a_i, b_j by $Z_2^i a_i, Z_2^j b_j$ respectively.

For $r > 2$, we obtain[5] (2) by applying (1) to the polynomials $f_1, \ldots, f_{r-1}$, for each *ansatz* $\Xi(F_i) = f_i$, $1 \le i < r$; denoting their roots by

$$(1, \lambda_2^{(i)}, \ldots, \lambda_r^{(i)}), \quad 1 \le i \le D_r = \prod_{i=1}^{r-1} d_i,$$

let us then define $R := \prod_{i=1}^{D_r} F_r(1, \lambda_2^{(i)}, \ldots, \lambda_r^{(i)}) \in \mathbb{D}$, which clearly satisfies $R = \Xi(\mathrm{Res}(F_1, \ldots, F_r))$ and which is the numerator of a symmetric function of the roots and thus can be expressed in terms of the coefficients of the f_i. Thus $\mathrm{Res}(F_1, \ldots, F_r)$ is isobaric of degree $d_r D_r = D$ and is homogeneous of degree D_r in the variables $a_{r,\tau}$.   ◉

41.2 Bézout's Generating Set

Let us denote by $\mathsf{J} := (F_1, \ldots, F_r) \subset \mathbb{D}[Z_1, \ldots, Z_r]$ the homogeneous ideal generated by the generic forms F_i and, for each $\delta \in \mathbb{N}$,

$$\mathsf{J}_\delta := \mathsf{J} \cap \mathcal{W}_\delta = \{F \in \mathsf{J} \text{ homogeneous}, \deg(F) = \delta\}.$$

We remark that[6]

$$B := \{\omega F_i : \omega \in \mathcal{W}_{\delta - d_i}, 1 \le i \le r\}$$

is a $\mathbb{D}$-generating set of J_δ so that, for each homogeneous polynomial $F \in \mathsf{J}_\delta$, there are homogeneous polynomials

$$S_i = \sum_{\omega \in \mathcal{W}_{\delta - d_i}} c(S_i, \omega)\omega \in \mathbb{D}[Z_1, \ldots, Z_r], \quad \deg(S_i) = \delta - d_i \text{ for each } i,$$

such that $F = \sum_{i=1}^{r} S_i F_i$; such representations are, of course, not unique.

---

[5] I limit myself here to sketching Poisson's proof of this result, which is never used in the argument which leads to Theorem 41.5.3. For a proof of Fact 41.1.2(2) I refer to Remark 41.5.1, below.

[6] With a slight abuse of notation, $\mathcal{W}_z = \emptyset$ for each $z \in \mathbb{Z}$, $z < 0$.

Uniqueness can be forced, however, by restricting the spans of the F_i: let us write, for each $i \leq r + 1$,

$$\mathcal{W}_\delta^{(i)} := \{Z_1^{a_1} \cdots Z_r^{a_r} \in \mathcal{W}_{\delta-d_i} : a_j < d_j \text{ for each } j < i\}$$

such that

- $\mathcal{W}_\delta^{(1)} = \mathcal{W}_{\delta-d_1}$;
- $\mathcal{W}_\delta^{(2)}$ consists of all terms of degree $\delta - d_2$ which are not divisible by $Z_1^{d_1}$;
- $\mathcal{W}_\delta^{(3)}$ consists of all terms of degree $\delta - d_3$ which are divisible by neither $Z_1^{d_1}$ nor $Z_2^{d_2}$;

  ...

- $\mathcal{W}_\delta^{(i)}$ consists of all terms τ of degree $\delta - d_i$ such that $\tau \notin (Z_1^{d_1}, \ldots, Z_{i-1}^{d_{i-1}})$;

  ...

- $\mathcal{W}_\delta^{(r+1)} := \{Z_1^{a_1} \cdots Z_r^{a_r} \in \mathcal{W}_\delta : a_j < d_j \text{ for each } j \leq r\}$.

Remark 41.2.1.　For each $\delta \geq d = 1 + \sum_{i=1}^r (d_i - 1)$, we have $\mathcal{W}_\delta^{(r+1)} = \emptyset$ since for $Z_1^{a_1} \cdots Z_r^{a_r} \in \mathcal{W}_\delta^{(r+1)}$ we have the contradiction

$$1 + \sum_i a_i \leq 1 + \sum_{i=1}^r (d_i - 1) = d \leq \delta = \sum_{i=1}^r a_i.$$

$\boxdot$

Lemma 41.2.2.　*With the notation above, for any $v \leq r$, $\mathcal{W}_\delta$ has the following partition:*

$$\mathcal{W}_\delta = \left\{\tau Z_1^{d_1}, \tau \in \mathcal{W}_\delta^{(1)}\right\} \sqcup \cdots \sqcup \left\{\tau Z_v^{d_v}, \tau \in \mathcal{W}_\delta^{(v)}\right\} \sqcup \mathcal{W}_\delta^{(v+1)}$$
$$= \mathcal{V}_\delta^{(1)} \sqcup \cdots \sqcup \mathcal{V}_\delta^{(i)} \sqcup \cdots \mathcal{V}_\delta^{(v)} \sqcup \mathcal{W}_\delta^{(v+1)},$$

where, for each $i \leq r$,

$$\mathcal{V}_\delta^{(i)} := \{Z_1^{a_1} \cdots Z_i^{a_i} \in \mathcal{W}_{\delta-d_i} : a_i \geq d_i, a_j < d_j \text{ for each } j < i\}.$$

$\boxdot$

Theorem 41.2.3 (Bézout).　*For any $v \leq r$, each homogeneous polynomial $F \in \mathbb{D}[Z_1, \ldots, Z_r]$ of degree δ can be uniquely expressed as*

$$\Delta F = \sum_{i=1}^v Q_i F_i + Q_{v+1},$$

where $\Delta \in \mathbb{D}, \Delta \neq 0$ and

$$Q_i = \sum_{\omega \in \mathcal{W}_\delta^{(i)}} c(Q_i, \omega)\omega \in \mathbb{D}[Z_1, \ldots, Z_r] \quad \text{for each } i.$$

Proof. Let us begin by remarking that, by Lemma 41.2.2 above,

$$\sum_{i=1}^{v+1} \#\mathcal{W}_\delta^{(i)} = \sum_{i=1}^{v} \#\mathcal{V}_\delta^{(i)} + \mathcal{W}_\delta^{(v+1)} = \#\mathcal{W}_\delta;$$

this counting argument can be also deduced by considering the polynomial $\sum_{i=1}^{v} Q_i Z_i^{d_i} + Q_{v+1}$, where each term of degree δ appears once and once only.

Equating, for each $\tau \in \mathcal{W}_\delta$, the coefficient of τ in F with that in $\sum_{i=1}^{v} Q_i F_i + Q_{v+1}$, we obtain $\#\mathcal{W}_\delta$ equations in the $\sum_{i=1}^{v+1} \#\mathcal{W}_\delta^{(i)} = \#\mathcal{W}_\delta$ unknowns $c(Q_i, \omega)$, $\omega \in \mathcal{W}_\delta^{(i)}$, $1 \le i \le v+1$.

The corresponding determinant Δ cannot vanish, otherwise we would obtain a non-trivial solution of the equation $\sum_{i=1}^{v} Q_i F_i + Q_{v+1} = 0$; it would then be sufficient to make the *ansatz* $F_i := Z_i^{d_i}$, $1 \le i \le v$, in order to deduce a contradictory identity $\sum_{i=1}^{v} Q_i Z_i^{d_i} + Q_{v+1} = 0$, where some Q_i is non-vanishing. Hence the theorem is proved.   ◉

Remark 41.2.4.   Denote by S_r the symmetric group of all the permutations π : $\{1, \ldots, r\} \to \{1, \ldots, r\}$ over $\{1, \ldots, r\}$.

The result of Theorem 41.2.3 being independent of the ordering of the variables, for each permutation $\pi \in \mathsf{S}_r$ we have that each homogeneous polynomial $F \in \mathbb{D}[Z_1, \ldots, Z_r]$ of degree δ can be uniquely expressed as $\Delta F = \sum_{i=1}^{v} \bar{Q}_i F_{\pi(i)} + \bar{Q}_{v+1}$, where $v \le r$, $\Delta \in \mathbb{D}$, $\Delta \ne 0$ and $\bar{Q}_i \in \mathrm{Span}_{\mathbb{K}}(\mathcal{W}_{\pi\delta}^{(i)})$ with

$$\mathcal{W}_{\pi\delta}^{(i)} := \{Z_1^{a_1} \cdots Z_r^{a_r} \in \mathcal{W}_{\delta-d_{\pi(i)}} : a_{\pi(j)} < d_{\pi(j)} \text{ for each } j < i\}.$$

◉

41.3 Macaulay's Matrix

Let us now represent the $\mathbb{D}$-generating set

$$B := \{\omega F_i : \omega \in \mathcal{W}_{d-d_i} 1 \le i \le r\}$$

of J_d by a matrix, *Macaulay's matrix*, whose columns are indexed by the $\binom{d+r-1}{r-1}$ terms $\tau \in \mathcal{W}_d$ and each of whose rows is indexed by one of the elements

$$\sum_{\tau \in \mathcal{W}_d} c(\omega F_i, \tau)\tau = \omega F_i \in B$$

and has as its τ-entry

$$c(\omega F_i, \tau) := \begin{cases} a_{i,v} & \text{if } \tau = v\omega, \\ 0 & \text{if } \omega \nmid \tau. \end{cases}$$

Example 41.3.1.   Let us set $Q = K[x, y, z]$, $d_1 = d_2 = 2, d_3 = 1$ and $\delta = 3$. In Figure 41.1 we represent such a matrix. Each 0 indicates that the entry is 0; each x

	x^3	x^2y	x^2z	xy^2	xyz	xz^2	y^3	y^2z	yz^2	z^3
xF_1	**X**	x	x	x	x	x	0	0	0	0
yF_1	0	**X**	0	x	x	0	x	x	x	0
zF_1	0	0	**X**	0	x	x	0	x	x	x
xF_2	x	x	x	**X**	x	x	0	0	0	0
yF_2	0	x	0	x	x	0	**X**	x	x	0
zF_2	0	0	x	0	x	x	0	**X**	x	x
x^2F_3	x	x	X	0	0	0	0	0	0	0
xyF_3	0	x	0	x	**X**	0	0	0	0	0
xzF_3	0	0	x	0	x	**X**	0	0	0	0
y^2F_3	0	0	0	x	0	0	x	X	0	0
yzF_3	0	0	0	0	x	0	0	x	**X**	0
z^2F_3	0	0	0	0	0	x	0	0	x	**X**

Figure 41.1. Macaulay's Matrix

indicates that the entry is a variable, $a_{i.\upsilon}$. The elements $a_{i.z_i^{d_i}}$ are represented by **X**. For the remaining notation, see Examples 41.3.3 and 41.4.6 below.   ⊙

Definition 41.3.2 (Macaulay). *The (Macaulay's) resultant of $F_1, \ldots, F_r$ is the greatest common divisor of all the determinants of the above matrix.*   ⊙

Denoting by

$$\mathsf{R} := \mathsf{R}(d_1, \ldots, d_r) := \mathsf{R}(F_1, \ldots, F_r) \in \mathbb{D} = \mathbb{Z}[a_{i.\tau}, 1 \le i \le r, \tau \in \mathcal{W}_{d_i}]$$

the Macaulay's resultant of $F_1, \ldots, F_r$, let us consider the determinant D of the square matrix[7] $\overline{\mathcal{D}}$ obtained by selecting the rows indexed by the polynomials in the basis

$$\mathsf{B} := \{\omega F_i : \omega \in \mathcal{W}_d^{(i)}, 1 \le i \le r\}$$

and let us study its properties.

Example 41.3.3. In Figure 41.1, removal of the two rows in sans serif (indexed by x^2F_3 and y^2F_3) leaves the matrix $\overline{\mathcal{D}}$.   ⊙

Proposition 41.3.4 (Macaulay). *Setting, for each $i \le r$,*

$$a_i := a_{i.z_i^{d_i}}, \quad c_i := a_{i.z_r^d}, \quad \mu_i := \#\mathcal{W}_d^{(i)},$$

the following hold:

(1) $c(\mathsf{D}, a_1^{\mu_1} \cdots a_r^{\mu_r}) = \pm 1;$

(2) $\Xi_{\mathbf{f}}(c_i) = \Xi_{\mathbf{f}}\left(a_{i.z_r^d}\right) = c(f_i, Z_r^d) = 0 \text{ for each } i \implies \Xi_{\mathbf{f}}(\mathsf{D}) = 0;$

(3) for each $\tau \in \mathcal{W}_d$, $\mathsf{D}\tau \in \mathsf{J};$

---

[7] The matrix is square as a consequence of Lemma 41.2.2 and Remark 41.2.1.

(4) $\mu_r := \#(\mathcal{W}_d^{(r)}) = \prod_{i=1}^{r-1} d_i = D_r;$

(5) D *is homogeneous of degree D_r in the coefficients of F_r, while*

(6) D *is homogeneous in the coefficients of F_j with degree $> D_j := D/d_j$ for each $j < r$;*

(7) D *is isobaric with weight $D := \prod_{i=1}^{r} d_i$.*

Proof.

(1) Clearly in each row there appears one and only one a_i; in order to prove the claim we must show that no two such a_i can appear in the same column. This is a trivial consequence of two remarks above, namely that $\mathcal{W}_d^{(r+1)} = \emptyset$ and that, making in the expression $\sum_{i=1}^{r} Q_i F_i + Q_{r+1} = \sum_{i=1}^{r} Q_i F_i$ the *ansatz* $F_i := a_i Z_i^{d_i}$, $1 \le i \le r$, we obtain $\sum_{i=1}^{r} Q_i a_i Z_i^{d_i}$, where each term of degree d appears once and once only.

(2) The column indexed by Z_r^d contains all zeros except in the rows indexed by $Z_r^{d-d_i} F_i$, where the value is $a_{i,Z_r^{d_i}}$.

(3) Denoting by $D_{i,\omega}$ the subdeterminant obtained by crossing the column indexed by τ and the row corresponding to the polynomial ωF_i, and solving for each τ the linear equations given by $\overline{\mathcal{D}}$, we have

$$\mathsf{D}\tau = \sum_{i,\omega} D_{i,\omega} c(\omega F_i, \tau) \equiv - \sum_{\substack{i,\omega}} \sum_{\substack{\upsilon \in \mathcal{W}_d \\ \upsilon \ne \tau}} D_{i,\omega} c(\omega F_i, \upsilon)\upsilon = 0 \bmod \mathsf{J}.$$

(4) We have $\deg(Q_r) = d - d_r = 1 + \sum_{i=1}^{r}(d_i - 1) - d_r = \sum_{i=1}^{r-1}(d_i - 1)$, and the terms of Q_r consist of the set of all terms in the expansion

$$\prod_{i=1}^{r-1} \left(Z_r^{d_i - 1} + Z_i Z_r^{d_i - 2} + \cdots + Z_i^{d_i - 2} Z_r + Z_i^{d_i - 1} \right),$$

whence $\mu_r = \#(\mathcal{W}_d^{(r)}) = \#\mathrm{supp}(Q_r) = \prod_{i=1}^{r-1} d_i = D_r$.

(5) This is a trivial consequence of the result above.

(6) In general, denoting by Ω the set of all terms ω in the expansion

$$\prod_{i=1}^{j-1} \left(1 + Z_i + \cdots + Z_i^{d_i - 1} \right),$$

and, for each $\omega \in \Omega$, writing

- $U_\omega := \{ Z_j^{a_j} \cdots Z_r^{a_r} : \sum_{i=j}^{r} a_r = d - d_j - \deg(\omega) \}$,
- $V_\omega := \{ Z_j^{a_j} \cdots Z_r^{a_r} : a_i < d_i, \text{ for each } i > j, a_j = \deg(\omega) - \sum_{i=j+1}^{r} a_r \}$,

the terms $\tau = \omega\upsilon$ of Q_j are each the product of a term $\omega \in \Omega$, and a term $\upsilon \in U_\upsilon$.

Note that since, for each $\upsilon \in U_\omega$,

$$\deg(\upsilon) = d - d_j - \deg(\omega) \ge 1 + \sum_{i=1}^{r}(d_i - 1) - d_j - \sum_{i=1}^{j-1}(d_i - 1) = \sum_{i=j+1}^{r}(d_i - 1),$$

we have $V_\omega \subseteq U_\omega$, with equality only for $\deg(\omega) = \sum_{i=1}^{r}(d_i - 1)$, so that

$$\mu_j = \#(\mathcal{W}_d^{(j)}) = \#\operatorname{supp}(Q_j) > \#\Omega\,\#V_\omega = \prod_{i=1}^{j-1} d_i \prod_{i=j+1}^{r} d_i = \prod_{\substack{i=1 \\ i \neq j}} d_i = D_i.$$

(7) For any element

$$\omega F_i = \sum_{\upsilon \in \mathcal{W}_{d-d_i}^{(r)}} a_{i,\upsilon}\omega\upsilon = \sum_{\tau \in \mathcal{W}_d} c(\omega F_i, \tau)\tau \in \mathsf{B}$$

we have, for each $c(\omega F_i, \tau) \neq 0$,

$$\operatorname{wt}(c(\omega F_i, \tau)) = \operatorname{wt}(a_{i,\upsilon}) = \deg_r(\upsilon) = \deg_r(\tau) - \deg_r(\omega);$$

hence, on expanding D, the weight of each term is

$$\sum_{\tau \in \mathcal{W}_d} \deg_r(\tau) - \sum_{i=1}^{r} \sum_{\omega \in \mathcal{W}_d^{(i)}} \deg_r(\omega).$$

Thus D is isobaric and, by (1), has weight

$$\operatorname{wt}(a_1^{\mu_1} \cdots a_r^{\mu_r}) = \mu_r \operatorname{wt}(a_r) = D_r d_r = D. \qquad \boxed{\odot}$$

Lemma 41.3.5 (Macaulay). *Any other determinant* $\mathsf{D}' = \det\left(\overline{\mathcal{D}}'\right)$ *of the above matrix has a common factor with* D *which is homogeneous of degree* D_r *in the coefficients of* F_r.

Proof. Let us denote by $H_j,\ 1 \leq j \leq \binom{d+r-1}{r-1}$, the polynomials corresponding to the rows of $\overline{\mathcal{D}}'$; according to Theorem 41.2.3, any arbitrarily $\mathbb{K}$-linear combination

$$\sum_{j} \alpha_j H_j = \sum_{i=1}^{r} A_i F_i \in \operatorname{Span}_{\mathbb{K}}(\mathcal{W}_d)$$

has a unique representation

$$\sum_{i=1}^{r} A_i F_i = \sum_{i=1}^{r} Q_i F_i, \quad Q_i \in \operatorname{Span}_{\mathbb{K}}(\mathcal{W}_d^{(i)}).$$

There is therefore a matrix $\mathcal{N} \in GL\left(\binom{d+r-1}{r-1}, \mathbb{K}\right)$ such that $\overline{\mathcal{D}}' = \mathcal{N}\overline{\mathcal{D}}$ and $\mathsf{D}\det(\mathcal{N}) = \mathsf{D}'$.

We can compute each $Q_i \in \operatorname{Span}_{\mathbb{K}}(\mathcal{W}_d^{(i)})$, and therefore the matrix $\mathcal{N}$, by the following recursive procedure.

- Compute the polynomials (whose existence is implied by Theorem 41.2.3) $Y_i^{(r)} \in$ $\mathrm{Span}_{\mathbb{K}}\left(\mathcal{W}_{d-d_r}^{(i)}\right)$, $i \le r$, such that $A_r = \sum_{i=1}^{r-1} Y_i^{(r)} F_i + Y_r^{(r)}$; since

$$\sum_{i=1}^{r} A_i F_i = \sum_{i=1}^{r-1} A_i F_i + A_r F_r$$

$$= \sum_{i=1}^{r-1} A_i F_i + \sum_{i=1}^{r-1} Y_i^{(r)} F_i F_r + Y_r^{(r)} F_r$$

$$= \sum_{i=1}^{r-1} (A_i + Y_i^{(r)} F_r) F_i + Y_r^{(r)} F_r$$

we can set $Q_r := Y_r^{(r)}$ and reduce the problem of solving the equation

$$\sum_{i=1}^{r} A_i F_i = \sum_{i=1}^{r} Q_i F_i$$

to that of solving $\sum_{i=1}^{r-1} (A_i + Y_i^{(r)} F_r) F_i = \sum_{i=1}^{r-1} Q_i F_i$.

- Q_{r-1} is then obtained by computing the polynomials[8] $Y_i^{(r-1)} \in \mathrm{Span}_{\mathbb{K}}\left(\mathcal{W}_{d-d_{r-1}}^{(i)}\right)$, $i \le r-1$, such that

$$A_{r-1} + Y_{r-1}^{(r)} F_r = \sum_{i=1}^{r-2} Y_i^{(r-1)} F_i + Y_{r-1}^{(r-1)},$$

thus allowing one to obtain, setting $Q_{r-1} := Y_{r-1}^{(r-1)}$,

$$\sum_{i=1}^{r} A_i F_i = Q_r F_r + \sum_{i=1}^{r-1} (A_i + Y_i^{(r)} F_r) F_i$$

$$= Q_r F_r + \sum_{i=1}^{r-2} (A_i + Y_i^{(r)} F_r) F_i + \sum_{i=1}^{r-2} Y_i^{(r-1)} F_i F_{r-1} + Y_{r-1}^{(r-1)} F_{r-1}$$

$$= \sum_{i=1}^{r-2} (A_i + Y_i^{(r)} F_r + Y_i^{(r-1)} F_{r-1}) F_i + \sum_{i=r-1}^{r} Q_i F_i.$$

- Inductively, we assume that

$$\sum_{i=1}^{r} A_i F_i = \sum_{i=1}^{r-j} \left(A_i + \sum_{l=r-j+1}^{r} Y_i^{(l)} F_l \right) F_i + \sum_{i=r-j+1}^{r} Q_i F_i$$

and compute the polynomials $Y_i^{(r-j)} \in \mathrm{Span}_{\mathbb{K}}\left(\mathcal{W}_{d-d_{r-j}}^{(i)}\right)$, $i \le r-j$, such that

$$A_{r-j} + \sum_{l=r-j+1}^{r} Y_{r-j}^{(l)} F_l = \sum_{i=1}^{r-j-1} Y_i^{(r-j)} F_i + Y_{r-j}^{(r-j)},$$

---

[8] Whose existence is implied by Theorem 41.2.3.

so that, setting $Q_{r-j} := Y_{r-j}^{(r-j)}$, we obtain

$$
\begin{aligned}
\sum_{i=1}^{r} A_i F_i &= \sum_{i=1}^{r-j}\left(A_i + \sum_{l=r-j+1}^{r} Y_i^{(l)} F_l\right) F_i + \sum_{i=r-j+1}^{r} Q_i F_i \\
&= \sum_{i=1}^{r-j-1}\left(A_i + \sum_{l=r-j+1}^{r} Y_i^{(l)} F_l\right) F_i + \sum_{i=1}^{r-j-1} Y_i^{(r-j)} F_i F_{r-j} \\
&\quad + Y_{r-j}^{(r-j)} F_{r-j} + \sum_{i=r-j+1}^{r} Q_i F_i \\
&= \sum_{i=1}^{r-j-1}\left(A_i + \sum_{l=r-j}^{r} Y_i^{(l)} F_l\right) F_i + \sum_{i=r-j}^{r} Q_i F_i
\end{aligned}
$$

- and so forth.

The point is that in this procedure all the polynomials $Y_i^{(l)}$ are independent of the coefficients $a_{r,\tau}$ of F_r, and the same is true for $\det(\mathcal{N}) \in \mathbb{K}$; whence the claim follows. $\quad\boxdot$

As the determinant D depends on the ordering of the variables, we can in fact choose $r!$ diffierent determinants D_π, $\pi \in \mathsf{S}_r$, each satisfying (cf. Remark 41.2.4) an appropriate reformulation of Theorem 41.2.3.

In particular we have

Corollary 41.3.6. *It holds that:*

(1) for each $i \le r$ and each $\pi \in \mathsf{S}_r$ such that $\pi(r) = i$, the determinant D_π is homogeneous of degree D_i in the coefficients of F_i;

(2) with such a determinant any other determinant D' has a common factor which is homogeneous of degree D_i in the coefficients of F_i. $\quad\boxdot$

Corollary 41.3.7. *The resultant R is homogeneous of degree $D_i = \mu_i$ in the coefficients of F_i for each i.* $\quad\boxdot$

Example 41.3.8. In Example 41.3.1, if we consider the permutation (132) the construction would return Figure 41.2. $\quad\boxdot$

Definition 41.3.9. *The element $\mathsf{A} \in \mathbb{D}$ such that $\mathsf{D} = \mathsf{A}\mathsf{R}$ is called the* extraneous factor *of D.* $\quad\boxdot$

Denoting by $\chi_v : \mathbb{D}[Z_1, \ldots, Z_r] \to \mathbb{D}[Z_1, \ldots, Z_{v-1}]$ the evaluation

$$
F(Z_1, \ldots, Z_r) \mapsto \chi_v(F) = F(Z_1, \ldots, Z_{v-1}, 0, \ldots, 0),
$$

we have:

	x^3	x^2y	x^2z	xy^2	xyz	xz^2	y^3	y^2z	yz^2	z^3
$x^2 F_3$	x	x	**X**	0	0	0	0	0	0	0
$xy F_3$	0	x	0	x	**X**	0	0	0	0	0
$xz F_3$	0	0	x	0	x	**X**	0	0	0	0
$y^2 F_3$	0	0	0	x	0	0	x	**X**	0	0
$yz F_3$	0	0	0	0	x	0	0	x	**X**	0
$z^2 F_3$	0	0	0	0	0	x	0	0	x	**X**
$x F_1$	**X**	x	x	x	x	x	0	0	0	0
$y F_1$	0	**X**	0	x	x	0	x	x	x	0
zF_1	0	0	**X**	0	x	x	0	x	x	x
$x F_2$	x	x	x	**X**	x	x	0	0	0	0
$y F_2$	0	x	0	x	x	0	**X**	x	x	0
zF_2	0	0	x	0	x	x	0	**X**	x	x

Figure 41.2. Macaulay's Matrix (2)

Corollary 41.3.10.   *The following holds:*

(1) $c(\mathsf{R}, a_1^{D_1} \cdots a_r^{D_r}) = \pm 1$;

(2) R *is homogeneous of degree* D_i *in the coefficients of* F_i *for each* i;

(3) R *is isobaric of degree* D;

(4) A *is isobaric of degree* 0;

(5) $c(\mathsf{A}, a_{i,\tau}) = 0$ *for each* $\tau \in \mathcal{W} : Z_r \mid \tau$;

(6) A *is independent of the coefficients* $a_{r,\tau}$ *of* F_r;

(7) A *depends only on the coefficients of the generic polynomials*

$$\chi_r(F_1), \ldots, \chi_r(F_{r-1}) \in \mathbb{Z}[a_{i,\tau}, 1 \le i < r, \tau \in \mathcal{W}_{d_i}][Z_1, \ldots, Z_{r-1}];$$

(8) $\Xi_{\mathbf{f}}(c_i) = \Xi_{\mathbf{f}}\left(a_{i,Z_r^d}\right) = c(f_i, Z_r^d) = 0$ *for each* $i \implies \Xi_{\mathbf{f}}(\mathsf{R}) = 0.$

Proof.

(1) This part follows from Proposition 41.3.4(1) and from the equality $D_i = \mu_i$ for each i.

(2) This is a direct consequence of Corollary 41.3.7.

(3) Each factor of an isobaric function is isobaric too; the weight of R is $\mathrm{wt}(a_1^{D_1} \cdots a_r^{D_r}) = D$.

(4) This is a direct corollary of (3), and (5) of (4).

(6) This is the statement at the end of the proof of Lemma 41.3.5.

(7) Here we have a résumé of (5) and (6).

(8) From Proposition 41.3.4(2) which implies that $\Xi_{\mathbf{f}}(\mathsf{D}) = 0$, and (5) which implies that A is independent of each $c_i = a_{i,Z_r^d}$.   ⊙

41.4 The Extraneous Factor

In order to give an explicit and effective representation of the extraneous factor A of D, Macaulay needed a deeper analysis and a convenient notation, starting from

Bézout's formula (Theorem 41.2.3). Thus, considering, for each $v \leq r$ and each $\delta \geq \min(d_1, \ldots, d_v)$, the *submatrix* obtained by selecting the columns indexed by the terms in

$$\mathcal{W}_\delta \setminus \mathcal{W}_\delta^{(v+1)} = \left\{ \tau Z_1^{d_1}, \tau \in \mathcal{W}_\delta^{(1)} \right\} \sqcup \cdots \sqcup \left\{ \tau Z_v^{d_v}, \tau \in \mathcal{W}_\delta^{(v)} \right\}$$

and the rows indexed by the elements

$$\omega F_i = \sum_{\tau \in \mathcal{W}_d} c(\omega F_i, \tau)\tau \in \mathsf{B}_v := \{\omega F_i : \omega \in \mathcal{W}_\delta^{(i)}, 1 \leq i \leq v\},$$

Macaulay denotes by $\mathsf{D}(v, \delta)$ its determinant and sets[9]

$$\mathsf{R}(v, \delta) := \gcd(\mathsf{D}_\pi(v, \delta) : \pi \in \mathsf{S}_v\}.$$

He also sets $\mathsf{D}(v, \delta) = \mathsf{R}(v, \delta) = 1$ for each $\delta < \min(d_1, \ldots, d_v)$.

Remark 41.4.1.   The following conditions are equivalent:

(1) $\Xi_{\mathbf{f}}(\mathsf{D}(v, \delta)) = 0$;
(2) $\sum_{i=1}^{v} Q_i \chi_v(f_i) = Q_{v+1}$ for suitable $Q_i = \sum_{\omega \in \mathcal{W}_\delta^{(i)}} c(Q_i, \omega)\omega \in \mathcal{Q}, i \leq v+1$.

Moreover, it is sufficient to repeat the same argument as that which led to the proof of Corollary 41.3.7, in order to obtain that $\mathsf{R}(v, \delta)$ is homogeneous for each $i \leq v$ in the coefficients of F_i with degree

$$\#\{Z_1^{a_1} \cdots Z_r^{a_r} \in \mathcal{W}_{\delta - d_i} : a_j < d_j \text{ for each } j \neq i\}. \qquad \boxed{\odot}$$

     This notation allows one to state

Theorem 41.4.2 (Macaulay).   *It holds that*

$$\left| \frac{\mathsf{D}(r, \delta)}{\mathsf{R}(r, \delta)} \right| = \left| \prod_{j=0}^{d_r - 1} \frac{\mathsf{D}(r-1, \delta - j)}{\mathsf{R}(r-1, \delta - j)} \right| \left| \prod_{j=d_r}^{\delta - 1} \mathsf{D}(r-1, \delta - j) \right|.$$

Proof.   The resultant $\mathsf{R}(r, \delta)$ is a factor of $\mathsf{D}(r, \delta)$ and, by Corollary 41.3.10(6), the other factors are independent of the coefficients of F_r.

     Let us consider an arbitrary combination

$$\sum_{i=1}^{r} Q_i F_i = 0, \qquad Q_i = \sum_{\omega \in \mathcal{W}_\delta^{(i)}} x(Q_i, \omega)\omega, \qquad 1 \leq i \leq r,$$

in terms of the unknowns $x(Q_i, \omega)$, where

$$\mathcal{W}_\delta^{(i)} := \{Z_1^{a_1} \cdots Z_r^{a_r} \in \mathcal{W}_{\delta - d_i} : a_j < d_j \text{ for each } j < i\},$$

and write $a := a_{r, Z_r^{d_r}}$ and $w := \#\mathcal{W}_\delta^{(r)}$.

---

[9]   In this construction the value v fixes the precise set $\{Z_1, \ldots, Z_v\}$ of the first v variables; therefore, in this setting Remark 41.2.4 is applicable to those variables only.

The element a appears w times in the matrix, namely in the positions satisfying $c(\omega F_r, \omega Z_r^{d_r}) = a = a_{r, Z_r^{d_r}}$, where ω runs over the elements of $\mathcal{W}_\delta^{(r)}$; more precisely, for each such ω, a appears in the position corresponding to the column indexed by $\omega Z_r^{d_r}$ and the row representing ωF_r.

The columns where a does not appear are those indexed by the terms

$$\left\{\tau Z_1^{d_1}, \tau \in \mathcal{W}_\delta^{(1)}\right\} \sqcup \cdots \sqcup \left\{\tau Z_{r-1}^{d_{r-1}}, \tau \in \mathcal{W}_\delta^{(r-1)}\right\}.$$

Hence the coefficient of a^w in the expansion of $\mathrm{D}(r, \delta)$ is the determinant whose vanishing is the condition that the identity

$$\sum_{i=1}^{r-1} Q_i F_i = Q_r, \quad Q_i = \sum_{\omega \in \mathcal{W}_\delta^{(i)}} x(Q_i, \omega)\omega, \quad 1 \le i \le r,$$

is satisfied.

In order to evaluate such a determinant, assume the identity is satisfied and set $Z_r = 0$, obtaining the identity

$$\sum_{i=1}^{r-1} \chi_r(Q_i)\chi_r(F_i) = \chi_r(Q_r).$$

Therefore, for each $\mathbf{f} := \{f_1, \ldots, f_r\}$, either

- such an identity is non-trivially satisfied, that is, $\Xi_{\mathbf{f}}(\mathrm{D}(r-1, \delta)) = 0$, or
- $\chi_r(Q_1) = \cdots = \chi_r(Q_r) = 0$, which means that $Z_r \mid Q_i$ for each i.

In the latter case we obtain a similar identity,

$$\sum_{i=1}^{r-1} Q_i' f_i = Q_r', \qquad Q_i' = \sum_{\omega \in \mathcal{W}_{\delta-1}^{(i)}} x(Q_i, Z_r\omega)\omega, \quad 1 \le i \le r.$$

Repeating the same argument we obtain that either $\Xi_{\mathbf{f}}(\mathrm{D}(r-1, \delta-1)) = 0$ or there is an identity

$$\sum_{i=1}^{r-1} Q_i'' f_i = Q_r'', \qquad Q_i'' = \sum_{\omega \in \mathcal{W}_{\delta-2}^{(i)}} x(Q_i, Z_r^2\omega)\omega, \quad 1 \le i \le r.$$

Since we have only the limitation

$$\max \left\{a_r : Z_1^{a_1} \cdots Z_r^{a_r} \in \cup_{i=1}^{r-1} \mathcal{W}_\delta^{(i)}\right\} = \delta,$$

by iteration we deduce that the required determinant is $\prod_{j=0}^{\delta-1} \mathrm{D}(r-1, \delta-j)$, so that

$$\mathrm{D}(r, \delta) = \left(\prod_{j=0}^{\delta-1} \mathrm{D}(r-1, \delta-j)\right) a^w + \cdots.$$

Let us therefore now evaluate the coefficient of a^w in $R(r, \delta)$; to do so we consider another permutation, say the cyclic one, $\pi(i) \equiv i-1 \pmod{r}$, and the corresponding identity

$$Q_1 F_r + Q_2 F_1 + \cdots + Q_r F_{r-1} = 0, \qquad Q_i = \sum_{\omega \in \mathcal{W}_{\pi\delta}^{(i)}} x(Q_i, \omega)\omega, \quad 1 \le i \le r,$$

where $\mathcal{W}_{\pi\delta}^{(i)} := \{ Z_1^{a_1} \cdots Z_r^{a_r} \in \mathcal{W}_{\delta - d_{\pi(i)}} : a_{\pi(j)} < d_{\pi(j)} \text{ for each } j < i \}$ and, in particular,

$$\mathcal{W}_{\pi\delta}^{(1)} := \{ Z_1^{a_1} \cdots Z_r^{a_r} \in \mathcal{W}_{\delta - d_r} : a_r < d_r \}.$$

The variable $a := a_{r, Z_r^{d_r}}$ appears $s := \# \mathcal{W}_{\pi\delta}^{(1)}$ times in the positions that satisfy $c(\omega F_r, \omega Z_r^{d_r}) = a$ and correspond to the column indexed by $\omega Z_r^{d_r}$ and the row representing ωF_r, where ω runs over the elements of $\mathcal{W}_{\pi\delta}^{(1)}$.

The remaining columns are those indexed by the terms

$$\left\{ \tau Z_1^{d_1}, \tau \in \mathcal{W}_{\pi\delta}^{(2)} \right\} \sqcup \cdots \sqcup \left\{ \tau Z_{r-1}^{d_{r-1}}, \tau \in \mathcal{W}_{\pi\delta}^{(r)} \right\}$$

and the coefficient of a^s in the expansion of $D_\pi(r, \delta)$ is the determinant whose vanishing is the condition that the identity

$$\sum_{i=1}^{n-1} Q_{i+1} F_i = Q_1, \qquad Q_i = \sum_{\omega \in \mathcal{W}_{\pi\delta}^{(i)}} x(Q_i, \omega)\omega, \, 1 \le i \le r,$$

is satisfied.

We can therefore reapply the same argument as above, setting $Z_r = 0$ and obtaining that, for each $\mathbf{f} := \{f_1, \ldots, f_r\}$, either $\Xi_{\mathbf{f}}(D_\pi(r-1, \delta)) = 0$ or each Q_i is divisible by Z_r; however, since we have the stricter limitation

$$\max \left\{ a_r : Z_1^{a_1} \cdots Z_r^{a_r} \in \cup_{i=1}^{r-1} \mathcal{W}_\delta^{(i)} \right\} = d_r,$$

we obtain only

$$D_\pi(r, \delta) = \left(\prod_{j=0}^{d_r - 1} D_\pi(r-1, \delta - j) \right) a^s + \cdots .$$

Since the same argument can be applied for each permutation π', $\pi'(1) = r$, we obtain

$$R(r, \delta) = \left(\prod_{j=0}^{d_r - 1} R(r-1, \delta - j) \right) a^s + \cdots .$$

Since $R(r, \delta)$ is a factor of $D(r, \delta)$ and the other factors are independent of the coefficients of F_r we thus obtain the claim. $\qquad \boxed{\odot}$

Definition 41.4.3. *A term* $Z_1^{a_1} \cdots Z_r^{a_r} \in \mathcal{W}$ *satisfying*

$$a_j < d_j \quad \text{for each } j \in \{i_1, \dots, i_h\} \subset \{1, \dots, r\}$$

is said to be reduced *in* $\{Z_{i_1}, \dots, Z_{i_h}\}$. $\qquad\qquad\qquad$ ⊙

Now let

$\mathsf{S} \subset \mathcal{W}$ be the semigroup ideal generated by $\{Z_i^{d_i}, 1 \le i \le r\}$,

$\mathsf{S}_\star \subset \mathcal{W}$ be the semigroup ideal generated by $\{Z_i^{d_i} Z_j^{d_j}, 1 \le i < j \le r\}$,

$\mathcal{W}_\star \subset \mathcal{W}$ be the set of terms which is divisible by a *single* term $Z_i^{d_i}$,

$\mathcal{W}_{\star\delta}^{(i)} := \{Z_1^{a_1} \cdots Z_r^{a_r} \in \mathcal{W}_\delta^{(i)} : a_j < d_j \text{ for each } j > i\}$,

$\mathcal{U}_\delta^{(i)} := \{Z_1^{a_1} \cdots Z_r^{a_r} \in \mathcal{W}_\delta^{(i)} : \text{ there exists } h > i : a_h \ge d_h\}$,

and set $\mathsf{S}_\delta := \mathsf{S} \cap \mathcal{W}_\delta$, $\mathcal{W}_{\star\delta} := \mathcal{W}_\star \cap \mathcal{W}_\delta$ and $\mathsf{S}_{\star\delta} := \mathsf{S}_\star \cap \mathcal{W}_\delta$. Then

Lemma 41.4.4. *It holds that:*

(1) $\mathcal{W} \setminus \mathsf{S} = \bigcup_\delta \mathcal{W}_\delta^{(r+1)}$ *is the set of all terms reduced in* $\{Z_i, 1 \le i \le r\}$;

(2) $\mathcal{W} = \mathcal{W}_\star \sqcup \mathsf{S}_\star \sqcup \bigcup_\delta \mathcal{W}_\delta^{(r+1)}$;

(3) $\mathcal{W}_\delta \setminus \mathcal{W}_\delta^{(r+1)} = \mathcal{W}_{\star\delta} \sqcup \mathsf{S}_{\star\delta}$;

(4) $\mathcal{W}_\delta = \mathcal{W}_{\star\delta} \sqcup \mathsf{S}_{\star\delta}$ *for each* $\delta \ge d$;

(5) $\mathcal{W}_\star = \bigcup_{i=1}^r \{Z_1^{a_1} \cdots Z_r^{a_r} \in \mathcal{W} \setminus \bigcup_\delta \mathcal{W}_\delta^{(r+1)} : a_j < d_j \text{ for each } j \ne i\}$;

(6) $\mathcal{W}_\star$ *is the set of all terms which are reduced in* $\{Z_j, j \ne i\}$ *for some i but are not reduced in* $\{Z_1, \dots, Z_r\}$;

(7) $\mathcal{W}_{\star\delta}^{(i)} = \{Z_1^{a_1} \cdots Z_r^{a_r} \in \mathcal{W}_{\delta - d_i} : a_j < d_j \text{ for each } j \ne i\}$ *is the set of terms of degree* $\delta - d_i$ *which are reduced in* $\{Z_1, \dots, Z_{i-1}, Z_{i+1}, \dots, Z_r\}$;

(8) $\overline{\mathcal{W}}_\delta := \mathcal{W}_{\star\delta} \cup \mathcal{W}_\delta^{(r+1)} = \bigcup_i \mathcal{W}_{\star\delta}^{(i)}$ *and* $\mathcal{W}_\delta^{(r+1)} = \bigcap_i \mathcal{W}_{\star\delta}^{(i)}$;

(9) $\mathcal{W}_\delta^{(i)} = \mathcal{W}_{\star\delta}^{(i)} \sqcup \mathcal{U}_\delta^{(i)}$;

(10) $\mathcal{U}_\delta^{(i)} = \{Z_1^{a_1} \cdots Z_r^{a_r} \in \mathcal{W}_{\delta - d_i} : a_j < d_j \text{ for each } j < i, \text{there exists } h > i : a_h \ge d_h\}$;

(11) $\mathcal{W}_{\star\delta}^{(r)} = \mathcal{W}_\delta^{(r)}$;

(12) *for each* $\tau \in \mathcal{W}_{\star\delta}$ *there is an* $i \le r$ *and an* $\omega \in \mathcal{W}_{\star\delta}^{(i)}$ *such that* $c(\omega F_i, \tau) = a_{i,Z_i^{d_i}}$;

(13) *for each* $i \le r$ *and* $\omega \in \mathcal{W}_{\star\delta}^{(i)}$, *we have that*

$$c(\omega F_i, \tau) = a_{i,Z_i^{d_i}} \implies \tau = \omega Z_i^{d_i} \in \mathcal{W}_{\star\delta}.$$
$\qquad\qquad$ ⊙

Definition 41.4.5. *The* extraneous factor *of* $\mathsf{D}(r, \delta)$, $\mathsf{A}(r, \delta)$, *is the determinant of the minor of* $\mathsf{D}(r, \delta)$ *obtained on removing the columns indexed by the terms in* $\mathcal{W}_{\star\delta}$ *and the rows indexed by the polynomials which contain the elements* $a_{i,Z_i^{d_i}}$, $1 \le i \le r$, *in the omitted columns, that is, the rows indexed by the set*

$$\{\omega F_i : \omega \in \mathcal{W}_{\star d}^{(i)}, 1 \le i \le r\}.$$

Alternatively, the surviving columns are those indexed by the terms $\tau \in S_{*\delta}$ *and the surviving rows are those relating to the elements in*

$$\{\omega F_i : \omega \in \mathcal{U}_d^{(i)}, 1 \le i \le r\}. \qquad \boxed{\odot}$$

Example 41.4.6.   In Figures 41.1 and 41.2 the elements of the extraneous factor are in boxes.

Note that in Figure 41.2 variables (respectively, polynomials) are ordered as z, x, y (respectively, F_3, F_1, F_2). $\boxed{\odot}$

Remark 41.4.7.   Since, in the construction of $D(\nu, \delta)$ and $R(\nu, \delta)$, the value ν fixes the precise set $\{Z_1, \ldots, Z_\nu\}$ of the first ν variables, the definition of the extraneous factor can be naturally extended to define $A(\nu, \delta)$. $\boxed{\odot}$

Theorem 41.4.8 (Macaulay).   *It holds that:*

(1) $|A(r, \delta)| = \left| \prod_{j=0}^{d_r - 1} A(r - 1, \delta - j) \right| \left| \prod_{j=d_r}^{\delta - 1} D(r - 1, \delta - j) \right|$;

(2) $D(2, \delta) = A(2, \delta)R(2, \delta)$;

(3) $D(r, \delta) = A(r, \delta)R(r, \delta)$.

Proof.   Since (2) requires only a trivial verification and allows the deduction of (3) from (1), we just need to prove (1).

The vanishing of $A(r, \delta)$ is the condition that the identity

$$\sum_{i=1}^{r-1} Q_i F_i = Q_r, \quad Q_i = \sum_{\omega \in \mathcal{U}_\delta^{(i)}} x(Q_i, \omega)\omega, \quad 1 \le i < r, \quad Q_r = \sum_{\omega \in \overline{\mathcal{W}}_\delta} x(Q_r, \omega)\omega$$

can be solved in terms of the unknowns $x(Q_i, \omega)$.

The number of linear equations and unknowns are equal and $A(r, \delta)$ is non-zero since, for the *ansatz* $\Xi_Z(F_i) := Z_i^{d_i}$, in the polynomial $\sum_{i=1}^{r-1} Q_i Z_i^{d_i} + Q_r$ each term in $\mathcal{W}_\delta$ occurs once and once only.

Assume, again, that the identity is satisfied and set $Z_r = 0$, obtaining the identity

$$\sum_{i=1}^{r-1} \chi_r(Q_i)\chi_r(F_i) = \chi_r(Q_r).$$

Now reapply the same argument as in Theorem 41.4.2, obtaining, for each $\mathbf{f} := \{f_1, \ldots, f_r\}$, that either $\Xi_\mathbf{f}(A(r - 1, \delta)) = 0$ or each Q_i is divisible by Z_r; repeating this argument we obtain that either $\Xi_\mathbf{f}(A(r - 1, \delta - i)) = 0, 0 \le i < d_r$ or there is an identity $\sum_{i=1}^{r-1} Q_i'' f_i = Q''$, where

$$Q_i'' = \sum_{\omega \in \mathfrak{U}_{\delta-d_r}^{(i)}} x(Q_i, Z_r^{d_r}\omega)\omega, \quad 1 \le i < r, \qquad Q'' = \sum_{\omega \in \mathfrak{W}_{\delta-d_r}} x(Q_i, Z_r^{d_r}\omega)\omega,$$

for $\mathfrak{U}_{\delta-d_r}^{(i)} := \{\omega : Z_r^{d_r}\omega \in \mathcal{U}_\delta^{(i)}\}$ and $\mathfrak{W}_{\delta-d_r} := \{\omega : Z_r^{d_r}\omega \in \overline{\mathcal{W}}_\delta\} = \mathcal{W}_{\delta-d_r}^{(r)}$.

We have a similar relation,

$$\mathfrak{U}^{(i)}_{\delta-d_r} := \{\omega : Z_r^{d_r}\omega \in \mathcal{U}^{(i)}_\delta\} = \mathcal{W}^{(i)}_{\delta-d_r},$$

also for $i < r$ since

$$Z_r^{d_r}\omega \in \mathcal{W}^{(i)}_{\star\delta} = \{Z_1^{a_1}\cdots Z_r^{a_r} \in \mathcal{W}_{\delta-d_i} : a_j < d_j \text{ for each } j \neq i\} \implies i = r,$$

whence

$$\begin{aligned}
\mathfrak{U}^{(i)}_{\delta-d_r} &= \{\omega : Z_r^{d_r}\omega \in \mathcal{U}^{(i)}_\delta\}\\
&= \{\omega : Z_r^{d_r}\omega \in \mathcal{W}^{(i)}_\delta \setminus \mathcal{W}^{(i)}_{\star\delta}\}\\
&= \{\omega : Z_r^{d_r}\omega \in \mathcal{W}^{(i)}_\delta\}\\
&= \mathcal{W}^{(i)}_{\delta-d_r}.
\end{aligned}$$

Thus, we can conclude that either $\Xi_{\mathbf{f}}(\mathsf{A}(r-1,\delta-i)) = 0$, $0 \leq i < d_r$, or there is an identity

$$\sum_{i=1}^{r-1} Q_i'' f_i = Q_r'', \qquad Q_i'' = \sum_{\omega \in \mathcal{W}^{(i)}_{\delta-d_r}} x(Q_i, X_n^{d_n}\omega)\omega, \quad 1 \leq i \leq r,$$

that is (by Theorem 41.4.2), $\prod_{j=d_n}^{\delta-1} \Xi_{\mathbf{f}}(\mathsf{D}(n-1,\delta-j)) = 0$. ◉

Corollary 41.4.9. *The* extraneous factor A *of* $\mathsf{D} = \mathsf{D}(n,d)$ *satisfying*

$$\mathsf{A} = \frac{\mathsf{D}}{\mathsf{R}} = \frac{\mathsf{D}(n,d)}{\mathsf{R}(n,d)}$$

is $\mathsf{A} := \mathsf{A}(n,d).$ ◉

41.5 Macaulay's Resultant

Remark 41.5.1. If the resultant $\mathrm{Res}(f_1,\ldots,f_r)$ vanishes then the f_i, have a common root $\alpha \in \mathbb{P}^{r-1}(K)$, and all polynomials in

$$\Xi_{\mathbf{f}}(B) := \{\omega f_i : \omega \in \mathcal{W}^{(i)}_d, 1 \leq i \leq r\}$$

vanish when evaluated at such a root; thus, setting $x_\upsilon := \upsilon(\alpha)$ for each $\upsilon \in \mathcal{W}$, $(x_\upsilon : \upsilon \in \mathcal{W}_d)$ is a common root of the linear equations

$$\sum_{\upsilon \in \mathcal{W}_d} x_{\omega\upsilon} c(f_i, \upsilon) = 0, \quad \omega \in \mathcal{W}^{(i)}_d, \quad 1 \leq i \leq r,$$

and $\Xi_{\mathbf{f}}(\mathsf{D}_\pi) = 0$ for each $\pi \in \mathsf{S}_r$.

Thus $\mathrm{Res}(d_1,\ldots,d_r)$ divides each D_π, $\pi \in \mathsf{S}_r$, and hence divides R; since both R (Corollary 41.3.10) and $\mathrm{Res}(d_1,\ldots,d_r)$ (Fact 41.1.2(2)) are isobaric of weight D, in principle we can conclude that R is the sought-after resultant. However, since we have not given a complete proof of Fact 41.1.2(2), we prefer to explicitly prove that R is the resultant and deduce Fact 41.1.2(2) from Theorem 41.5.3 below. ◉

Lemma 41.5.2 (Macaulay). *The coefficients of a generic member of* J_{d-1} *satisfy one and only one identical linear relation.*[10] ⊙

Proof. We need to prove that $\dim_{\mathbb{K}}(\mathsf{J}_{d-1}) = \#\mathcal{W}_{d-1} - 1$.

In fact, the equation

$$\sum_{i=1}^{r} A_i F_i = \sum_{i=1}^{r} Q_i F_i, \quad Q_i \in \operatorname{Span}_{\mathbb{K}}(\mathcal{W}_{d-1}^{(i)}),$$

can be solved by the method used in Lemma 41.3.5 for arbitrary given polynomials A_i; thus $\dim_{\mathbb{K}}(\mathsf{J}_{d-1})$ is less than or equal to the number of coefficients in the expression $\sum_{i=1}^{r} Q_i Z_i^{d_i}$, which is $\#\mathcal{W}_{d-1} - 1$ since each term in $\#\mathcal{W}_{d-1}$ except $\Omega := \prod_{i=1}^{r} Z_i^{d_i-1}$ occurs once and only once in that expression.

In order to prove that this equality is strict, it is sufficient to show that it is satisfied by at least a specific *ansatz*. Macaulay considers the *ansatz* $\Xi_1(F_i) := f_i$, $1 \le i \le r$, where

$$f_i := (Z_i - Z_{i+1})Z_i^{d_i-1}, \quad 1 \le i < r, \qquad f_r := (Z_r - Z_1)Z_r^{d_r-1};$$

clearly $\Xi_1(\mathsf{R}) = 0$ since the system $f_1 = \cdots = f_r = 0$ has the common root $(1, 1, \ldots, 1)$.

In order to prove that $\dim_K(\Xi_1(\mathsf{J}_{d-1})) = \#\mathcal{W}_{d-1} - 1$, Macaulay shows that, for each term $\tau := Z_1^{a_1} \cdots Z_r^{a_r}$, $\tau - \Omega \in (f_1, \ldots, f_r) = \Xi_1(\mathsf{J})$. He does this by proposing an interesting rewriting procedure, which it is worthwhile to quote. Given τ, set $\iota := 1$ and repeatedly perform the following transformation:

- if $a_\iota \ge d_\iota$, set $\tau := Z_1^{a_1} \cdots Z_{\iota-1}^{a_{\iota-1}} Z_\iota^{d_\iota-1} Z_{\iota+1}^{a_{\iota+1}+a_\iota-d_\iota+1} \cdots Z_r^{a_r}$, which is equivalent to transforming τ to

$$\tau - (Z_\iota^{a_\iota} - Z_\iota^{d_\iota-1} Z_{\iota+1}^{a_\iota-d_\iota+1})\frac{\tau}{Z_\iota^{a_\iota}} = \tau - (Z_\iota^{a_\iota-d_\iota+1} - Z_{\iota+1}^{a_\iota-d_\iota+1})Z_\iota^{d_\iota-1}\frac{\tau}{Z_\iota^{a_\iota}}$$

$$= \tau - \frac{Z_\iota^{a_\iota-d_\iota+1} - Z_{\iota+1}^{a_\iota-d_\iota+1}}{Z_\iota - Z_{\iota+1}}\frac{\tau}{Z_\iota^{a_\iota}}(Z_\iota - Z_{\iota+1})Z_\iota^{d_\iota-1}$$

$$= \tau - \frac{Z_\iota^{a_\iota-d_\iota+1} - Z_{\iota+1}^{a_\iota-d_\iota+1}}{Z_\iota - Z_{\iota+1}}\frac{\tau}{Z_\iota^{a_\iota}}\Xi_1(f_\iota),$$

- set $\iota := \iota + 1 \bmod r$,

going round the cycle[11] $Z_1, Z_2, \ldots, Z_r, Z_1$ as many times as needed until we obtain the term Ω. ⊙

---

[10] Both this result and the theorem below require $d \ge 2$, that is, the existence of an at least non-linear polynomial. However, if $d_i = 1$ for each i, so that $d = 1$, this lemma is empty but the theorem below then claims that the determinant of a system of r linear equations in r variables vanishes if and only if the system has a common root.

[11] For $r = 4$, $d_i = 4$, and $d - 1 = 12$ we have e.g.

$$Z_1^7 Z_4^5 \quad \rightarrow \quad Z_1^3 Z_2^4 Z_4^5 \quad \rightarrow \quad Z_1^3 Z_2^3 Z_3 Z_4^5 \quad \rightarrow$$

$$Z_1^5 Z_2^3 Z_3 Z_4^3 \quad \rightarrow \quad Z_1^3 Z_2^5 Z_3 Z_4^3 \quad \rightarrow \quad Z_1^3 Z_2^3 Z_3^3 Z_4^3$$

Theorem 41.5.3 (Macaulay). $\mathsf{R} = \mathrm{Res}(F_1, \ldots, F_r)$.

Proof.   We have (Proposition 41.3.4(3)) that $\mathsf{A}\mathsf{R}Z_r^d \in \mathsf{J}$; setting $Z_r := 1$ and applying *Kronecker substitution*, which changes $c_i := a_{i,Z_r^d}$ to $c_i - F_i$, $1 \leq i \leq r$, then A is not changed (as a consequence of Corollary 41.3.10(5)) while R is changed to $\mathsf{R} - \sum_{i=1}^r A_i F_i$; as a consequence $\mathsf{R} \in (F_1, \ldots, F_r, Z_r - 1)$.

Hence $\Xi_{\mathbf{f}}(\mathsf{R}) = 0$ if the equations $f_1 = \cdots = f_r = 0$ have a common solution $(z_1, \ldots, z_{r-1}, 1) \in \mathbb{P}^{r-1}(K)$.

Let us assume that $\Xi_{\mathbf{f}}(\mathsf{R}) = 0$ gives a relation among the coefficients of $f_1, \ldots, f_r$, so that there are fewer than $\#\mathcal{W}_d$ linearly independent members in $\Xi_{\mathbf{f}}(\mathsf{B}) := \{\omega f_i : \omega \in \mathcal{W}_\delta^{(i)}, 1 \leq i \leq r\}$.

Hence the coefficients x_τ of the generic element

$$\sum_{\tau \in \mathcal{W}_d} x_\tau \tau \in \Xi_{\mathbf{f}}(\mathsf{J}_d) = (f_1, \ldots, f_r)_d$$

must satisfy the linear relation $\sum_{\tau \in \mathcal{W}_d} x_\tau c_\tau = 0$.

Moreover, by Lemma 41.5.2, the generic element

$$f := \sum_{\upsilon \in \mathcal{W}_{d-1}} x_\upsilon \upsilon \in \Xi_{\mathbf{f}}(\mathsf{J}_{d-1}) = (f_1, \ldots, f_r)_{d-1}$$

also satisfies a linear relation, $\sum_{\upsilon \in \mathcal{W}_{d-1}} x_\upsilon c_\upsilon = 0$.

Since each $Z_i f = \sum_{\upsilon \in \mathcal{W}_{d-1}} x_\upsilon Z_i \upsilon \in (f_1, \ldots, f_r)_d$, the unknowns x_υ must satisfy the r equations

$$\sum_{\upsilon \in \mathcal{W}_{d-1}} x_\upsilon c_{\upsilon Z_i} = 0,$$

which are necessarily equivalent, so that for each $\upsilon \in \mathcal{W}_{d-1}$ the continued ratio $c_{\upsilon Z_1} : c_{\upsilon Z_2} : \cdots : c_{\upsilon Z_r}$ is the same. Denoting it by $\alpha_1 : \alpha_2 : \cdots : \alpha_r$, it follows that, for each $\tau := Z_1^{a_1} \cdots Z_r^{a_r} \in \mathcal{W}_d$, c_τ is proportional to $\alpha_1^{a_1} \cdots \alpha_r^{a_r}$, that is, $\alpha := (\alpha_1, \ldots, \alpha_r) \in \mathbb{P}^{r-1}(K)$ satisfies $f_1(\alpha) = \cdots = f_n(\alpha) = 0$.   ◉

Corollary 41.5.4.   *With the notation of Corollary 41.3.10 and writing*

$$R_\rho := \mathrm{Res}\left(\chi_\rho(F_1), \ldots, \chi_\rho(F_{\rho-1})\right),$$

the following also hold:

(9) $c(\mathsf{R}, a_1^{D_1} \cdots a_{r-1}^{D_{r-1}}) = a_r^{D_r}$;

(10) $c(\mathsf{R}, a_r^{D_r}) = R_r^{d_r}$;

(11) $c(\mathsf{R}, a_\rho^{D_\rho} \cdots a_r^{D_r}) = R_\rho^{d_\rho \cdots d_r}$.

Proof.   Part (9) is obvious and (11) is a repreated application of (10). For (10): if we consider an *ansatz* $\Xi : \Xi(a_{r,\tau}) = 0$ for each $\tau \neq Z_r^{d_r}$ then $\Xi(F_r) = Z_r^{d_r}$ and $\Xi(\mathsf{D}) = a_n^{D_n} \bar{R}$, where $\bar{R}$ is the subdeterminant whose rows correspond to the basis elements $\{\omega F_i : \omega \in \mathcal{W}_\delta^{(i)}, 1 \leq i < r\}$ and whose columns are labeled by the terms

$$\left\{\tau Z_1^{d_1}, \tau \in \mathcal{W}_\delta^{(1)}\right\} \sqcup \cdots \sqcup \left\{\tau Z_{r-1}^{d_{r-1}}, \tau \in \mathcal{W}_\delta^{(r-1)}\right\},$$

then the vanishing of such a determinant under an *ansatz* $\Xi : \Xi(F_r) = Z_r^{d_r}$ is the condition that the identity

$$\sum_{i=1}^{r-1} Q_i \, \Xi(F_i) = Q_r Z_r^{d_r}, \qquad Q_i = \sum_{\omega \in \mathcal{W}_\delta^{(i)}} x(Q_i, \omega)\omega, \qquad 1 \le i \le r,$$

is satisfied.

In order to evaluate such a determinant, assume that the identity is satisfied and set $Z_r = 0$, obtaining the identity

$$\sum_{i=1}^{r-1} \chi_r(Q_i)\chi_r(F_i) = 0.$$

Thus $\bar{R}$ is necessarily a multiple of $R_r^{d_r}$; evaluating the weight gives that the multiplicity is d_r. $\qquad\qquad\boxed{\odot}$

Proposition 41.5.5. *It holds that:*

(1) $\mathrm{Res}(F_1, \ldots, F_{r-1}, F_r'F_r'') = \mathrm{Res}(F_1, \ldots, F_{r-1}, F_r')\,\mathrm{Res}(F_1, \ldots, F_{r-1}, F_r'')$;
(2) $\mathrm{Res}(F_1, F_2, \ldots, F_{r-1}, F_r)$ *is irreducible.*

Proof.

(1) Since, for each *ansatz*, $\mathrm{Res}(f_1, \ldots, f_{r-1}, f_r')\,\mathrm{Res}(f_1, \ldots, f_{r-1}, f_r'')$ vanishes if and only if $f_1 = f_2 = \cdots = f_r'f_r'' = 0$ have a common root, we obtain

$$\mathrm{Res}(F_1, \ldots, F_{r-1}, F_r'F_r'') \mid \mathrm{Res}(F_1, \ldots, F_{r-1}, F_r')\,\mathrm{Res}(F_1, \ldots, F_{r-1}, F_r'');$$

equality is then obtained since both are isobaric of the same weight.

(2) By contradiction let d_r be the least value for which $\mathrm{Res}(d_1, \ldots, d_r) = \mathrm{Res}(F_1, \ldots, F_r)$ has a non-trivial factorization $\mathrm{Res}(d_1, \ldots, d_r) = R_1 R_2$. Choose any two positive values d_r' and d_r'' such that $d_r = d_r' + d_r''$. By the minimality of d_r, both $\mathrm{Res}(d_1, \ldots, d_r') = \mathrm{Res}(F_1, \ldots, F_r')$ and $\mathrm{Res}(d_1, \ldots, d_r'') = \mathrm{Res}(F_1, \ldots, F_r'')$ are irreducible. Denote $a_\tau := c(\tau, F_r)$, $a_\omega' := c(\omega, F_r')$ and $a_\upsilon'' := c(\upsilon, F_r'')$ for each $\tau \in \mathcal{W}_{d_r}$, $\omega \in \mathcal{W}_{d_r'}$, $\upsilon \in \mathcal{W}_{d_r''}$.

From the *ansatz*

$$\Xi(a_\tau) = \sum_{\omega \in \mathcal{T}_{d_r'}, \, \omega \mid \tau} a_\omega' a_{\frac{\tau}{\omega}}'',$$

which implies that $\Xi(F_r) = F_r'F_r''$, we have by (1)

$$\Xi(R_1)\Xi(R_2) = \mathrm{Res}(F_1, \ldots, F_r'F_r'') = \mathrm{Res}(F_1, \ldots, F_r')\,\mathrm{Res}(F_1, \ldots, F_r'');$$

the irreducibility of $R_1, R_2, \mathrm{Res}(F_1, \ldots, F_r')$ and $\mathrm{Res}(F_1, \ldots, F_r'')$, which is a consequence of the minimality of d_r, implies that, say,

$$\Xi(R_1) = \mathrm{Res}(F_1, \ldots, F_r'), \qquad \Xi(R_2) = \mathrm{Res}(F_1, \ldots, F_r'').$$

Since both R_1 and R_2 depend on the a_τ, $\Xi(R_1) = \mathrm{Res}(F_1, \ldots, F_r')$ depends not only on a_ω' but also on a_υ''. This clearly leads to a contradiction.   ⊡

Remark 41.5.6 (Lazard).   Given $h \geq r$ "generic" forms

$$F_1, \ldots, F_h \in \mathbb{D}[Z_1, \ldots, Z_r], \quad \deg(F_i) := d_i, \quad \mathbb{D} := \mathbb{Z}[a_{i,\tau}, 1 \leq i \leq h, \tau \in \mathcal{W}_{d_i}],$$

with

$$d_1 \geq d_2 \geq \cdots \geq d_r \geq \cdots \geq d_h, \quad d = 1 - r + \sum_{i=1}^{r} d_i,$$

the construction of Macaulay's matrix $\mathcal{M}$, discussed at the start of Section 41.3, can be generalized to obtain h blocks, the ith block consisting of the $\binom{d-d_i+r-1}{r-1}$ rows related to the K-generators ωf_i, $\omega \in \mathcal{W}_{d-d_i}$.

Consequently, the notion of Macaulay's resultant (Definition 41.3.2) generalizes naturally to this setting as still the greatest common divisor of all determinants of the $\binom{d+r-1}{r-1} \times \sum_{i=1}^{h} \binom{d-d_i+r-1}{r-1}$ Macaulay's matrix. It is clear that such a Macaulay's resultant $\mathsf{R} := \mathsf{R}(F_1, \ldots, F_h)$ is the greatest common divisor of the $\binom{h}{r}$ original Macaulay's resultants $\mathsf{R}(F_{j_1}, F_{j_2}, \ldots, F_{j_r})$ obtained by choosing any subset $\{j_1, \ldots, j_r\} \subset \{1, \ldots, h\}$ of r indices.

Equally trivially, denoting, for each set of forms $\mathbf{f} := \{f_1, \ldots, f_h\} \in \mathcal{Q}$, as $\Xi_{\mathbf{f}}$ the *ansatz* $\Xi_{\mathbf{f}}(F_i) = f_i$, $\mathbf{f}$ has a common root if and only if $\Xi_{\mathbf{f}}(\mathsf{R}) = 0$.[12]   ⊡

Remark 41.5.7.   In the non-homogeneous setting[13]

the resultant of n given non-homogeneous polynomials in $n - 1$ variables is the resultant of the corresponding homogeneous polynomials of the same degree obtained by introducing a variable x_0 of homogeneity.

In other words, given $h \geq r$ non-homogeneous polynomials

$$f_1, \ldots, f_h \in K[Z_1, \ldots, Z_r], \quad d_i := \deg(f_i)$$

and introducing the homogenizing variable Z_0, we consider the generic forms

$$F_i \in \mathbb{D}[Z_0, Z_1, \ldots, Z_r], \quad \deg(F_i) := d_i,$$

and the *ansatz*

$$\Xi : \mathbb{D}[Z_0, Z_1, \ldots, Z_r] \to K[Z_0][Z_1, \ldots, Z_r] : \Xi(F_i) := Z_0^d f_i\left(\frac{Z_1}{Z_0}, \ldots, \frac{Z_r}{Z_0}\right).$$

More precisely, we consider the $\binom{d+r}{r} \times \sum_{i=1}^{h} \binom{d-d_i+r}{r}$ Macaulay's matrix whose columns are indexed by the set $\mathcal{W}(d)$ of all terms of degree bounded by d and whose rows represent the polynomials

---

[12] Macaulay's statements consider only the case $h = r$ but such a result is already implicit in the introduction of the determinants $\mathrm{D}(\nu, \delta)$ (Section 41.4). Moreover, the construction of the u-resultants, in the next section, freely uses these implicit definitions and constructions.

[13] Macaulay F.S., *The Algebraic Theory of Modular Systems*, Cambridge University Press (1916), p. 3.

$$\omega Z_0^d f_i \left(\frac{Z_1}{Z_0}, \ldots, \frac{Z_r}{Z_0} \right), \qquad \omega \in \mathcal{W}(d - d_i), \qquad 1 \le i \le h,$$

the corresponding resultant being an element of $K[Z_0]$. ◉

41.6 Macaulay: The *u*-Resultant

Let us consider $r \le n$ homogeneous polynomials

$$\mathbf{f} := \{f_1, \ldots, f_r\} \subset k[x_1, \ldots, x_n]$$

of degrees respectively $d_1 \le \cdots \le d_r$.

One can therefore expect that the ideal $M := (f_1, \ldots, f_r)$ has rank r, in which case Macaulay considers its extension and contraction ideal[14]

$$M^{(r)} := Mk(x_{r+1}, \ldots, x_n)[x_1, \ldots, x_r] \cap k[x_1, \ldots, x_n].$$

He is aware that if a "generic" change of coordinates has been already performed then each $f_i \in M^{(r)}$ is homogeneous of degree d_i in the variables[15] $x_1, \ldots, x_r$ and that the assumption on the rank is satisfied iff the resultant $F \in k[x_{r+1}, \ldots, x_n][x_r]$ of the r polynomials f_i w.r.t. the $r - 1$ variables $x_1, \ldots, x_{r-1}$ does not vanish, thus granting the existence of a root.

In this context and under these assumptions, adapting the notation of Chapters 31, 32 and 39 we set

$$k[x_1, \ldots, x_n] = k[Z_1, \ldots, Z_r, V_1, \ldots, V_d] = k[x_{r+1}, \ldots, x_n][x_1, \ldots, x_r],$$

denote by[16]

$$\pi : k[x_1, \ldots, x_n] = k[V_1, \ldots, V_d][Z_1, \ldots, Z_r] \to k[Z_1, \ldots, Z_r]$$

the projection defined by $\pi(F) = F(Z_1, \ldots, Z_r, 0, \ldots, 0]$ for each

$$F(x_1, \ldots, x_r, x_{r+1}, \ldots, x_n) = F(Z_1, \ldots, Z_r, V_1, \ldots, V_d),$$

$K = k(V_1, \ldots, V_d)$, $\mathcal{R} = k[V_1, \ldots, V_d]$, and consider

$$(f_1, \ldots, f_r) = M^{(r)} \subset \mathcal{R}[Z_1, \ldots, Z_r],$$

remarking that with this new notation we have $F \in \mathcal{R}[Z_r]$.

We begin by remarking that the polynomials

$$\bar{f}_i := f_i(Z_0 V_1, \ldots, Z_0 V_d, Z_1, \ldots, Z_{r-1}, Z_0 Z_r) \in \mathcal{R}[Z_0, Z_1, \ldots, Z_{r-1}, Z_r]$$

are homogeneous in the variables $Z_1, \ldots, Z_{r-1}, Z_0$ and that the resultant $\overline{F} \in \mathcal{R}[Z_r]$ w.r.t. $Z_1, \ldots, Z_{r-1}, Z_0$ of $\bar{\mathbf{f}} := \{\bar{f}_1, \ldots, \bar{f}_r\}$ is a homogeneous polynomial, in the

---

[14] Compare the discussion in Section 30.5.
[15] The variable x_r being chosen as the variable of homogeneity.
[16] Compare Section 30.5, n. 48.

variables $V_1, \ldots, V_d, Z_r$, of degree $D := \prod_{i=1}^{r} d_i$ and such that $\mathbf{T}(\overline{F}) = R_{r+1} Z_r^D$, where, by the assumption on the rank,

$$R_{r+1} = \mathrm{Res}(\pi(f_1), \ldots, \pi(f_r)) \neq 0.$$

Instead of solving for one of the unknown variables Z_i, we solve for their *Liouville substitution*

$$Z = U_1 Z_1 + U_2 Z_2 + \cdots + U_r Z_r,$$

setting $f_u := Z - U_1 Z_1 - U_2 Z_2 - \cdots - U_r Z_r$, considering the polynomial set $\mathbf{f}^{(u)} := \{f_1, \ldots, f_r, f_u\}$ as a subset of $\mathcal{R}[U_1, \ldots, U_r][Z, Z_1, \ldots, Z_r]$ and computing their resultant $F^{(u)} \in \mathcal{R}[U_1, \ldots, U_r][Z]$ w.r.t. $Z_1, \ldots, Z_r$.

Definition 41.6.1 (Macaulay). $F^{(u)}$ *is called the u-resultant of* $\mathbf{f}$.   ⊙

With an argument similar to that we gave for $\overline{F}$, setting

$$\hat{f}_i := f_i(Z_0 V_1, \ldots, Z_0 V_d, Z_1, \ldots, Z_r), \quad 1 \le i \le r, \quad \text{and} \quad \hat{f}_u := Z_0 Z - \sum_{i=1}^{r} U_i Z_i$$

and writing

$$\hat{\mathbf{f}}^{(u)} := \{\hat{f}_1, \ldots, \hat{f}_r, \hat{f}_u\} \subset \mathcal{R}[U_1, \ldots, U_r][Z_0, Z_1, \ldots, Z_r, Z],$$

we have that $F^{(u)}$ is the resultant of $\hat{\mathbf{f}}^{(u)}$ w.r.t. $Z_1, \ldots, Z_r, Z_0$ and, being homogeneous in the variables $V_1, \ldots, V_d, Z$ and of degree $D := \prod_{i=1}^{r} d_i$, we have $\mathbf{T}(F^{(u)}) = R'_{r+1} Z^D$, where $R'_{r+1} = \mathrm{Res}\,(\pi(f_1), \ldots, \pi(f_r), \pi(f_u))$.

If we consider in the expansion of $F^{(u)}$ the indeterminate coefficient a, representing the coefficient $a := c(Z, f_u)$ of Z in f_u, degree considerations allow us to deduce that $a^D \mid R'_{r+1}$ and $R'_{r+1} = a^D R_{r+1}$, whence $R'_{r+1} = R_{r+1}$ since the *ansatz* evaluates a as $c(Z, f_u) = 1$.

With the same kind of argument as in the proof of Theorem 41.5.3, we can deduce that to each root $\alpha_r^{(j)}$ of $\overline{F}$ there corresponds a root $(\alpha_1^{(j)}, \ldots, \alpha_r^{(j)})$ of $\bar{\mathbf{f}}$; there are D solutions altogether, all of which are "finite"[17] since $R_{r+1} \neq 0$. Similarly, to each of the D roots $z^{(j)}$ of $F^{(u)}$ there corresponds a root $(\beta_1^{(j)}, \ldots, \beta_r^{(j)}, z^{(j)})$ of $\hat{\mathbf{f}}^{(u)}$; clearly, up to a re-enumeration we have

$$(\alpha_1^{(j)}, \ldots, \alpha_r^{(j)}) = (\beta_1^{(j)}, \ldots, \beta_r^{(j)})$$

and, since $\hat{f}_u(\beta_1^{(j)}, \ldots, \beta_r^{(j)}, z^{(j)}) = 0$, we have $z^{(j)} = \sum_{i=1}^{r} U_i \alpha_i^{(j)}$, so that

$$F^{(u)} = R'_{r+1} \prod_{i=1}^{D} \left(Z - \sum_{i=1}^{r} U_i \alpha_i^{(j)} \right).$$

In conclusion we have

---

[17] That is, they are affine points $(\alpha_1^{(j)}, \ldots, \alpha_r^{(j)}) \in K^r$ corresponding to the projective point $(1, \alpha_1^{(j)}, \ldots, \alpha_r^{(j)}) \in \mathbb{P}^r(K)$.

Proposition 41.6.2 (Macaulay). *The u-resultant $F^{(u)}$ is a product of D factors which are linear in $Z, U_1, \ldots, U_r$, and the coefficients of $U_1, \ldots, U_r$ in each factor supply a solution of the system* **f**.

Also, the number of solutions is either $D = \prod_{i-1}^{r} d_i$ or infinite, the latter being the case when $\overline{F}$ vanishes identically. ◉

Remark 41.6.3 (Macaulay).

(1) Denoting by D the determinant for the generic forms $\hat{F}_1, \ldots, \hat{F}_r, \hat{F}_u$, regarding $Z_1, \ldots, Z_r, Z_0$ as variables and letting A be its extraneous factor, we have that $\mathsf{D} = \mathsf{A}\,\mathrm{Res}(\hat{F}_1, \ldots, \hat{F}_r, \hat{F}_u)$ and, setting $Z_0 = 0$, that (see Corollary 41.3.10(7)) A depends only on the coefficients of the generic polynomials $\chi_{r+1}(F_1), \ldots, \chi_{r+1}(F_r)$. Hence A is independent of $V_1, \ldots, V_d$ and $U_1, \ldots, U_r$.

(2) In the case of non-homogeneous polynomials, the preliminary generic change of coordinates does not affect the homogeneity variable; thus it is possible for R_{r+1} to vanish identically. The consequence is a diminution in the number of finite solutions for Z but not in the number of linear factors of $\overline{F}$; such factors can be expressed as $\sum_{i=1}^{r} U_i \alpha_i$ and correspond to an *infinite solution*[18] in the set of ratios $\alpha_1 : \alpha_2 : \cdots : \alpha_r$. ◉

In the generalized setting of Remarks 41.5.6 and 41.5.7, Macaulay's result can be read as follows:

Proposition 41.6.4 (Lazard). *Given h (non-homogeneous) polynomials*

$$f_1, \ldots, f_h \in \mathcal{Q}, \quad \deg(f_i) := d_i, \quad h \geq r, d_1 \geq d_2 \geq \cdots \geq d_h, \quad d = 1 - r + \sum_{i=1}^{r} d_i,$$

and setting

- $f_{h+1} := U_0 + \sum_{i-1}^{r} U_i Z_i,$
- $\mathcal{M} \in K[Z_0, U_0, U_1, \ldots, U_r]$ *as the Macaulay's matrix constructed, according to Remarks 41.5.6 and 41.5.7, via $f_1, \ldots, f_h, f_{h+1}$,*
- R *as the corresponding Macaulay's resultant,*
- $G := \mathsf{R}(1, U_0, U_1, \ldots, U_r) \in K[U_0, U_1, \ldots, U_r],$

we have that

(1) *$f_1, \ldots, f_h$ have a finite number of common roots iff $\mathcal{M}$ has rank $\binom{d+r}{r}$, that is, $G \neq 0$;*

(2) *if $G \neq 0$, $\deg(G)$ is the number of common roots of $f_1, \ldots, f_h$, counting multiplicity and zeros at infinity;*

(3) *G is homogeneous and, in $\mathsf{K}[U_0, U_1, \ldots, U_n]$, is a product of linear polynomials;*

---

[18] That is, a projective point $(0, \alpha_1, \alpha_2, \ldots, \alpha_r)$.

(4) *if $\alpha_0 U_0 + \alpha_1 U_1 + \cdots + \alpha_n U_n$ is a linear factor of G then*
  - *if $\alpha_0 \neq 0$ then $(\alpha_1/\alpha_0, \ldots, \alpha_n/\alpha_0) \in \mathsf{k}^n$ is a root of the f_i;*
  - *if $\alpha_0 = 0$ then $(\alpha_0, \alpha_1, \ldots, \alpha_n)$ is a common zero at infinity.* $\boxed{\odot}$

41.7 Kronecker's Resolvent

Let us consider a finite set

$$F_n := \{f_1^{(n)}, \ldots, f_{s_n}^{(n)}\} \subset \mathcal{P} = k[X_1, \ldots, X_n] = k[X_1, \ldots, X_{n-1}][X_n]$$

of non-homogeneous polynomials, generating an ideal $\mathsf{I} := \mathbb{I}(F_n)$, which we assume to be in a sufficiently "generic" position, the variables having been subjected to a change of coordinates beforehand. As a consequence, in particular each $f_i^{(n)}$ is regular in X_n.[19]

Also, I is in *allgemeine* position (Definition 34.4.3) so that, for each primary component q of I, $\dim(\mathsf{q}) = d$, we have $\mathsf{I} \cap k[X_1, \ldots, X_d] = (0)$, and the construction at the start of Chapter 39 can be compacted and extended: we can introduce the fields $K_d := k(X_1, \ldots, X_d)$ and their algebraic closures $\mathsf{K}_d \subset \Omega(k)$ knowing that, for each primary component q of rank $r = n - d$, the corresponding roots are $(X_1, \ldots, X_d, \beta_1, \ldots, \beta_r)$, $\beta_i \in \mathsf{K}_d$.

We can iteratively, for $v := n, n - 1, \ldots, 1$, compute:[20]

- $D_v := \gcd(F_v) \in k[X_1, \ldots, X_{v-1}][X_v]$,
- $g_i^{(v)} := f_i^{(v)}/D_v \in k[X_1, \ldots, X_{v-1}][X_v], 1 \leq i \leq s_v$,
- $G_v := \{g_i^{(v)}, 1 \leq i \leq s_v\} \subset k[X_1, \ldots, X_{v-1}][X_v]$,
- $f := \sum_{i=1}^{s_v} U_i g_i^{(v)} \in k[X_1, \ldots, X_{v-1}][U_1, W_1, \ldots, U_{s_v}, W_{s_v}][X_v]$,
- $g := \sum_{i=1}^{s_v} W_i g_i^{(v)} \in k[X_1, \ldots, X_{v-1}][U_1, W_1, \ldots, U_{s_v}, W_{s_v}][X_v]$,
- $R_v := \mathrm{Res}(f, g) =: \sum_{v \in \mathcal{U}^{(s_v)}} f_v v \in k[X_1, \ldots, X_{v-1}][U_1, W_1, \ldots, U_{s_v}, W_{s_v}]$,
  where, for each value $j \in \mathbb{N}$, we use the notation

$$\mathcal{U}^{(j)} := \left\{ U_1^{a_1} \cdots U_j^{a_j} W_1^{b_1} \cdots W_j^{b_j} : (a_1, \ldots, a_j, b_1, \ldots, b_j) \in \mathbb{N}^{2j} \right\},$$

- $F_{v-1} := \left\{ f_1^{(v-1)}, \ldots, f_{s_{v-1}}^{(v-1)} \right\} := \left\{ f_v, v \in \mathcal{U}^{(s_v)} \right\} \subset k[X_1, \ldots, X_{v-1}]$

and remark that

(1) $1 = \gcd(G_v) \in k[X_1, \ldots, X_{v-1}][X_v]$, so that
(2) $R_v \neq 0$,
(3) each $f^{(v)}$ is regular in X_v since we are assuming that each variable has been subjected to a generic change of coordinates,

---

[19] A polynomial $f = \sum_{\tau \in \mathcal{T}} c(f, \tau)$, $\tau \in \mathcal{P}$, $\deg(f) = d$, is *regular* in X_i iff $c(f, X_i^d) \neq 0$.
[20] Compare Section 20.4.

(4) each common root of F_ν is either a root of D_ν or a common root of G_ν and

(5) each common root α of G_ν is a common root of $F_{\nu-1}$ since $R_\nu \in \mathbb{I}(f, g)$,

(6) if, however, $\beta := (X_1, \ldots, X_{\nu-1}, \beta_1)$, $\beta_1 \in \mathsf{K}_{\nu-1}$ is such that $D_\nu(X_1, \ldots, X_{\nu-1}, \beta_1) = 0$ then $R_{\nu+1}(X_1, \ldots, X_{\nu-1}, \beta_1) = 0$ and

$$f := \sum_{i=1}^{s_{\nu+1}} U_i\, g_i^{(\nu+1)}(\beta_1, X_{\nu+1}) \quad \text{and} \quad g := \sum_{i=1}^{s_{\nu+1}} W_i\, g_i^{(\nu+1)}(\beta_1, X_{\nu+1})$$

have a common root $\beta_2 \in \mathsf{K}_{\nu-1}$, so that, for each i, $1 \le i \le s_{\nu+1}$, we have

$$g_i^{(\nu+1)}(X_1, \ldots, X_{\nu-1}, \beta_1, \beta_2) = 0 \quad \text{and} \quad f_i^{(\nu+1)}(X_1, \ldots, X_{\nu-1}, \beta_1, \beta_2) = 0,$$

(7) and, by iterating this argument, each root $(X_1, \ldots, X_{\nu-1}, \beta_1)$ of D_ν lifts to a root (of rank $n - \nu + 1$ and dimension $\nu - 1$) of I.

Definition 41.7.1. *The polynomial $\prod_1^n D_\nu$ is called the* complete (total) resolvent *of* $\mathbb{I}(F_n)$*; each factor D_ν is called the* complete partial resolvent *of* $\mathbb{I}(F_n)$ *of dimension $\nu - 1$ and rank $n - \nu + 1$.* ◎

Proposition 41.7.2. *The complete resolvent of $\mathbb{I}(F_n)$ is a member of $\mathbb{I}(F_n)$.*

Proof. In fact, for each ν there are

$$p_\nu, q_\nu \in k[X_1, \ldots, X_{\nu-1}][U_1, W_1, \ldots, U_{s_\nu}, W_{s_\nu}][X_\nu],$$

such that $R_\nu = p_\nu \left(\sum_{i=1}^{s_\nu} U_i g_i^{(\nu)} \right) + q_\nu \left(\sum_{i=1}^{s_\nu} W_i g_i^{(\nu)} \right)$. Thus $F_{\nu-1} \subset \mathbb{I}(G_\nu)$ and $D_\nu F_{\nu-1} \subset \mathbb{I}(F_\nu)$, whence, by an inductive argument, $F_1 \prod_{\nu=1}^n D_\nu \subset \mathbb{I}(F_n)$, where either $F_1 \in k$ or $s_1 > 1$ and $F_1 = \{f_1, \ldots, f_{s_1}\} \in k[X_1]$, with $\gcd(F_1) = 1$, so that there are polynomials $q_i(X_1) \in k[X_1]$ such that $1 = \sum_{i=1}^{s_n} q_i f_i$ and $\prod_{\nu=1}^n D_\nu = \sum_{i=1}^{s_n} \left(q_i \prod_{\nu=1}^n D_\nu \right) f_i \in \mathbb{I}(F_n)$. ◎

Remark 41.7.3 (Macaulay).

(1) As a direct consequence we have Hilbert's Nullstellensatz: if J has no root, the complete resolvent is 1 so that $1 \in \mathsf{I}$.

(2) Suppose that we are given n forms f_i in n variables, each of degree l, which have no common solution, so that the complete resolvent is $D_1 = X_1^{\mu}$.[21] Since the elements of F_n all have degree l, the elements of F_{n-1} all have degree l^2 and those of F_{n-2} have degree $(l^2)^2$; in general the terms of F_ν all have degree $(l^{2^{\nu-1}})^2 = l^{2^\nu}$, so that $\mu = l^{2^{n-1}}$. Macaulay remarks that[22]

We should arrive at a similar result if we change x_i to $x_i + a_i$ ($i = 1, 2, \ldots, n$) beforehand, thus making the polynomials non-homogeneous. The complete resolvent would be $(x_n + a_n)^{l^{2^{n-1}}}$. The resultant would be $(x_n + a_n)^{l^n}$. The difference in the two results is explained by the fact that the resultant is obtained by a process applying uniformly to all the variables, and the resolvent by a process applied to the variables in succession. ◎

---

[21] The construction reads the forms as polynomials and the sought roots are considered affine, so that if there is no proper solution then the origin is to be considered a root with appropriate multiplicity.

[22] Macaulay F.S., *The Algebraic Theory, op. cit.*, pp. 21–22.

41.8 Kronecker: The u-Resolvent

Given a basis $F := \{f_1, \ldots, f_s\} \subset k[X_1, \ldots, X_n]$, let us consider new variables $X, \Lambda_1, \ldots, \Lambda_n$ such that

$$X = \Lambda_1 X_1 + \cdots + \Lambda_n X_n$$

and perform the Liouville substitution[23]

$$X_i = \begin{cases} \dfrac{X - \Lambda_1 X_1 + \cdots + \Lambda_{n-1} X_{n-1}}{\Lambda_n}, & \text{if } i = n, \\ X_i & \text{otherwise,} \end{cases}$$

thus obtaining

- the polynomials[24]

$$f_i' := \Lambda_n^{l_n} F_i\left(X_1, \ldots X_{n-1}, \frac{X - \sum_1^{n-1} \Lambda_i X_i}{\Lambda_n}\right), \quad l_n := \deg_n(F_i),$$

- the basis $F' := \{f_1', \ldots, f_s'\} \subset k[\Lambda_1, \ldots, \Lambda_n][X_1, \ldots, X_{n-1}][X]$,
- the ideal $\mathsf{I}' := \mathbb{I}(F') \subset k[\Lambda_1, \ldots, \Lambda_n][X_1, \ldots, X_{n-1}, X]$.

Clearly there is a one-to-one correspondence between

- the roots $(\xi_1, \ldots, \xi_n) \in \mathcal{Z}(F)$ and
- the roots $(\xi_1, \ldots, \xi_{n-1}, \xi) \in \mathcal{Z}(F')$,

the relation being given by $\xi = \Lambda_1 \xi_1 + \cdots + \Lambda_n \xi_n$.

Definition 41.8.1. *The complete resolvent* $F_u := \prod_1^n D_v$ *of* $\mathbb{I}(F')$ *is called the complete* u*-resolvent of* $\mathbb{I}(F)$.   ⊙

We have $F_u(X_1, \ldots, X_{n-1}, \Lambda_1 X_1 + \cdots + \Lambda_n X_n) \in \mathbb{I}(F)$, since (by Proposition 41.7.2) $F_u \in \mathbb{I}(F')$; moreover F_u, considered as a univariate polynomial,

$$F_u \in k[\Lambda_1, \ldots, \Lambda_n][X_1, \ldots, X_{n-1}][X],$$

factors into linear factors; those of dimension $v - 1$, that is, the factors of the component D_v, can be expressed as

$$X - \Lambda_1 X_1 - \cdots - \Lambda_{v-1} X_{v-1} - \Lambda_v \xi_r - \cdots - \Lambda_n \xi_1. \tag{41.2}$$

---

[23] Clearly

$$x := \Lambda_1 x_1 + \cdots + \Lambda_n x_n \in k(\Lambda_1, \ldots, \Lambda_n)[x_1, \ldots, x_n]$$
$$:= k(\Lambda_1, \ldots, \Lambda_n)[X_1, \ldots, X_n]/\mathbb{I}(F)^e$$

is a primitive element in $k[x_1, \ldots, x_n] := k[X_1, \ldots, X_n]/\mathbb{I}(F)$ for any "generic" evaluation of the Λ_i.

[24] The multiplier $\Lambda_n^{l_n}$ "being introduced to make $|f_i'|$ integral in $|\Lambda_n|$."
Macaulay F.S., *The Algebraic Theory, op. cit.*, p. 24.
This grants that each $f_i' \in k[\Lambda_1, \ldots, \Lambda_n][X_1, \ldots, X_n]$.

Remark 41.8.2.   The linear factors (41.2) of the complete *partial* resolvent D_ν are related to components of dimension $d := \nu - 1$ and rank $r := n - \nu + 1$.

According to our notation the irreducible components

$$R \in k[\Lambda_1, \ldots, \Lambda_n][X_1, \ldots, X_{n-1}][X]$$

of D_ν should be read as elements $R \in k[\Lambda_1, \ldots, \Lambda_n][V_1, \ldots, V_d, Z_1, \ldots, Z_r][X]$ and the linear factors (41.2) as

$$X - \Lambda_1 V_1 - \cdots - \Lambda_{\nu-1} V_d - \Lambda_{d+1} \xi_1 - \cdots - \Lambda_n \xi_r.$$

According to the notation introduced by Macaulay[25] and reported in Sections 30.5 and 41.6, R must be read as an element

$$R \in k[\Lambda_1, \ldots, \Lambda_n][x_n, \ldots, x_{r+1}][x_r, \ldots, x_1][X]$$

and the linear factors (41.2) as

$$X - \Lambda_1 x_n - \cdots - \Lambda_d x_{r+1} - \Lambda_{d+1} \xi_r - \cdots - \Lambda_n \xi_1.$$

$\boxed{\odot}$

41.9 Kronecker Parametrization

In general the splitting factorization of D_ν can contain linear factors (41.2) in which some ξ_i depends on the Λs.

Definition 41.9.1.   *A linear factor (41.2) of D_ν where each ξ_i is independent of the Λs is called* true.[26] $\boxed{\odot}$

Remark 41.9.2 (Macaulay).   Kronecker stated, without proving it, that each factor (41.2) is true: "whether this is so or not must be considered doubtful."[27]

It could however be proved that:

- a solution supplied by a non-true factor is necessarily embedded;
- any irreducible component R of a partial resultant D_ν either factors in true linear factors only or has no true factor. $\boxed{\odot}$

Historical Remark 41.9.3.   Today the natural way to restrict ourselves to *true* factors is to get rid of embedded components via a radical computation; the more so since Macaulay gave a procedure (Algorithm 30.7.3) for recovering embedded components and their multiplicity.

---

[25] To be more precise, the Liouville substitution performed by Macaulay was

$$x = u_1 x_1 + \cdots + u_n x_n.$$

As we pointed out in Section 36.3, footnote 15, in order to adapt Macaulay's notation to the current usage of choosing the *first* variables as parameters, one has to set $x_i := X_{n-i}$.

[26] In this case, in relation to Remark 41.8.2, the ξ_i are elements, with our notation, of the algebraic closure of $K = k(V_1, \ldots, V_d)$ and with Macaulay's notation, of the algebraic closure of $k(x_n, \ldots, x_{r+1})$.

[27] Macaulay F.S., *The Algebraic Theory, op. cit.*, p. 26.

But I guess that Seidenberg's Algorithm (Corollary 35.2.3) was the first procedure proposed for radical computation. ⊙

So, let us consider an irreducible component

$$R(X) \in k[\Lambda_1, \ldots, \Lambda_n][X_1, \ldots, X_{\nu-1}][X]$$

of the partial resultant D_ν having a linear factorization into true factors of dimension $d := \nu - 1$ and rank $r := n - \nu + 1$:

$$R(X) = \prod_{j=1}^{\delta} \left(X - \Lambda_1 X_1 - \cdots - \Lambda_{\nu-1} X_{\nu-1} - \Lambda_\nu \xi_{1j} - \cdots - \Lambda_n \xi_{rj} \right)$$

$$= \prod_{j=1}^{\delta} \left(X - \Lambda_1 V_1 - \cdots - \Lambda_{\nu-1} V_d - \Lambda_{d+1} \xi_{1j} - \cdots - \Lambda_{d+r} \xi_{rj} \right),$$

where $\xi_{ij} \in K_{\nu-1}$. Let us evaluate it at

$$\Lambda_1 X_1 + \cdots + \Lambda_n X_n = \Lambda_1 V_1 + \cdots + \Lambda_d V_d + \Lambda_{d+1} Z_1 + \Lambda_{d+r} Z_r,$$

obtaining

$$R' := R(\Lambda_1 V_1 + \cdots + \Lambda_d V_d + \Lambda_{d+1} Z_1 + \Lambda_{d+r} Z_r)$$

$$= \prod_{j=1}^{\delta} \left(\Lambda_{d+1}(Z_1 - \xi_{1j}) + \cdots + \Lambda_{d+i}(Z_i - \xi_{ij}) + \cdots + \Lambda_n(Z_r - \xi_{rj}) \right),$$

so that $R' \in k[\Lambda_\nu, \ldots, \Lambda_n, Z_1, \ldots, Z_r] = k[Z_1, \ldots, Z_r][\Lambda_\nu, \ldots, \Lambda_n]$.

To $[R]$ corresponds what is called an *irreducible spread*, viz. the spread of all points $[\xi_n, \ldots, \xi_{r+1}, \xi_{rj}, \ldots, \xi_{1j}]$ in which $[\xi_n, \ldots, \xi_{r+1}]$ take all finite values, and $[\xi_{rj}, \ldots, \xi_{1j}]$ the $[\delta]$ sets of values supplied by the linear factors of $[R]$ which vary as $[\xi_n, \ldots, \xi_{r+1}]$ vary. [...]

No linear factor of $[R]$ can be repeated, unless $[X_1, \ldots, X_{\nu-1}]$ are given special values; for otherwise $[R]$ and $[\partial R/\partial X]$ would have an H.C.F. involving $[X]$, and $[R]$ would be the product of two factors.[28]

In other words, to the irreducible component R of the partial resultant D_ν there corresponds a *prime* component $\mathfrak{f} := \mathfrak{f}_R$ of $\mathbb{I}(F)$ and an associated variety $\mathcal{Z}(R) := \mathcal{Z}(\mathfrak{f}_R)$ of dimension $d := \nu - 1$ and rank $r := n - d$. Moreover, in the expansion of $R' \in k[Z_1, \ldots, Z_r][\Lambda_\nu, \ldots, \Lambda_n]$ the coefficients of the terms in

$$\left\{ \Lambda_\nu^{a_\nu} \cdots \Lambda_n^{a_n} : (a_\nu, \ldots, a_n) \in \mathbb{N}^r \right\}$$

all vanish at every point of the spread $[\mathcal{Z}(R)]$ and do not all vanish at any other point.[29]

In particular the coefficient of

- Λ_{d+1}^{δ} is $q(V_1, \ldots, V_d, Z_1) := \prod_{j=1}^{\delta}(Z_1 - \xi_{1j}) \in k[V_1, \ldots, V_d][Z_1]$,

[28] Macaulay F.S., *The Algebraic Theory, op. cit.*, p. 27.
[29] Macaulay F.S., *The Algebraic Theory, op. cit.*, p. 27.

- $\Lambda_{d+i}\Lambda_{d+1}^{\delta-1}$, $1 < i \leq r$, is

$$q(V_1, \ldots, V_d, Z_1)\sum_j \frac{Z_i - \xi_{ij}}{Z_1 - \xi_{1j}} = \frac{\partial q}{\partial Z_1}(V_1, \ldots, V_d, Z_1)Z_i - w_i(V_1, \ldots, V_d, Z_1),$$

where $w_i(V_1, \ldots, V_d, Z_1) = q(V_1, \ldots, V_d, Z_1)\sum_j \frac{\xi_{ij}}{Z_1 - \xi_{1j}}$;

moreover, we also have

$$q(V_1, \ldots, V_d, Z_1) = \frac{\partial q}{\partial Z_1}(V_1, \ldots, V_d, Z_1)Z_1 - w_1(V_1, \ldots, V_d, Z_1),$$

$$w_1(V_1, \ldots, V_d, Z_1) = q(V_1, \ldots, V_d, Z_1)\sum_j \frac{\xi_{1j}}{Z_1 - \xi_{1j}}.$$

Thus the roots $(\xi_1, \ldots, \xi_r) \in \mathcal{Z}(R)$ satisfy the parametrization[30]

$$
\begin{cases}
q(V_1, \ldots, V_d, T) &= 0, \\
Z_1 &= \dfrac{w_1(V_1, \ldots, V_d, T)}{\dfrac{\partial q}{\partial T}(V_1, \ldots, V_d, T)}, \\
&\vdots \\
Z_r &= \dfrac{w_r(V_1, \ldots, V_d, T)}{\dfrac{\partial q}{\partial T}(V_1, \ldots, V_d, T)}.
\end{cases}
\tag{41.3}
$$

Definition 41.9.4. *A parametrization (41.3) of a prime ideal*

$$\mathsf{I} \subset \mathcal{P}, \qquad \dim(\mathsf{I}) = \nu - 1,$$

in "generic" position is called a Kronecker parametrization *of* I.   ⊙

41.10 Historical Intermezzo: From Bézout to Cayley

In connection with Sylvester's resultant, both Sylvester and Cayley quote[31] *Bézout's abridged method to obtain the resultant* or[32] *Bézout's abbreviated method of elimination*. Without pretending to give a complete survey of Bézout's result, I think it would be helpful to give some pointers to it for the interested reader.

Bézout's method for computing resultants is preliminarily described[33] in the case of three homogeneous linear equations[34] $\sum_{j=1}^4 a_{ij}X_j$, $1 \leq i \leq 3$: he considers the

---

[30] Here we simply substitute Z_1 by T.

[31] Sylvester J.J., On a theory of the syzygietic relations of two rational integral functions, comprising an application to the theory of Sturm's functions, and that of the greatest algebraic common measure, *Phil. Trans. Royal Soc. London* **CXLIII** (1853), 407–548.

[32] Cayley A., A fourth memoir upon quantics, *Phil. Trans. Royal Soc. London* **CXLVIII** (1858), 415–427.

[33] Bézout E., *Théorie Generale des Èquations Algébriques*, Pierres, Paris (1771), §200–203 pp.174–176.

[34] I consider it more suitable *not* to follow the original notation but to adapt it appropriately.

product $\prod_{j=1}^{4} X_j$ and successively, for $i = 1 \ldots 3$, substitutes each X_j with a_{ij}, *observing the sign rule.*[35] We thus obtain

$$a_{11}X_2X_3X_4 - a_{12}X_1X_3X_4 + a_{13}X_1X_2X_4 - a_{14}X_1X_2X_3 \qquad\qquad i=1$$

$$\begin{aligned}
&(a_{11}a_{22} - a_{21}a_{12})X_3X_4 - (a_{11}a_{23} - a_{21}a_{13})X_2X_4\\
&+ (a_{11}a_{24} - a_{21}a_{14})X_2X_3 + (a_{12}a_{23} - a_{22}a_{13})X_1X_4 \qquad\qquad i=2\\
&- (a_{12}a_{24} - a_{22}a_{14})X_1X_3 + (a_{13}a_{24} - a_{23}a_{14})X_1X_2
\end{aligned}$$

$$\begin{aligned}
&[(a_{11}a_{22} - a_{21}a_{12})a_{33} - (a_{11}a_{23} - a_{21}a_{13})a_{32} + (a_{12}a_{23} - a_{22}a_{13})a_{31}]\,X_4\\
&- [(a_{11}a_{22} - a_{21}a_{12})a_{34} - (a_{11}a_{24} - a_{21}a_{14})a_{32} + (a_{12}a_{24} - a_{22}a_{14})a_{31}]\,X_3\\
&+ [(a_{11}a_{23} - a_{21}a_{13})a_{34} - (a_{11}a_{24} - a_{21}a_{14})a_{33} + (a_{13}a_{24} - a_{23}a_{14})a_{31}]\,X_2 \quad i=3\\
&- [(a_{12}a_{23} - a_{22}a_{13})a_{34} - (a_{12}a_{24} - a_{22}a_{14})a_{33} + (a_{13}a_{24} - a_{23}a_{14})a_{32}]\,X_1
\end{aligned}$$

whence he deduces that

$$
\begin{cases}
X_1 = X_4 \dfrac{-[(a_{12}a_{23} - a_{22}a_{13})a_{34} - (a_{12}a_{24} - a_{22}a_{14})a_{33} + (a_{13}a_{24} - a_{23}a_{14})a_{32}]}{(a_{11}a_{22} - a_{21}a_{12})a_{33} - (a_{11}a_{23} - a_{21}a_{13})a_{32} + (a_{12}a_{23} - a_{22}a_{13})a_{31}}\,,\\[2.2em]
X_2 = X_4 \dfrac{[(a_{11}a_{23} - a_{21}a_{13})a_{34} - (a_{11}a_{24} - a_{21}a_{14})a_{33} + (a_{13}a_{24} - a_{23}a_{14})a_{31}]}{(a_{11}a_{22} - a_{21}a_{12})a_{33} - (a_{11}a_{23} - a_{21}a_{13})a_{32} + (a_{12}a_{23} - a_{22}a_{13})a_{31}}\,,\\[2.2em]
X_3 = X_4 \dfrac{-[(a_{11}a_{22} - a_{21}a_{12})a_{34} - (a_{11}a_{24} - a_{21}a_{14})a_{32} + (a_{12}a_{24} - a_{22}a_{14})a_{31}]}{(a_{11}a_{22} - a_{21}a_{12})a_{33} - (a_{11}a_{23} - a_{21}a_{13})a_{32} + (a_{12}a_{23} - a_{22}a_{13})a_{31}}\,,
\end{cases}
$$

that is, Cramer's formula.

As Muir put it,[36]

the unreal product $\prod_{j=1}^{4} X_j$ at the very outset must have been a sore puzzle to students. [...]

To throw light upon the process, let us compare the above solution of a set of three linear equations with the following solution, which from one point of view may be looked upon as an improvement on the ordinary determinantal modes of solution as presented to modern readers.

[...] The numerators of the values of X_1, X_2, X_3 and the common denominator are [...] the coefficients of X_1, X_2, X_3, X_4 in the determinant

$$
\begin{vmatrix}
a_{11} & a_{12} & a_{13} & a_{14}\\
a_{21} & a_{22} & a_{23} & a_{24}\\
a_{31} & a_{32} & a_{33} & a_{34}\\
X_1 & X_2 & X_3 & X_4
\end{vmatrix} := [X_1X_2X_3X_4]\,.
$$

More precisely, Muir explains, if we write

$$
[X_iX_j] := \begin{vmatrix} a_{3i} & a_{3j}\\ X_i & X_j \end{vmatrix}
\quad\text{and}\quad
[X_iX_jX_h] := \begin{vmatrix} a_{2i} & a_{2j} & a_{2h}\\ a_{3i} & a_{3j} & a_{3h}\\ X_i & X_j & X_h \end{vmatrix}
$$

<hr>

[35] The reference being to Cramer's rule of signs.

[36] Muir T., *The Theory of Determinants in the Historical Order of Development*, MacMillan (1906), p. 44.

then we have (by developing along the first row)

$$a_{11}\,[X_2X_3X_4] - a_{12}\,[X_1X_3X_4] + a_{13}\,[X_1X_2X_4] - a_{14}\,[X_1X_2X_3]$$

and, again developing, along the first row, the four determinants $\left[X_i X_j X_h\right]$:

$$(a_{11}a_{22} - a_{21}a_{12})\,[X_3X_4] - (a_{11}a_{23} - a_{21}a_{13}))\,[X_2X_4]$$
$$+ (a_{11}a_{24} - a_{21}a_{14}))\,[X_2X_3] + (a_{12}a_{23} - a_{22}a_{13}))\,[X_1X_4]$$
$$- (a_{12}a_{24} - a_{22}a_{14})\,[X_1X_3] + (a_{13}a_{24} - a_{23}a_{14})\,[X_1X_3]\,.$$

Finally, expanding these six determinants and collecting the terms together, we have

$$\begin{vmatrix} a_{11} & a_{12} & a_{13} \\ a_{21} & a_{22} & a_{23} \\ a_{31} & a_{32} & a_{33} \end{vmatrix} X_4 - \begin{vmatrix} a_{11} & a_{12} & a_{14} \\ a_{21} & a_{22} & a_{24} \\ a_{31} & a_{32} & a_{34} \end{vmatrix} X_3$$
$$+ \begin{vmatrix} a_{11} & a_{13} & a_{14} \\ a_{21} & a_{23} & a_{24} \\ a_{31} & a_{33} & a_{34} \end{vmatrix} X_2 - \begin{vmatrix} a_{12} & a_{13} & a_{14} \\ a_{22} & a_{23} & a_{24} \\ a_{32} & a_{33} & a_{34} \end{vmatrix} X_1\,.$$

The same method, that is, an expansion of the appropriate determinants expressed using a similar notation and process, is then applied by Bézout to *resolve* different systems of polynomial equations,[37] including a computation of the resultant in $k[X_1]$ of two polynomials in $k[X_1, X_2]$, and is then specialized to the computation of the resultant in k of two polynomials

$$\phi := \sum_{i=0}^{n} a_{i+1} X^{n-i} \quad \text{and} \quad \phi' := \sum_{i=0}^{m} a'_{i+1} X^{m-i} \in k[X].$$

The specialized method, which is the one called by the English School *Bézout's abridged* or *abbreviated method*, consists of the following:

(1) multiplying ϕ and ϕ' respectively by the polynomials

$$\Phi := \sum_{i=0}^{\nu} A_{i+1} X^{\nu-i} \quad \text{and} \quad \Phi' := \sum_{i=0}^{\mu} A'_{i+1} X^{\mu-i},$$

[37] Among the instances discussed (see the publication mentioned in footnote 33), we can list the following resultants:

- in $k[X]$ of a quadratic and a linear polynomial in $k[X, Y]$ (§278–§280, pp. 215–229);
- in $k[X]$ of two polynomials $XY - aX - bY - c$ (§281–§284, pp. 230–235);
- in $k[X]$ of two quadratic polynomials in $k[X, Y]$ (§285–§291, pp. 235–243; pp. 303–305, pp. 252–255);
- in $k[X]$ of a quadratic and two linear polynomials in $k[X, Y, Z]$ (§292, pp. 244–245);
- in $k[X]$ of three quadratic polynomials

$$aX^2 + bXY + cXZ + dX + eY + fZ + g$$

in $k[X, Y, Z]$ (§320, pp. 269–271);
- in k of three polynomials $XY - aX - bY - c$ (§373–§374, pp. 235–236);
- in k of three quadratic polynomials in $k[X, Y]$ (§375, pp. 326–328);
- in k of three quadratic polynomials in $k[X]$ (§462, pp. 389–390);
- in k of three cubic polynomials in $k[X]$ (§463-4, pp. 390–392).

whose degrees (actually the values are $\mu := n - 1$ and $\nu := m - 1$) are deduced by means of results similar to Theorem 41.2.3;[38]

(2) summing two such products, considering the linear system whose equations are the coefficients of the resulting polynomial and whose unknowns are the A_i and A'_i; and

(3) solving this system by the method discussed above, that is, via determinant expansion.

Example 41.10.1.　Let us illustrate the easiest case, that of two quadratic univariate polynomials

$$ax^2 + bx + c, a'x^2 + b'x + c' \in k[x];$$

multiplying them by, respectively, $Ax + B$ and $A'x + B'$ we obtain[39]

une équation de cette forme

$$Aax^3 + (Ab + Ba)x^2 + (Ac + Bb)x + Bc = 0.$$

Egalant à zero le coefficient total de x^3, celui de x^2, &c. je procède au calcul de $AA'BB'$, comme il suit:

Première ligne $aA'BB'$
Seconde ligne $(ab')BB' - aA'aB'$
Troisiéme ligne $(ab')bB' - (ac')aB'$

en rejettant le terme où resteroit A' qui n'étant point dans la dernière équation, ne peut plus influer sur l'equation finale.
　　Quatriéme ligne $(ab')(bc') - (ac')^2$.
　　On a donc pour équation finale $(ab')(bc') - (ac')^2 = 0$.[40]

If we expand Bézout's computation using the same notation as above, we have

$$aA'BB' - Aa'BB',$$
$$(ab' - a'b)BB' - aA'aB' + aA'Ba' + Aa'aB' - Aa'Ba',$$
$$(ab' - a'b)(bB' - Bb') - (ac' - ca')(aB' - Ba') + (aA' - Aa')(ab' - ba'),$$
$$(ab' - a'b)(bc' - cb') - (ac' - ca')^2,$$

which can be interpreted as the expansion of the determinant

$$\begin{vmatrix} a & a' & 0 & 0 \\ b & b' & a & a' \\ c & c' & b & b' \\ 0 & 0 & c & c' \\ \hline A & A' & B & B' \end{vmatrix}$$

---

[38] The point is to reach a degree at which, equating the appropriate coefficients of the terms in the equation $\phi\Phi + \phi'\Phi' = 0$, one obtains at least as many equations as unknowns. If the difference is positive, leaving some freedom, alternatively one can either equate some unknown to 0 or, in order to preserve symmetry, add equations of the type $a_i A'_j - a'_i A_j = 0$ for convenient i, j (for an illustration compare Example 41.10.2 below).

[39] We remark that Bézout uses the shorthand (ab') to denote the determinant $\begin{vmatrix} a & b \\ a' & b' \end{vmatrix}$.

[40] Bézout E., *op cit.*, §347, p. 300.

corresponding to the linear system

$$\begin{cases} Aa + A'a' &= 0, \\ Ab + Ba + A'b' + B'a' &= 0, \\ Ac + Bb + A'c' + B'b' &= 0, \\ Bc + B'c' &= 0. \end{cases}$$

Note that in the last expansion the terms in A and A' have been substituted by the corresponding coefficient 0 that annihilates the last summand of the third expansion, justifying Bézout's comment that the terms containing A' (and A) can be removed since in the fourth equations they do not appear and thus do not influence the expansion.

Finally, we remark that the matrix whose determinant has been computed is equivalent to Sylvester's matrix

$$\begin{vmatrix} a & b & c & 0 \\ 0 & a & b & c \\ a' & b' & c' & 0 \\ 0 & a' & b' & c' \end{vmatrix}.$$

Example 41.10.2. Let us now consider the system[41]

$$\begin{cases} ax^2 + bxy + cy^2 + dx + ey + f &= 0, \\ d'x + e'y + f' &= 0, \\ d''x + e''y + f'' &= 0, \end{cases}$$

where the three equations are multiplied, respectively, by C, $A'x + B'y + C'$ and $A''x + B''y + C''$; considering in an orderly way the coefficients of x^2 and xy, the *équation arbitraire* $B'd' + B''d'' = 0$ and the coefficients of $y^2, x, y, 1$ we have seven equations in connection with seven variables, and we thus obtain the resultant

$$\begin{vmatrix} d' & d'' & 0 & 0 & a & 0 & 0 \\ e' & e'' & d' & d'' & b & 0 & 0 \\ 0 & 0 & d' & d'' & 0 & 0 & 0 \\ 0 & 0 & e' & e'' & c & 0 & 0 \\ f' & f'' & 0 & 0 & d & d' & d'' \\ 0 & 0 & f' & f'' & e & e' & e'' \\ 0 & 0 & 0 & 0 & f & f' & f'' \end{vmatrix},$$

which Bézout presents as

$$(d'e'')\left(c(d'f'')^2 + (d'e'')(de'f'') - b(e'f'')(d'f'') + a(e'f'')^2 \right),$$

where he uses the shorthand notation $(de'f'') = \begin{vmatrix} d & e & f \\ d' & e' & f' \\ d'' & e'' & f'' \end{vmatrix}.$

[41] Bézout E., *op. cit.*, §369, pp. 319–20.

Historical Remark 41.10.3.

(1) A similar method is applied by Bézout to compute the resultants in $k[X]$ of two polynomials in $k[X, Y]$. The main differences are that

(2) the linear system is obtained considering only the coefficients divisible by Y,

(3) the solution of the system returns the coefficients A_i, A_i' as rational functions in the a_i and a_i',

(4) the resolvent is obtained by setting, in the polynomial obtained in step (1), $Y = 0$, that is, removing the coefficients used in step (2) and substituting each A_i, A_i' with their expression in terms of the a_i and a_i'.

$\boxed{\odot}$

Example 41.10.4.   Let us illustrate Bézout's approach by considering[42] the polynomials $ax^2 + bxy + cy^2 + dx + ey + f$ and $d'x + e'y + f'$, which are respectively multiplied by F and $D'x + E'y + F'$, giving

$$(Fa + D'd')x^2 + (Fb + D'e' + E'd)xy + (Fc + E'e')y^2$$
$$+ (Fd + D'f' + F'd')x + (Fe + E'f + F'e')y + fF + f'F'. \qquad (41.4)$$

We then consider the equations (relating to the coefficients of y^2, xy and y)

$$\begin{cases} Fc + E'e' &= 0, \\ Fb + D'e' + E'd &= 0, \\ Fe + E'f' + F'e' &= 0, \end{cases}$$

and we expand the expression $D'E'FF'$, obtaining[43]

$$- D'e'FF' + D'E'cF',$$
$$- e'e'FF' + D'e'bF' + e'E'cF' - D'd'cF',$$
$$(cd'e' - be'e')D' - ce'e'E' + e'e'e'F + (ce'f' - ee'e')F,$$

and the solution (sic!) $D' = cd'e' - be'e'$, $E' = ce'e'$, $F = e'e'e'$, $F' = ce'f' - ee'e'$; substituting it into (41.4) and setting $y = 0$ we obtain

$$(e'e'e'a + cd'e'd' - be'e'd')x^2 + (e'e'e'd + 2cd'e'f' - be'e'f'$$
$$- ee'e'd')x + (fe'e'e' + ce'f'f' - ee'e'f')$$

$$= e' \begin{vmatrix} c & bx + e & ax^2 + dx + f \\ e' & d'x + f' & 0 \\ 0 & e' & d'x + f' \end{vmatrix};$$

this is, up to the extraneous factor e', the expected Sylvester resultant.

$\boxed{\odot}$

---

[42] Bézout E., op. cit., §278 pp. 225–228.

[43] Compare the corresponding determinant, $\begin{vmatrix} 0 & e' & c & 0 \\ e' & d' & b & 0 \\ 0 & f' & e & e' \\ D' & E' & F & F' \end{vmatrix}$.

The Sylvester resultant was, at least implicitly, introduced by Euler.[44] His approach is essentially a variation of the approach illustrated in Example 41.10.1: given

$$\phi := \sum_{i=0}^{m} a_{i+1}(X) Y^{m-i} \quad \text{and} \quad \phi' := \sum_{i=0}^{n} a'_{i+1}(X) Y^{n-i} \in k(X)[Y],$$

he multiplies them, respectively, by

$$\Phi := a'_1 Y^{n-1} + \sum_{i=0}^{n-2} A_{i+1} Y^{n-2-i} \quad \text{and} \quad \Phi' := a_1 Y^{m-1} + \sum_{i=0}^{m-2} A'_{i+1} Y^{m-2-i}$$

and subtracts the result, obtaining a polynomial of degree $m + n - 2$, the coefficient of Y^{m+n-1} being 0; thus, equating the coefficients of $Y, Y^2, \ldots, Y^{m+n-2}$, we obtain $m + n - 2$ linear equations in $(m - 1) + (n - 1)$ variables. The solution is then substituted into the constant coefficient $A_{n-1}a_{m+1} - A'_{m-1}a'_{n+1}$, giving $E(X) \in k(X)$.

Thus the equation system can be expressed as

$$M \cdot (a'_1, A_1, \ldots, A_{n-1}, a_1, A'_1, \ldots, A'_{m-1})^T = (0, \ldots, 0, E(X))^T,$$

where M is Sylvester's matrix.

It is clear that the procedure proposed by Euler is equivalent to the computation of the Sylvester resultant, so that $E(X) = \mathrm{Res}(\phi, \phi')$ is the required resultant.

In order to present the English School's view of Bézout's abridged method, I refer to Salmon's *Higher Algebra*.[45]

Given two polynomials of the same degree,

$$U := \sum_{i=0}^{m} a_{i+1} X^{m-i}, \quad V := \sum_{i=0}^{n} a'_{i+1} X^{n-i}, m = n,$$

and writing

$$V_\rho := \sum_{i=0}^{\rho} a'_{i+1} X^{\rho-i}, \quad U_\rho := \sum_{i=0}^{\rho} a_{i+1} X^{\rho-i} \qquad \text{for each } \rho, 0 \le \rho < n,$$

we compute the n polynomials, of degree bounded by $m - 1$,

$$F_\rho := V_{\rho-1} U - U_{\rho-1} V := \sum_{\sigma=1}^{m} \alpha_{\rho\sigma} X^{m-\sigma}, \qquad 1 \le \rho \le n;$$

---

[44] Euler L., *Introductio in Analysin Infinitorum*. Tom. 2 (1748), Lausanne Chapter XIX, §483–§485.

[45] Salmon G., *Lessons Introductory to the Modern Higher Algebra*, Fifth Edn, Chelsea Publishing Company (1885), §84–§86, pp. 81–83.

    The method can be found in Bézout E., Recherches sur le degré des équations résultantes de l'évanouissement des inconnues, et sur les moyens qu'il convient d'employer pour trouver ses équations, *Mém. Acad. Roy. Sci. Paris* (1964), 288–338. I was unable to read this paper, so I rely on the description by Salmon and that by Wimmer H.K., On the History of the Bezoutian and the resultant matrix. *Lin. Alg. Appl.* **128** (1990), 27–34.

we thus obtain the square matrix $(\alpha_{\rho\sigma})$ whose determinant is the required resultant. In order to extend this construction of a square matrix to the case $m > n$, we need $e := m - n$ more polynomials of degree bounded by $m - 1$; the choice is to take $F_\rho := X^{n-\rho}V$, $n < \rho \le m$, so that the resultant

is, therefore, as it ought to be, of the nth degree in the coefficients of $|U|$, and of the mth in those of $|V|$.[46]

Example 41.10.5.   It is sufficient to apply this recipe to the case of Example 41.10.1 to realize the equivalence with Bézout's result. We have

$$F_1 = a(a'x^2 + b'x + c') - a'(ax^2 + bx + c)$$
$$= (a, b')x + (a, c'),$$
$$F_2 = (ax + b)(a'x^2 + b'x + c') - (a'x + b')(ax^2 + bx + c)$$
$$= (a, c')x + (b, c'),$$

whence the required determinant is $\begin{vmatrix} (ab') & (ac') \\ (ac') & (bc') \end{vmatrix}$.   ⊡

The first to study the *Bezoutic matrix* $\mathsf{B} := (\alpha_{\rho\sigma})$ was Jacobi,[47] who, among other comments, remarked that

- $\sum_{\rho\sigma} X^\rho \alpha_{\rho\sigma} X^\sigma = U(X)\dfrac{\partial V(X)}{\partial X} - V(X)\dfrac{\partial U(X)}{\partial X}$ (p. 103),[48]
- B is symmetric (p. 102) and
- has no *factore superfluo* (extraneous factor) (p. 104),
- if $\det(\mathsf{B}) \ne 0$ then the inverse of B is a Hankel matrix[49] (p. 104),
- if $\det(\mathsf{B}) = 0$ then the common roots α of U and V are related to the linear solutions $(1, \alpha, \dots, \alpha^{n-1})$ of B (p. 104).

---

[46] Salmon G., *op. cit.*, §86, p. 83.

[47] Jacobi C.G.I., De eliminatione variabilis e duabus aequationibus algebraicas, *J. Reine Ang. Math.* **XV** (1836), 101–124.

[48] With the present notation, Jacobi remarked that

$$\sum_{\rho=1}^{m}\sum_{\sigma=1}^{m} X^{m-\rho}\alpha_{\rho\sigma}X^{m-\sigma} = \sum_{\rho=1}^{m} X^{m-\rho}F^\rho$$
$$= \left(\sum_{\rho=1}^{m} X^{m-\rho}\sum_{i=0}^{\rho-1} a'_{i+1}X^{\rho-1-i}\right)U - \left(\sum_{\rho=1}^{m} X^{m-\rho}\sum_{i=0}^{\rho-1} a_{i+1}X^{\rho-1-i}\right)V$$
$$= \left(\sum_{i=0}^{m-1} (m-i)a'_{i+1}X^{m-i-1}\right)U - \left(\sum_{i=0}^{m-1} (m-i)a_{i+1}X^{m-i-1}\right)V$$
$$= \frac{\partial V(X)}{\partial X}U - \frac{\partial U(X)}{\partial X}V.$$

[49] That is, the matrix $\det(\mathsf{B})^{-1}\mathsf{B}^{-1} := (a_{\rho\sigma})$ satisfies, for each ρ, σ, $a_{\rho\sigma} = A_{\rho+\sigma-2}$ for suitable A_i, $0 \le i \le 2n$.

Sylvester[50] gave in 1842 a formula[51] to express the coefficients of the F_ρ (in the case $n = m$) in terms of determinants,

$$(i, j) := (a_i a'_j) := \begin{vmatrix} a_i & a_j \\ a'_i & a'_j \end{vmatrix}, \quad 1 \le i < j \le n + 1,$$

as follows:

[Conceive] a number of cubic blocks each of which has two numbers, termed its *characteristics*, inscribed upon one of its faces, upon which the values of such a block (itself called an *element*) depend.

For instance, the value of the *element*, whose *characteristics* are r, s, is the difference between two products: the one of the coefficient rth in order occurring in the polynomial U, by that which comes sth in order of the polynomial V; the other product is that of the coefficient sth in order of the polynomial U, by that rth in order of V; so that if the degree of each equation be n, there will be altogether $\frac{1}{2}n(n + 1)$ such elements.

The blocks are formed into squares or flats (*plafonds*) of which the number is $n/2$ or $(n + 1)/2$, according as n is even or odd. The first of these contains n blanks in a side, the next $(n - 2)$, the next $(n - 4)$, till finally we reach a square of four blocks or of one, according as n is even or odd. These flats are laid upon one another so as to form a regularly ascending pyramid, of which the two diagonal planes are termed the planes of separation and symmetry respectively. The former divides the pyramid into two halves, such that no element on the one side of it is the same as that of any block in the other. The plan of symmetry, as the name denotes, divides the pyramid into two exactly *similar* parts; it being a rule, that *all elements lying in any given line of a square (platfond) parallel to the plane of separation are identical*; moreover the sum of the characteristics is the same for *all* elements lying *anywhere* in a *plane* parallel to that of separation.

The formula behind this rule is

$$\alpha_{\rho\sigma} = \alpha_{\sigma\rho} = \sum_{j=\sigma+1}^{n+1} (a_{\rho+\sigma+1-j} a'_j). \tag{41.5}$$

Example 41.10.6. For $n = 2$ we have the same formula as that produced by Bézout (Example 41.10.1): $(ab' - a'b)(bc' - cb') - (ac' - ca')^2$.

We illustrate Sylvester's formula and construction for $n = 3$:[52] we have

<table>
<tr><td rowspan="3">| 2, 3 |</td><td>1, 2</td><td>1, 3</td><td>1, 4</td></tr>
<tr><td>1, 3</td><td>1, 4</td><td>2, 4</td></tr>
<tr><td>1, 4</td><td>2, 4</td><td>3, 4</td></tr>
</table>

[50] Sylvester J.J., Memoir on the dialytic method of elimination. Part I. *Phil. Mag.* **XXXI** (1842), 534–539.

[51] Sylvester's matrix was given two years before, in Sylvester J.J., A method of determining by mere inspection the derivatives from two equations of any degree. *Phil. Mag.* **XVI** (1840), 132–135.

[52] The cases $n = 4, 5$ can be found in Salmon, *op. cit.*, §84–§85, pp. 81–82; all cases up to 6 are in Sylvester J.J., *Memoir, op. cit.*; see footnote 31.

giving the determinant

$$
\begin{vmatrix}
(a_1 a_2') & (a_1 a_3') & (a_1 a_4') \\
(a_1 a_3') & (a_1 a_4') + (a_2 a_3') & (a_2 a_4') \\
(a_1 a_4') & (a_2 a_4') & (a_3 a_4')
\end{vmatrix}
$$

$$
= \begin{vmatrix}
\begin{vmatrix} a_1 & a_2 \\ a_1' & a_2' \end{vmatrix} & & \begin{vmatrix} a_1 & a_3 \\ a_1' & a_3' \end{vmatrix} & & \begin{vmatrix} a_1 & a_4 \\ a_1' & a_4' \end{vmatrix} \\[2ex]
\begin{vmatrix} a_1 & a_3 \\ a_1' & a_3' \end{vmatrix} & \begin{vmatrix} a_1 & a_4 \\ a_1' & a_4' \end{vmatrix} + \begin{vmatrix} a_2 & a_3 \\ a_2' & a_3' \end{vmatrix} & \begin{vmatrix} a_2 & a_4 \\ a_2' & a_4' \end{vmatrix} \\[2ex]
\begin{vmatrix} a_1 & a_4 \\ a_1' & a_4' \end{vmatrix} & & \begin{vmatrix} a_2 & a_4 \\ a_2' & a_4' \end{vmatrix} & & \begin{vmatrix} a_3 & a_4 \\ a_3' & a_4' \end{vmatrix}
\end{vmatrix},
$$

which is to be compared with Sylvester's determinant

$$
\begin{vmatrix}
a_1 & a_2 & a_3 & a_4 & 0 & 0 \\
0 & a_1 & a_2 & a_3 & a_4 & 0 \\
0 & 0 & a_1 & a_2 & a_3 & a_4 \\
a_1' & a_2' & a_3' & a_4' & 0 & 0 \\
0 & a_1' & a_2' & a_3' & a_4' & 0 \\
0 & 0 & a_1' & a_2' & a_3' & a_4'
\end{vmatrix}.
$$

If $m > n$ the same formula is applied simply by expressing V as follows:

$$
V := \sum_{i=0}^{n} a_{i+1}' X^{n-i} = \sum_{i=0}^{m} b_{i+1} X^{m-i},
$$

with

$$
b_{i+1} = \begin{cases} 0 & 0 \le i < m - n \\ a_{i+1-m+n}' & m - n \le i \le m. \end{cases}
$$

In 1857 Cayley[53] gave in *Crelle*

la forme la plus simple sous laquelle on peut présenter cette méthode.
Pour éliminer [x] entre deux équations du n^{ième} *degré*[54]

$$
\sum_{i=0}^{n} a_{i+1} x^{n-i} = 0, \qquad \sum_{i=0}^{n} a_{i+1}' x^{n-i} = 0
$$

---

[53] Cayley A., Note sur la méthode d'élimination de Bezout, *J. Reine Ang. Math.* **LIII** (1857) 366–367.
    The result is, however, earlier. It was explicitly reported in 1853 by Sylvester in On a theory, op. cit., §62.
[54] Cayley gives the formula for two homogeneous forms; I have adapted his formulas to the non-homogeneous case and I use a modern notation instead of the one introduced by him.

on n'a qu'a former l'équation identique

$$\frac{\sum_{i=0}^{n} a_{i+1} x^{n-i} \sum_{i=0}^{n} a'_{i+1} y^{m-i} - \sum_{i=0}^{n} a'_{i+1} x^{m-i} \sum_{i=0}^{n} a_{i+1} y^{n-i}}{x - y}$$

$$= (y^{n-1}, y^{n-2}, \ldots, 1) \begin{pmatrix} a_{0,0} & a_{1,0} & \cdots & a_{n-1,0} \\ a_{0,1} & a_{1,1} & \cdots & a_{n-1,1} \\ \vdots & \vdots & \ddots & \vdots \\ a_{0,n-1} & a_{1,n-1} & \cdots & a_{n-1,n-1} \end{pmatrix} \begin{pmatrix} x^{n-1} \\ x^{n-2} \\ \vdots \\ 1 \end{pmatrix}$$

oú l'expression qui forme le second membre représente la fonction suivant:

$$\left(a_{0,0} x^{n-1} + a_{1,0} x^{n-2} + \cdots + a_{n-1,0} \right) y^{n-1}$$

$$+ \left(a_{0,1} x^{n-1} + a_{1,1} x^{n-2} + \cdots + a_{n-1,1} \right) y^{n-2}$$

$$+ \cdots$$

$$+ \left(a_{0,n-1} x^{n-1} + a_{1,n-1} x^{n-2} + \cdots + a_{n-1,n-1} \right):$$

le résultat de l'élimination sera

$$\begin{vmatrix} a_{0,0} & a_{1,0} & \cdots & a_{n-1,0} \\ a_{0,1} & a_{1,1} & \cdots & a_{n-1,1} \\ \vdots & \vdots & \ddots & \vdots \\ a_{0,n-1} & a_{1,n-1} & \cdots & a_{n-1,n-1} \end{vmatrix}.$$

Example 41.10.7. For $n = 2$ we have

$$\frac{(ax^2 + bx + c)(a'y^2 + b'y + c') - (a'x^2 + b'x + c')(ay^2 + by + c)}{x - y}$$

$$= (ab' - ba')xy + (ac' - ca')x + (ac' - ca')y + (bc' - cb'),$$

again returning the matrix $\begin{pmatrix} (ab') & (ac') \\ (ac') & (bc') \end{pmatrix}$.

For $n = 3$ and $U(x) = ax^3 + bx^2 + cx + d$, $V(x) = a'x^3 + b'x^2 + c'x + d'$ we have

$$\frac{U(x)V(y) - V(x)U(y)}{x - y} = (ab')x^2 y^2 + (ac')(x^2 y + xy^2) + (ad')x^2$$

$$+ \big((ad') + (bc')\big) xy + (ad')y^2 + (bd')x + (bd')y + (cd'),$$

giving the determinant $\begin{vmatrix} (ab') & (ac') & (ad') \\ (ac') & (ad') + (bc') & (bd') \\ (ad') & (bd') & (cd') \end{vmatrix}.$ ◎

Sylvester,[55] in the case $n = m$, denotes the polynomials F_ρ, $1 \leq \rho \leq n$, the *Bezoutians* of U and V and remarks that

---

[55] Sylvester J.J., On a theory, *op. cit.*, §5.

The determinant formed by arranging in a square the n sets of coefficients of the n Bezoutians, and which I shall term the Bezoutian matrix,[56] gives, as is well known, the Resultant (meaning thereby the Result in its simplest form of eliminating the variables out) of U and V.

Eliminating dialytically, first X^{n-1} between the first and the second, then X^{n-1} and X^{n-2} between the first, second and the third, and so on, and finally, all the powers of X between the first, second, third, ...,nth of these Bezoutians, and repeating the first of them, we obtain a derived set of n equations, the right-hand members of which I shall term the secondary Bezoutians to U and V.

The "dialytical elimination" performed by Sylvester on the expressions

$$V_{\rho-1}U - U_{\rho-1}V = F_\rho, \quad 1 \le \rho \le n,$$

returns

$$
\begin{aligned}
V_0 U - U_0 V &= F_1 &=:\ & B_1, \\
(\alpha_{21}V_0 - \alpha_{11}V_1)U - (\alpha_{21}U_0 - \alpha_{11}U_1)V &= \alpha_{21}F_1 - \alpha_{11}F_2 &=:\ & B_2, \\
&\vdots \\
S_\rho U - T_\rho V & &=:\ & B_\rho, \\
&\vdots \\
S_{n-1}U - T_{n-1}V & &=:\ & B_{n-1},
\end{aligned}
$$

where we have $\deg(S_\rho) = \deg(T_\rho) = \rho - 1$, $\deg(B_\rho) = n - \rho$; thus, assuming U, V to be

perfectly unrelated, and each the most general function that can be formed of the same degree.[57]

In the case $m = n$,[58] if we repeatedly perfom the Division Algorithm and *change the sign* of each remainder, as in Section 13.3, we obtain the polynomial

---

[56] But he also uses the term *Bezoutic square*. The term then stabilizes as *Bezoutic matrix* (compare Cayley's entry Mathematics, recent terminology in, in the *English Cyclopædia*, vol. V (1860), pp. 534–542). In particular, Cayley (A fourth memoir, *op.cit.*, §88) labels the *bezoutic emanant* of U and V the polynomial $(U(x)V(y) - V(x)U(y))/(x - y)$ introduced by him.

The term *Bezoutiant*, in fact, as we will see below, has been already associated with the quadratic function which (A. Cayley, A fourth memoir, *op. cit.*, §91):

Professor Sylvester forms with the matrix of the Bezoutic emanant and a set of m facients $\{u, v, ...\}$ an m-ary quadratic function, which he terms the Bezoutiant.

[57] Sylvester J.J., On a theory, *op. cit.*, §1.

[58] This argument and construction is in §5. In the sequent §6, Sylvester explains how to extend it to the case $m = n + e, e > 0$.

He defines

$$V_\rho := \sum_{i=0}^{\rho} a'_{i+1} X^{\rho-i}, \quad U_{\rho+e} := \sum_{i=0}^{\rho+e} a_{i+1} X^{\rho+e-i} \quad \text{for each } \rho, 0 \le \rho < n,$$

and computes the n polynomials (all of degree bounded by $m - 1$)

$$F_\rho := V_{\rho-1}U - U_{\rho-1}V := \sum_{\sigma=1}^{m} \alpha_{\rho\sigma} X^{m-\sigma}, \quad 1 \le \rho \le n.$$

sequence[59] $U, V, R_2, \ldots, R_m$, where each R_i necessarily satisfies $\deg(R_i) = m - i + 1$. Thus, the[60]

> n successive *Secondary Bezoutians* to the system U, V [...] will (saving at least a numerical factor of a magnitude and algebraic sign to be determined, but which, when proper conventions are made, will be subsequently proved to be $+1$) represent the simplified [...] residue[s] to U/V.

That is, $B_\rho = R_{\rho+1}$ for each ρ.

Once he has obtained the Bezoutic square of two polynomials f, ϕ of the same degree m, Sylvester remarks that[61]

> this square [...] is symmetrical about one of its diagonals, and corresponds therefore (as every symmetrical matrix must do) to a homogeneous quadratic function of m variables of which it expresses the determinant. This quadratic function, which plays a great part in [...] the theory of real roots, I term the Bezoutiant.
>
> [...]
>
> In Section V, Arts. 56, 57, I show that the *total* number of effective intercalations between the roots of two functions of the same degree is given by the *inertia* of that quadratic form[62] which we agreed to term the Bezoutiant to f and ϕ; and in the following article (58) the result

Next he introduces the e polynomials $X^\mu V, 0 \le \mu < e$, and, for $\rho, 1 \le \rho \le n$, using these e polynomials and the ρ polynomials $F_r, 1 \le r \le \rho$, he produces the relation

$$S_\rho U - T_\rho V = B_\rho, \quad \deg(S_\rho) = \rho - 1, \quad \deg(T_\rho) = e + \rho - 1, \quad \deg(B_\rho) = m - \rho:$$

thus the argument given in the case $n = m$ applies *verbatim*.

[59] which is the Sturm sequence (Definition 13.3.1) if $V = U'$.

[60] Sylvester J.J., On a theory, *op. cit.*, §5.

[61] Sylvester J.J., On a theory, *op. cit.*, Introduction.

[62] Sylvester introduced the notion of *inertia* and proved a *Law of Inertia* in *On a theory*, §44–45.
Recall that, given a quadratic form

$$f(x_1, \ldots, x_n) = \sum_i \sum_j \beta_{ij} x_i x_j, \quad \beta_{ij} = \beta_{ji}.$$

and writing, for each two vectors $u = (c_1, \ldots, c_n), v = (d_1, \ldots, d_n)$ in k^n,

$$f(u, v) := \sum_i \sum_j \beta_{ij} c_i d_j$$

then the vector space $N := \{ w \in k^n : f(w, u) = 0 \text{ for each } u \in k^n \}$ is invariant for linear transformations and such is also its dimension $n - r$.
Thus k^n has an orthogonal basis $v_1, \ldots, v_r, v_{r+1}, \ldots, v_n$ such that

$$N = \mathrm{Span}_k(v_{r+1}, \ldots, v_n) \quad \text{and} \quad f(v_i, v_j) = \begin{cases} 0 & i \ne j. \\ \gamma_i \ne 0 & i = j \le r. \\ 0 & i = j > r. \end{cases}$$

so that for each $u = \sum_i c_i v_i \in k^n$ we have $f(u, u) := \sum_i c_i^2 \gamma_i$.
Sylvester's Law of Inertia states that, if $k = \mathbb{R}$, the "number of integers in the excess of positive over negative signs which adheres to a quadratic form expressed as the sum of positive and negative squares" (which Sylvester names the *inertia* of the quadratic form) is "unchangeable notwithstanding any real linear transformation impressed upon such form", that is, the inertia is the invariant

$$\#\{\gamma_i > 0, 1 \le i \le r\} - \#\{\gamma_i < 0, 1 \le i \le r\}.$$

is extended to embrace the case contemplated in M. Sturm's theorem; that is to say, I show, that on replacing the function of x by a homogeneous function of x and y, the Bezoutiant of the two functions, which are respectively the differential derivates of f with respect to x and with respect to y, will serve to determine by its form or *inertia* the total number of real roots and of *equal* roots in $f(x)$.[63] The subject is pursued in the following Arts., 59, 60. [...] In Arts. 61, 62, 63, it is proved that the Bezoutiant is an invariative function of the functions from which it is derived; and in Art. 64 the important remark is added, that it is an invariant of that particular class to which I have given the name of Combinants, which have the property of remaining unaltered, not only for linear transformations of the variables, but also for linear combinations of the functions containing the variables,[64] possessing thus a character of double invariability. In Arts. 65, 66 I consider the relation of the Bezoutiant to the differential determinant, so called by Jacoby, but which for greater brevity I call the Jacobian. On proper substitutions being made in the Bezoutiant for the m variables which it contains [...], the Bezoutiant becomes identical with the Jacobian of f and Φ.

To illustrate the "proper substitution to be done" I again call on Sylvester:[65]

[The Bezoutiant] $B(u_1, \ldots, u_m)$ being a covariant of the system f and ϕ [...], on making $u_1, \ldots, u_m$ equal to $[x^{m-1}, x^{m-2}y, \ldots, y^{m-1}]$, B will become [...] what I am in the habit of calling the Jacobian (after the name of the late but ever-illustrious Jacobi), a term capable of application to any number of homogeneous functions of as many variables. In the case before us, where we have two functions of two variables, the Jacobian

$$
J(f, \phi) = \begin{vmatrix} \dfrac{df}{dx}, & \dfrac{d\phi}{dx} \\[2mm] \dfrac{df}{dy}, & \dfrac{d\phi}{dy} \end{vmatrix} = \frac{df}{dx}\frac{d\phi}{dy} - \frac{df}{dy}\frac{d\phi}{dx}.
$$

---

[63] In other words, Sylvester

- considers the polynomial $f(x) = \sum_{i=0}^{m} a_i x^{m-i}$ and its derivate $f'(x)$,
- performs the Division Algorithm, obtaining

$$
f_1(x) = mf(x) - xf'(x) = \sum_{i=1}^{m} i a_i x^{m-i}.
$$

- computes the Bezoutian secondaries of f_1 and f', $B_1, \ldots, B_{m-1}$, which in this case are exactly the Sturm sequence and
- evaluates "the number of pairs of imaginary roots in $f(x)$" by counting "the number of *variations* of sign betwen consecutive terms" obtained on evaluating $f_1, f', B_1, \ldots, B_{m-1}$ at $+\infty$.

Remark that, setting $g(x, y) = \sum_{i=0}^{m} a_i x^{m-i} y^i$ we actually have

$$
\frac{\partial g}{\partial y} = \sum_{i=1}^{m} i a_i x^{m-i} y^{i-1}, \qquad f_1(x) = \frac{\partial g}{\partial y}(x, 1),
$$

justifying Sylvester's reference to the derivate with respect to y.

[64] That is, he considers the Bezoutiant of the functions $kf + i\phi$, and $k'f + i'\phi$ and remarks that each entry on the Bezoutic matrix is multiplied by $ki' - k'i$, so that the Bezoutiant (§64) becomes increased in the ratio of $(ki' - k'i)^m$, that is remains always unaltered in point of form and absolutely immutable, provided that $ki' - k'i$ be taken, as we may always suppose to be the case, equal to 1.

[65] Sylvester J.J., On a theory, *op. cit.*, §65.

[...] So in the case of a single function F of the degree m, the Bezoutoid, that is the Bezoutiant to $dF/dx, dF/dy$, on making the $(m-1)$ variables which it contains identical with $x^{m-2}, x^{m-3}y, \ldots, y^{m-2}$ respectively, becomes identical with the Jacobian to $dF/dx, dF/dy$, that is the Hessian of F, namely

$$\begin{vmatrix} \dfrac{d^2F}{dx^2}, & \dfrac{d^2F}{dxdy} \\[2ex] \dfrac{d^2F}{dxdy}, & \dfrac{d^2F}{dy^2} \end{vmatrix}.$$

As an example of this property of the Bezoutiant, suppose

$$f = ax^3 + bx^2y + cxy^2 + dy^3,$$
$$\phi = \alpha x^3 + \beta x^2y + \gamma xy^2 + \delta y^3.$$

The Bezoutiant matrix becomes

$$\begin{matrix} a\beta - b\alpha, & a\gamma - c\alpha, & a\delta - d\alpha, \\[1ex] a\gamma - c\alpha, & \left(\begin{array}{c} a\delta - d\alpha \\ + \\ b\gamma - c\beta \end{array}\right), & b\gamma - c\beta, \\[1ex] a\delta - d\alpha, & b\gamma - c\beta, & c\delta - d\gamma. \end{matrix}$$

The Bezoutiant accordingly will be the quadratic function

$$(a\beta - b\alpha)u_1^2 + \{(a\delta - d\alpha) + (b\gamma - c\beta)\}u_2^2 + (c\delta - d\gamma)u_3^2$$
$$+ 2(a\gamma - c\alpha)u_1u_2 + 2(a\delta - d\alpha)u_3u_1 + 2(b\gamma - c\beta)u_2u_3,$$

which on making

$$u_1 = x^2, \quad u_2 = xy, \quad u_3 = y^2$$

becomes

$$Lx^4 + Mx^3y + Nx^2y^2 + Pxy^3 + Qy^4,$$

where L, M, N, P, Q respectively will be the sum of the terms lying in the successive bands drawn parallel to the sinister diagonal of the Bezoutiant matrix, that is

$$L = (a\beta - b\alpha),$$
$$M = 2(a\gamma - c\alpha),$$
$$N = 3(a\delta - d\alpha) + (b\gamma - c\beta),$$
$$P = 2(b\gamma - c\beta),$$
$$Q = (c\delta - d\gamma).$$

The biquadratic function in x and y [...] will be found on computation to be identical in point of form with the Jacobian to f, ϕ, namely

$$(3ax^2 + 2bxy + cy^2)(\beta x^2 + 2\gamma xy + 3\delta y^2) - (3\alpha x^2 + 2\beta xy + \gamma y^2)(bx^2 + 2cxy + 3dy^2)$$

this latter being in fact

$$3Lx^4 + 3Mx^3y + 3Nx^2y^2 + 3Pxy^3 + 3Qy^4.$$

He concludes by commenting

The remark is not without some interest, that in fact the Bezoutiant, which is capable (as has been shown already) of being mechanically constructed, gives the best and readiest means of calculating the Jacobian; for in summing the sinister bands transverse to the axis of symmetry the only numerical operation to be performed is that of addition of positive integers, whereas the direct method involves the necessity of numerical subtractions as well as additions, inasmuch as the same terms will be repeated with different signs.

He further remarks, in a different example, that, unlike the computation via the Bezoutiant, a direct evaluation requires one to effectively employ also *division* in order to reduce the Jacobian, being divisible by $\deg(f) = \deg(\phi)$, to its simplest form.

41.11 Dixon's Resultant

The computation of a resultant of r forms in r variables was already solved by Bézout as an instance of this general approach.

An alternative proposal was put forward by Cayley based on what today we could call a solution via linear syzygies.[66] He assumes m_1 variables connected by m_2 linear equations, not being all independent but connected by m_3 linear equations, again not necessarily linearly independent; we thus obtain s matrices $M_\sigma = \left(a_{ij}^{(\sigma)}\right)$, the σth matrix having m_σ columns and $m_{\sigma+1}$ rows, the m_σ being related by $\sum_{\sigma=1}^{s+1}(-1)^\sigma m_\sigma = 0$:

the number of quantities $|m_1|$ will be equal to the number of really independent equations connecting them, and we may obtain by the elimination of these quantities a result $\Delta = 0$.

The approach, denoting $\mu_\varrho := \sum_{\sigma=\varrho}^{s+1}(-1)^{\sigma-\varrho} m_\sigma = 0$, consists in:

- selecting $\mu_{s+1} = m_{s+1}$ indexes $I_s \subset \{1, \ldots, m_s\}$ and computing the determinant Q_s of the μ_{s+1}–square minor of M_s obtained by selecting the rows indexed by I_s;
- selecting $\mu_s = m_s - \mu_{s+1}$ indexes $I_{s-1} \subset \{1, \ldots, m_{s-1}\}$ and computing the determinant Q_{s-1} of the μ_s–square minor of M_{s-1} obtained by selecting the rows indexed by I_{s-1} and the columns indexed by $\{1, \ldots, m_s\} - I_s$;
- …;
- selecting $\mu_\varrho = m_\varrho - \mu_{\varrho+1}$ indexes $I_{\varrho-1} \subset \{1, \ldots, m_{\varrho-1}\}$ and computing the determinant $Q_{\varrho-1}$ of the μ_ϱ–square minor of $M_{\varrho-1}$ obtained by selecting the rows indexed by $I_{\varrho-1}$ and the columns indexed by $\{1, \ldots, m_\varrho\} - I_\varrho$;
- …;
- selecting $\mu_3 = m_3 - \mu_4$ indexes $I_2 \subset \{1, \ldots, m_2\}$ and computing the determinant Q_2 of the μ_2–square minor of M_2 obtained by selecting the rows indexed by I_2 and the columns indexed by $\{1, \ldots, m_3\} - I_3$;
- computing, on the basis of the remark that $m_1 = \mu_2 = m_2 - \mu_3$, the determinant Q_1 of the μ_2–square minor of M_1 obtained by selecting the columns indexed by $\{1, \ldots, m_2\} - I_2$.

[66] Cayley A., On the theory of elimination, *Camb. Dublin Math. J.* **III** (1848), 116–120.

Finally, if each Q_i is non-zero, one obtains Δ by computing

$$\Delta = Q_1 Q_2^{-1} Q_3 Q_4^{-1} \cdots = \prod_{\sigma=1}^{s} Q_\sigma^{(-1)^{\sigma-1}}.$$

The application considers a set of forms $\{f_1, \ldots, f_u\}$ and, fixing an appropriate degree $d \in \mathbb{N}$, aims to eliminate all terms of degree d among the equations $F = 0$, where F runs over the forms in the set

$$\mathsf{F} := \{\tau f_i, 1 \le i \le u, \tau \in \mathcal{T}, \deg(\tau f_i) = d\};$$

it consists in computing a linear resolution of the elements in F and applying to it the computation suggested above. Cayley, however, remarks that

I am not in possession of any method of arriving *at once* at the final result in its more simplified form; my process, on the contrary, leads me to a result encumbered by an extraneous factor, which is only got rid of by a number of successive divisions.

The first solution, apart from that of Bèzout, for computing the resultant of more than two polynomials is due to A. L. Dixon;[67] it generalized Cayley's interpretation of the Bezoutic/Bezoutian matrix in terms of the Bezoutic emanant, proposing such an emanant for three polynomials in two variables and remarking that the construction easily generalizes to polynomials in any number of variables.

Given three polynomials

$$\phi(X_1, X_2) = \sum_{r=1}^{n} \sum_{s=1}^{m} A_{rs} X_1^r X_2^s,$$

$$\psi(X_1, X_2) = \sum_{r=1}^{n} \sum_{s=1}^{m} B_{rs} X_1^r X_2^s,$$

$$\chi(X_1, X_2) = \sum_{r=1}^{n} \sum_{s=1}^{m} C_{rs} X_1^r X_2^s,$$

Dixon considers the determinant

$$\Delta := \begin{vmatrix} \phi(X_1, X_2) & \psi(X_1, X_2) & \chi(X_1, X_2) \\ \phi(X_1, Y_2) & \psi(X_1, Y_2) & \chi(X_1, Y_2) \\ \phi(Y_1, Y_2) & \psi(Y_1, Y_2) & \chi(Y_1, Y_2) \end{vmatrix}$$

and, remarking that it vanishes if we put $X_1 = Y_1$ and also if we put $X_2 = Y_2$ and so is divisible by $(X_1 - Y_1)(X_2 - Y_2)$, he considers the polynomial

$$D(X_1, X_2, Y_1, Y_2) = \frac{\Delta(X_1, X_2, Y_1, Y_2)}{(X_1 - Y_1)(X_2 - Y_2)}$$

---

[67] Dixon A.L., The eliminant of three quantics in two independent variables, *Proc. London Math. Soc.* **7** (1908), 49–69.

which is of degree

$$\begin{cases} 2n - 1 & \text{in } X_1, \\ m - 1 & \text{in } X_2, \\ n - 1 & \text{in } Y_1, \\ 2m - 1 & \text{in } Y_2, \end{cases}$$

so that

equating to zero the cofficients of $|Y_1^r Y_2^s|$, for all values of r and s, $|D = 0|$ is equivalent to $2mn$ equations in $|X_1, X_2|$ and the number of terms in these equations is also $2mn$. Thus the eliminant[68] can be at once written down as a determinant of order $2mn$, each constituent of which is the sum of determinants of the third order of the type $\Delta := \begin{vmatrix} A_{pq} & A_{rs} & A_{tu} \\ B_{pq} & B_{rs} & B_{tu} \\ C_{pq} & C_{rs} & C_{tu} \end{vmatrix}$.

In other words, writing $\mathfrak{a} := \{X_1^r X_2^s; r < 2n, s < m\}$ and $\mathfrak{b} := \{Y_1^r Y_2^s; r < n, s < 2m\}$, we have

$$D(X_1, X_2, Y_1, Y_2) = \sum_{\tau \in \mathfrak{a}} \sum_{\upsilon \in \mathfrak{b}} d_{\tau \upsilon} \tau \upsilon, \qquad d_{\tau \upsilon} \in \mathbb{Z}[A_{pq}, B_{rs}, C_{tu}].$$

Clearly the vanishing of the determinant of the matrix $(d_{\tau \upsilon})$ is equivalent to the existence of a common root of ϕ, ψ, χ.

Finally, Dixon remarks that such a method is

applicable to the problem of elimination when the number of variables is greater than two.

Write, for each $i, 0 \le i \le n$,

$$g(\mathbf{X}_i) := g(Y_1, \ldots, Y_i, X_{i+1}, \ldots, X_n) \quad \text{for each } g(X_1, \ldots, X_n) \in \mathcal{P},$$

so that, in particular, $g(\mathbf{X}_0) = g(X_1, \ldots, X_n)$ and $g(\mathbf{X}_n) = g(Y_1, \ldots, Y_n)$.

Given $n + 1$ polynomials $f_1, \ldots, f_{n+1} \in k[X_1, \ldots, X_n]$ each of degree n_i in the variable X_i, one can consider the determinant[69]

$$\Delta := \begin{vmatrix} f_1(\mathbf{X}_0) & \cdots & f_{n+1}(\mathbf{X}_0) \\ f_1(\mathbf{X}_1) & \cdots & f_{n+1}(\mathbf{X}_1) \\ \vdots & \ddots & \vdots \\ f_1(\mathbf{X}_i) & \cdots & f_{n+1}(\mathbf{X}_i) \\ \vdots & \ddots & \vdots \\ f_1(\mathbf{X}_n) & \cdots & f_{n+1}(\mathbf{X}_n) \end{vmatrix}. \tag{41.6}$$

---

[68] Salmon used the term *eliminant* to denote what we call the *resultant*.

[69] It was Dixon himself who reversed the order in which the variables are transformed from X to Y; for two variables he transformed from right to left; in the final remark he makes the example of four polynomials in three variables and transforms them from left to right.

It is divisible by $\prod_{i=1}^{n}(X_i - Y_i)$, giving a polynomial

$$D(X_1, X_2, \ldots, X_n, Y_1, \ldots, Y_n)$$

of degree $m_i := (n + 1 - i)n_i - 1$ in X_i and $\mu_i := in_i - 1$ in Y_i so that

$$D(X_1, X_2, \ldots, X_n, Y_1, \ldots, Y_n) = \sum_{\tau \in \mathfrak{a}} \sum_{\upsilon \in \mathfrak{b}} d_{\tau\upsilon}\tau\upsilon,$$

where $\mathfrak{a} := \{X_1^{a_1} \cdots X_n^{a_n} : a_i \leq m_i\}$, $\mathfrak{b} := \{Y_1^{\alpha_1} \cdots Y_n^{\alpha_n} : \alpha_i \leq \mu_i\}$ and

$$\#\mathfrak{a} = \#\mathfrak{b} = n! \prod_{i=1}^{n} n_i := s.$$

Definition 41.11.1 (Kapur–Saxena–Yang). *The polynomial D is called the* Dixon *polynomial of* $f_1, \ldots, f_{n+1}$.

The matrix $\mathsf{D} := (d_{\tau\upsilon})$ *is called the* Dixon *matrix and its determinant the* Dixon *resultant.* ◉

Remark 41.11.2 (Kapur–Saxena–Yang). Let $\mathsf{D} := (c_{\tau\upsilon})$ be the Dixon matrix of $f_1, \ldots, f_{n+1}$, and let us enumerate the elements of $\mathfrak{a}$ as

$$\tau_1 := 1, \quad \tau_2 := X_1, \quad \ldots, \quad \tau_{n+1} := X_n, \quad \tau_{n+2}, \quad \ldots, \quad \tau_s.$$

If $\alpha := (a_1, \ldots, a_n) \in \mathcal{Z}(f_1, \ldots, f_{n+1})$ then

$$(1, a_1, \ldots, a_n, \tau_{n+2}(\alpha), \ldots, \tau_s(\alpha))$$

is a solution of the linear system $\mathsf{D}(\tau_1, \ldots, \tau_s)^T$. ◉

Remark 41.11.3. If $n = 1$, the Dixon matrix and polynomial coincide with what Sylvester and Cayley called the Bezoutic (or Bezoutian) matrix and Bezoutic emanant. ◉

Remark 41.11.4 (Kapur–Saxena–Yang). Dixon considered "generic" polynomials all having the same degree in each variable; as a consequence the Dixon matrix is square and one can define its determinant and so introduce the Dixon resultant.

In specific instances, the default approach is the classical one already used by Sylvester and Cayley, namely, an approach using the assumption that the missing terms have coefficient zero.

For an alternative solution based on a restriction to complete intersection ideals, see below. ◉

In a previous paper[70] Dixon gave another interesting computational approach to evaluate the resultant[71] in terms of Cayley's formula: given two polynomials of the

---

[70] Dixon A.L. On a form of the eliminant of two quantics, *Proc. London Math. Soc.* **6** (1908), 468–478.
[71] Which he called *Bezout's determinant*.

same degree[72]

$$U := \sum_{i=0}^{n} a_{i+1} X^{n-i}, \quad V := \sum_{i=0}^{n} a'_{i+1} X^{n-i},$$

and writing

$$C(X, Y) := \frac{U(X)V(Y) - V(X)U(Y)}{X - Y}$$

he stated that

Lemma 41.11.5.   *It holds that*

$$\operatorname{Res}(U, V) = \Delta := \begin{vmatrix} d_{11} & \cdots & d_{1n} \\ \vdots & \ddots & \vdots \\ d_{n1} & \cdots & d_{nn} \end{vmatrix},$$

where

$$d_{i.j} = \frac{1}{i!j!} \frac{\partial^{i+j} C(X, Y)}{\partial X^i \partial Y^j}\bigg|_{X=Y=0}.$$

Proof.   One has

$$C(X, Y) = \sum_{p>q} (a_{n-p+1} a'_{n-q+1} - a'_{n-p+1} a_{n-q+1}) \frac{X^p Y^q - Y^p X^q}{X - Y}$$

$$= \sum_{p>q} (a_{n-p+1} a'_{n-q+1} - a'_{n-p+1} a_{n-q+1}) \left(\sum_{i=0}^{p-q-1} X^{p-1-i} Y^{q+i} \right);$$

thus $d_{i.j}$, which is the coefficient of $X^i Y^j$ in $C(X, Y)$, satisfies $d_{i.j} = \alpha_{ij}$ where α_{ij} is the result of Sylvester's construction (41.5).   ⊙

Dixon then fixes "two sets of arbitrary quantities" $x_1, \ldots, x_n$ and $y_1, \ldots, y_n$ and states

Proposition 41.11.6 (Dixon).   *It holds that*

$$\operatorname{Res}(U, V) = \frac{\begin{vmatrix} C(x_1, y_1) & \cdots & C(x_1, y_n) \\ \vdots & \ddots & \vdots \\ C(x_n, y_1) & \cdots & C(x_n, y_n) \end{vmatrix}}{\prod_{i>l}(x_i - x_j) \prod_{i>l}(y_i - y_j)}.$$

---

[72] A polynomial of lower degree is forced, as usual, to reach the highest degree by adding $0X^{m+1} + \cdots + 0X^n$.

Proof. If we expand each $\mathcal{C}(x_i, y_i)$ "in ascending powers of $x_i - X$, using *Taylor's theorem*", and the result in ascending powers of $y_j - Y$ then we have

$$\begin{vmatrix} \mathcal{C}(x_1, y_1) & \cdots & \mathcal{C}(x_1, y_n) \\ \vdots & \ddots & \vdots \\ \mathcal{C}(x_n, y_1) & \cdots & \mathcal{C}(x_n, y_n) \end{vmatrix}$$

$$= \begin{vmatrix} 1 & (y_1 - Y) & \cdots & (y_1 - Y)^{n-1} \\ \vdots & \vdots & \ddots & \vdots \\ 1 & (y_n - Y) & \cdots & (y_n - Y)^{n-1} \end{vmatrix} \begin{vmatrix} 1 & (x_1 - X) & \cdots & (x_1 - X)^{n-1} \\ \vdots & \vdots & \ddots & \vdots \\ 1 & (x_n - X) & \cdots & (x_n - X)^{n-1} \end{vmatrix} \Delta$$

$$= \prod_{i>l}(x_i - x_j) \prod_{i>l}(y_i - y_j) \operatorname{Res}(U, V).$$

$\boxdot$

Example 41.11.7. Let us consider

$$U = (X - 1)X(X - a) \quad \text{and} \quad V := (X + 1)(X + 2)(X - b)$$

and choose

$$x_1 = 1, \quad x_2 = -1, \quad x_3 = -2, \quad y_1 = 1, \quad y_2 = -2, \quad y_3 = 2,$$

so that

$$\left| \mathcal{C}(x_i, y_j) \right|$$

$$= \begin{vmatrix} 6(ab - a - b + 1) & -12(ab - a + 2b - 2) & 12(ab - a - 2b + 2) \\ 6(ab - a + b - 1) & 0 & 8(ab - 2a + b - 2) \\ 24(ab - a + 2b - 2) & -12(ab + 2a + 2b + 4) & 36(ab - 2a + 2b - 4) \end{vmatrix}$$

$$= \begin{vmatrix} 6(a - 1)(b - 1) & -12(a + 2)(b - 1)) & 12(a - 2)(b - 1) \\ 6(a + 1)(b - 1) & 0 & 8(a + 1)(b - 2) \\ 24(a + 2)(b - 1) & -12(a + 2)(b + 2) & 36(a + 2)(b - 2) \end{vmatrix}$$

$$= 2^6 3^3 (b - 1)b(a - b)(a + 1)(a + 2).$$

$\boxdot$

41.12 Toward Cardinal's Conjecture

Cayley's formulation of the Bezoutic matrix in terms of

$$\frac{U(X)V(Y) - V(X)U(Y)}{X - Y},$$

which was interpreted by Dixon in matrix terms as

$$\frac{\begin{vmatrix} U(X) & V(X) \\ U(Y) & V(Y) \end{vmatrix}}{X - Y}$$

was expressed by Cardinal in different (but equivalent) ways as follows:

$$\frac{U(X)V(Y) - V(X)U(Y)}{X - Y} = \begin{vmatrix} \dfrac{U(X) - U(Y)}{X - Y} & U(Y) \\ \dfrac{V(X) - V(Y)}{X - Y} & V(Y) \end{vmatrix} = \begin{vmatrix} \dfrac{U(X) - U(Y)}{X - Y} & U(X) \\ \dfrac{V(X) - V(Y)}{X - Y} & V(X) \end{vmatrix}.$$

In a similar way, given a set of n polynomials

$$\mathcal{F} := \{f_1, \ldots, f_n\} \in k[X_1, \ldots, X_n]$$

and denoting, for each polynomial $g \in k[X_1, \ldots, X_n]$,

- $D(g, \mathcal{F})$ and $\mathsf{D}(g, \mathcal{F})$ respectively the Dixon polynomial and matrix of $f_1, \ldots, f_n, g$,

- $g(\mathsf{X}_i) := g(Y_1, \ldots, Y_i, X_{i+1}, \ldots, X_n)$, for each i, $0 \le i \le n$,

- $\delta_i(g) := \dfrac{g(\mathsf{X}_i) - g(\mathsf{X}_{i-1})}{X_i - Y_i}$, for each i, and

- $\delta_i(g, h) := \dfrac{g(\mathsf{X}_i)h(\mathsf{X}_{i-1}) - h(\mathsf{X}_i)g(\mathsf{X}_{i-1})}{X_i - Y_i}$, for each i and each $h \in k[X_1, \ldots, X_n]$.

Cayley's interpretation by Cardinal was extended by the latter in order to give an alternative representation of the Dixon polynomials $D(1, \mathcal{F})$ and $D(X_i, \mathcal{F})$.

Lemma 41.12.1 (Cardinal). *We have*

$$D(1, \mathcal{F}) := \begin{vmatrix} \delta_1(f_1) & \cdots & \delta_1(f_n) \\ \vdots & \ddots & \vdots \\ \delta_n(f_1) & \cdots & \delta_n(f_n) \end{vmatrix}$$

and, for each i, $1 \le i \le n$,

$$D(X_i, \mathcal{F}) := \begin{vmatrix} \delta_1(f_1) & \cdots & \delta_1(f_n) \\ \vdots & \ddots & \vdots \\ \delta_{i-1}(f_1) & \cdots & \delta_{i-1}(f_n) \\ \delta_i(X_i f_1) & \cdots & \delta_i(X_i f_n) \\ \delta_{i+1}(f_1) & \cdots & \delta_{i+1}(f_n) \\ \vdots & \ddots & \vdots \\ \delta_n(f_1) & \cdots & \delta_n(f_n) \end{vmatrix}.$$

Proof. If in (41.6) we set $f_{n+1} = 1$ and then subtract the jth row from the $(j+1)$th row and divide it by $X_j - Y_j$ for each j, $1 \le j \le n$, we obtain

$$D(1, \mathcal{F}) = \begin{vmatrix} f_1(X_1, \ldots, X_n) & \cdots & f_n(X_1, \ldots, X_n) & 1 \\ \delta_1(f_1) & \cdots & \delta_1(f_n) & 0 \\ \vdots & \ddots & \vdots & \vdots \\ \delta_i(f_1) & \cdots & \delta_i(f_n) & 0 \\ \vdots & \ddots & \vdots & \vdots \\ \delta_n(f_1) & \cdots & \delta_n(f_n) & 0 \end{vmatrix}.$$

Now, in (41.6) we set $f_{n+1} = X_i$ and, after subtracting the ith row from the $(i+1)$th row and dividing it by $X_i - Y_i$ for each i, $1 \le i \le n$, we multiply the $(i+1)$th row by Y_i and subtract from it the ith row of (41.6). Since

$$\begin{vmatrix} f_1(X_0) & \cdots & f_{n+1}(X_0) \\ \vdots & \ddots & \vdots \\ f_1(X_{i-1}) & \cdots & f_{n+1}(X_{i-1}) \\ f_1(X_{i-1}) & \cdots & f_{n+1}(X_{i-1}) \\ f_1(X_{i+1}) & \cdots & f_{n+1}(X_{i+1}) \\ \vdots & \ddots & \vdots \\ f_1(X_n) & \cdots & f_{n+1}(X_n) \end{vmatrix} = 0$$

and

$$Y_i \frac{f_j(X_i) - f_j(X_{i-1})}{X_i - Y_i} - f_j(X_{i-1}) = \frac{Y_i f_j(X_i) - X_i f_j(X_{i-1})}{X_i - Y_i} = \delta_i(X_i, f_j),$$

we obtain

$$Y_i D(X_i, \mathcal{F}) = \begin{vmatrix} \delta_1(f_1) & \cdots & \delta_1(f_n) & 0 \\ \vdots & \ddots & \vdots & \vdots \\ \delta_{i-1}(f_1) & \cdots & \delta_{i-1}(f_n) & 0 \\ \delta_i(X_i f_1) & \cdots & \delta_i(X_i f_n) & 0 \\ \delta_{i+1}(f_1) & \cdots & \delta_{i+1}(f_n) & 0 \\ \vdots & \ddots & \vdots & \vdots \\ \delta_n(f_1) & \cdots & \delta_n(f_n) & 0 \\ f_1(Y_1, \ldots, Y_n) & \cdots & f_n(Y_1, \ldots, Y_n) & Y_i \end{vmatrix},$$

and we are through. $\boxed{\odot}$

With a similar proof we also have

Lemma 41.12.2 (Cardinal–Mourrain). *For each polynomial g it holds that*

$$
D(g, \mathcal{F}) = \frac{
\begin{vmatrix}
f_1(\mathbf{X}_0) & \cdots & f_n(\mathbf{X}_0) & g(\mathbf{X}_0) \\
f_1(\mathbf{X}_1) & \cdots & f_n(\mathbf{X}_1) & g(\mathbf{X}_1) \\
\vdots & \ddots & \vdots & \vdots \\
f_1(\mathbf{X}_n) & \cdots & f_n(\mathbf{X}_n) & g(\mathbf{X}_n)
\end{vmatrix}
}{\prod_{j=1}^{n} X_j - Y_j}
$$

$$
=
\begin{vmatrix}
f_1(\mathbf{X}_0) & \cdots & f_n(\mathbf{X}_0) & g(\mathbf{X}_0) \\
\delta_1(f_1) & \cdots & \delta_1(f_n) & \delta_1(g) \\
\vdots & \ddots & \vdots & \vdots \\
\delta_n(f_1) & \cdots & \delta_n(f_n) & \delta_n(g)
\end{vmatrix}
\tag{41.7}
$$

$$
=
\begin{vmatrix}
\delta_1(f_1) & \cdots & \delta_1(f_n) & \delta_1(g) \\
\vdots & \ddots & \vdots & \vdots \\
\delta_n(f_1) & \cdots & \delta_n(f_n) & \delta_n(g) \\
f_1(\mathbf{X}_n) & \cdots & f_n(\mathbf{X}_n) & g(\mathbf{X}_n)
\end{vmatrix}.
\tag{41.8}
$$

$\odot$

Remark 41.12.3 (Becker, Cardinal *et al.*). Jacobi's interpretation of the Bezoutic matrix in terms of Jacobians (Section 41.10) can be extended to more than two forms; with the present notation, since

$$
\frac{Y_i^\nu - X_i^\nu}{Y_i - X_i} = \sum_{j=0}^{\nu-1} Y_i^j X_i^{\nu-1-j},
$$

for the polynomial

$$
\delta_i(h)(X_i, Y_i) = \frac{h(\mathbf{X}_i) - h(\mathbf{X}_{i-1})}{X_i - Y_i} \in k[Y_1, \ldots, Y_{i-1}, X_{i+1}, \ldots, X_n][X_i, Y_i]
$$

we have

$$
\delta_i(h)(X_i, X_i) = \frac{\partial h(Y_1, \ldots, Y_{i-1}, X_i, X_{i+1}, \ldots, X_n)}{\partial X_i}
$$

so that if in $D(1, \mathcal{F})$ we substitute each Y_i by X_i we obtain the Jacobian matrix of $\mathcal{F}$.

$\odot$

Given the polynomial ring $\mathcal{P} := k[X_1, \ldots, X_n]$ and its monomial k-basis $\mathcal{T}$, we introduce n further variables $Y_1, \ldots, Y_n$ and we denote $\mathcal{P}_Y := k[Y_1, \ldots, Y_n]$, $\mathcal{T}_Y$ the monomial k-basis of $\mathcal{P}_Y$ and $\mathcal{P}_\otimes$ the ring

$$
\mathcal{P}_\otimes := \mathcal{P} \otimes \mathcal{P}_Y = k[X_1, \ldots, X_n, Y_1, \ldots, Y_n],
$$

whose k basis is $\{\tau \otimes \omega : \tau \in \mathcal{T}, \omega \in \mathcal{T}_Y\}$.

Let us consider a set of n polynomials $\mathcal{F} := \{f_1, \ldots, f_n\} \in \mathcal{P}$ generating an ideal I and denote $\mathsf{A} := \mathcal{P}/\mathsf{I}$; with a slight abuse of notation we will also denote as I the ideal in $\mathcal{P}_Y$ generated by $\{f_1(Y_1, \ldots, Y_n), \ldots, f_n(Y_1, \ldots, Y_n)\}$ and $\mathsf{A} := \mathcal{P}_Y/\mathsf{I}$.

With this notation we have

$$\mathsf{A} \otimes_k \mathsf{A} = \mathcal{P}_\otimes / \mathbb{I}\left(f_i(X_1, \ldots, X_n), f_i(Y_1, \ldots, Y_n), 1 \leq i \leq n\right);$$

in connection with this we also write $\mathsf{I}_X := \mathsf{I} \otimes \mathcal{P}_Y \subset \mathcal{P}_\otimes$ and $\mathsf{I}_Y := \mathcal{P} \otimes \mathsf{I} \subset \mathcal{P}_\otimes$.

For each $g \in \mathcal{P}$ set $D(g) := D(g, \mathcal{F}) \in \mathcal{P}_\otimes$, $D_0 := D(1)$ and $D_i := D(X_i)$, $1 \leq i \leq n$.

Let $\mathfrak{a} \subset \mathcal{T}$ and $\mathfrak{b} \subset \mathcal{T}_Y$ be suitable ordered finite sets such that we can express each D_i, $0 \leq i \leq n$, as follows:

$$D_i := \sum_{\tau \in \mathfrak{a}} \sum_{\omega \in \mathfrak{b}} d_{\tau\omega}^{(i)} \tau \otimes \omega = \sum_{\tau \in \mathfrak{a}} \sum_{\omega \in \mathfrak{b}} d_{\tau\omega}^{(i)} \tau\omega \in \mathcal{P}_\otimes.$$

Denote as $\mathsf{D}_i := \left(d_{\tau\omega}^{(i)}\right)$ the corresponding Dixon matrix.

Lemma 41.12.4. $D(f) = 0$ *for each* $f \in \mathcal{F}$. $\qquad\qquad$ ⊡

Lemma 41.12.5 (Cardinal). *For each* $g \in \mathcal{P}$ *it holds that*

$$D(g) - g(X_1, \ldots, X_n)D_0 \in \mathsf{I}_X \quad and \quad D(g) - g(Y_1, \ldots, Y_n)D_0 \in \mathsf{I}_Y.$$

Proof. By expanding $D(g, \mathcal{F})$ along the first (respectively, last) row of (41.7) (respectively, (41.8)) we obtain $D(g) - g(X_1, \ldots, X_n)D_0 \in \mathsf{I}_X$ (respectively, $D(g) - g(Y_1, \ldots, Y_n)D_0$). $\qquad$ ⊡

Corollary 41.12.6 (Cardinal). *For each* $g \in \mathcal{P}$, *writing*

$$D(g) := \sum_{\tau \in \mathfrak{a}} \sum_{\omega \in \mathfrak{b}} d_{\tau\omega} \tau \otimes \omega,$$

we have

$$\sum_{\tau \in \mathfrak{a}} d_{\tau\omega} \tau \equiv g(X_1, \ldots, X_n) \sum_{\tau \in \mathfrak{a}} d_{\tau\omega}^{(0)} \tau \mod \mathsf{I}_X \quad for \ each \ \omega \in \mathfrak{b}$$

and

$$\sum_{\omega \in \mathfrak{b}} d_{\tau\omega} \omega \equiv g(Y_1, \ldots, Y_n) \sum_{\omega \in \mathfrak{b}} d_{\tau\omega}^{(0)} \omega \mod \mathsf{I}_Y \quad for \ each \ \tau \in \mathfrak{a}.$$

$\qquad\qquad$ ⊡

Corollary 41.12.7. *For each* $g \in \mathcal{P}$ *it holds that*

$$\left(g(Y_1, \ldots, Y_n) - g(X_1, \ldots, X_n)\right)D_0 \in \mathbb{I}\left(f_i(\mathsf{X}_0), f_i(\mathsf{X}_n), 1 \leq i \leq n\right).$$

$\qquad\qquad$ ⊡

Corollary 41.12.8. *For each i, $1 \leq i \leq n$, $\omega \in \mathfrak{b}$ and $\tau \in \mathfrak{a}$, there are polynomials $k_\omega^{(i)}(X_1, \ldots, X_n) \in \mathcal{P}$ and $h_\tau^{(i)}(Y_1, \ldots, Y_n) \in \mathcal{P}_Y$ such that*

$$
(1) \quad D_i - X_i D_0 = \begin{vmatrix} f_1(\mathbf{X}_0) & \ldots & f_n(\mathbf{X}_0) & 0 \\ \delta_1(f_1) & \ldots & \delta_1(f_n) & 0 \\ \vdots & \ddots & \vdots & \vdots \\ \delta_i(f_1) & \ldots & \delta_i(f_n) & 1 \\ \vdots & \ddots & \vdots & \vdots \\ \delta_n(f_1) & \ldots & \delta_n(f_n) & 0 \end{vmatrix} = \sum_{\omega \in \mathfrak{b}} k_\omega^{(i)} \omega;
$$

$(2) \quad k_\omega^{(i)}(X_1, \ldots, X_n) \in \mathsf{I}$ *for each* $\omega \in \mathfrak{b}$;

$$
(3) \quad D_i - Y_i D_0 = \begin{vmatrix} \delta_1(f_1) & \ldots & \delta_1(f_n) & 0 \\ \vdots & \ddots & \vdots & \vdots \\ \delta_i(f_1) & \ldots & \delta_i(f_n) & 1 \\ \vdots & \ddots & \vdots & \vdots \\ \delta_n(f_1) & \ldots & \delta_n(f_n) & 0 \\ f_1(\mathbf{X}_n) & \ldots & f_n(\mathbf{X}_n) & 0 \end{vmatrix} = \sum_{\tau \in \mathfrak{a}} \tau h_\tau^{(i)};
$$

$(4) \quad h_\tau^{(i)}(X_1, \ldots, X_n) \in \mathsf{I}$ *for each* $\tau \in \mathfrak{a}$.

Let us now assume that $\mathsf{I} := \mathbb{I}(\mathcal{F})$ is an affine complete intersection (Definition 36.1.1) and thus a Gorenstein Algebra; as a consequence we have[73]

Fact 41.12.9. *With the present notation and under the assumption that $\mathsf{I} := \mathbb{I}(\mathcal{F})$ is an affine complete intersection, it holds that:*

(1) A *is a finite k-dimensional vector space;*

(2) *the morphism $D_0| : \mathrm{Hom}(\mathsf{A}, k) \to \mathsf{A}$ which associates with the functional $\Lambda : \mathsf{A} \to k$ the element*

$$
D_0|(\Lambda) := \sum_{\tau \in \mathfrak{a}} \sum_{\omega \in \mathfrak{b}} d_{\tau\omega}^{(0)} \tau \Lambda(\omega) \in \mathsf{A}
$$

    is actually an A-isomorphism;

(3) $\mathrm{Hom}(\mathsf{A}, k)$ *is a free A-module whose basis element $\ell \in \mathrm{Hom}(\mathsf{A}, k)$ satisfies $D_0|\ell = 1$;*

(4) $D_0 \equiv \sum_{i=1}^{D} a_i \otimes b_i$ *in $\mathsf{A} \otimes \mathsf{A}$, where $D = \dim_k(\mathsf{A})$ and $\{a_i, 1 \leq i \leq D\}$ and $\{b_i, 1 \leq i \leq D\}$ are suitable k-bases of A;*

(5) *both $\mathfrak{a}$ and $\mathfrak{b}$ are k-generating sets of A.*

⊙

---

[73] Compare Mourrain B., Bezoutian and quotient ring structure, *J. Symb. Comp.* **39** (2005), 397–415.

41.13 Cardinal–Mourrain Algorithm

Let $\mathcal{P}, \mathcal{P}_Y, \mathcal{P}_\otimes, \mathcal{T}, \mathcal{T}_Y, \mathcal{F} := \{f_1, \ldots, f_n\} \in \mathcal{P}$ be a set of n polynomials generating the affine complete intersection ideal I and let A, I_X and I_Y, $D_i, 0 \le i \le r$, $\mathfrak{a}, \mathfrak{b}$ and $\mathsf{D}_i := \left(d_{\tau\omega}^{(i)}\right), 0 \le i \le r$, be as defined after Remark 41.12.3.

Let us moreover write $V := \mathrm{Span}_k(\mathfrak{a})$, $W := \mathrm{Span}_k(\mathfrak{b})$,

$$K_0 := \mathrm{Span}_k\{k_\omega^{(i)} : 1 \le i \le n, \omega \in \mathfrak{b}\} \cap V \subset V \cap \mathsf{I} \subset \mathcal{P}$$

and

$$H_0 := \mathrm{Span}_k\{h_\tau^{(i)} : 1 \le i \le n, \tau \in \mathfrak{a}\} \cap W \subset W \cap \mathsf{I} \subset \mathcal{P}_Y,$$

where $k_\omega^{(i)}, h_\tau^{(i)}$ are the polynomials whose existence is stated in Corollary 41.12.8.

We present an algorithm which iteratively extends the vector spaces K_0, H_0 and returns, at termination,

- vectorspaces $K, H, K_0 \subseteq K = \mathsf{I} \cap V, H_0 \subseteq H = \mathsf{I} \cap W$,
- the supplementary vector spaces $A, B, A \otimes K = V, B \otimes H = W$, which satisfy the relation $\dim_k(A) = \dim_k(B) =: \delta$,
- the bases $\mathbf{a} := \{a_1, \ldots, a_\delta\}, \{b_1, \ldots, b_\delta\}$ of A and B respectively,
- δ-square matrices $M_i, 0 \le i \le n$,

such that

(1) M_0 is invertible,

(2) the $\overline{M}_p := M_0^{-1} M_p := \left(m_{ji}^{(p)}\right)$ satisfy

$$X_p a_i \equiv \sum_{j=1}^{\delta} m_{ji}^{(p)} a_j \bmod \mathsf{I} \quad \forall i, p, 1 \le i \le \delta, 1 \le p \le n,$$

so that, in particular,

(3) $A = \mathcal{P}/\mathsf{I} \cong V/K \cong \mathrm{Span}_k(\mathbf{a})$ and

(4) the assignment of the k-basis $\mathbf{a}$ and the square matrices

$$\overline{M}_p := M([X_p], \mathbf{a}), \quad 1 \le p \le n,$$

is a Gröbner representation of I.

Recall that the Dixon polynomials D_i are decomposed with respect to the same ordered sets of polynomials $\mathfrak{a} \subset \mathcal{P}$ and $\mathfrak{b} \subset \mathcal{P}_Y$ as those which index the rows and columns of the Dixon matrices D_i, so that

$$D_i = \mathfrak{a}^T \mathsf{D}_i \mathfrak{b}.$$

$i := 0$

Repeat

    ★ Apply the *saturation step* on K_i returning $K' : K_i \subseteq K' \subseteq I \cap V$,
       Perform the *quartering step* on the matrices D_i,
       Perform the *column reduction step* returning $K_{i+1} : K' \subseteq K_{i+1} \subset I \cap V$,
       Perform the *diagonalization step*,
       Perform the *row step* returning $H_{i+1} : H_i \subseteq H_{i+1} \subseteq I \cap W$,
until $H_{i+1} = H_i$ and $K_{i+1} = K_i$.

Figure 41.3. Cardinal–Mourrain Algorithm

At each step the algorithm performs simultaneously the *same* operation on the $n + 1$ matrices D_i, that is, it applies invertible transformations P, Q to their rows and columns, thus returning

$$PD_0Q, \quad PD_1Q, \quad \ldots, \quad PD_nQ$$

and transforming the bases of V and W in such a way that the indexes of the common rows and columns of the matrices PD_iQ are respectively $P^{-1^T}\mathfrak{a}$ and $Q^{-1}\mathfrak{b}$.

We present in Figure 41.3 the general scheme of the algorithm, whose operations we discuss here in terms of an example.

Example 41.13.1 (Cardinal). We will consider the ideal generated by $\mathcal{F} = \{f_1, f_2\} \in k[X_1, X_2]$, where

$$f_1 = X_1^2 + X_1X_2^2 - 1, \qquad f_2 = X_1^2X_2 + X_1.$$

We thus have

$$D_0 = -X_1X_2^2Y_1 - X_1X_2Y_1Y_2 - X_2Y_1^2Y_2 + X_1Y_1^2 + Y_1^3 - X_2Y_1 - Y_1Y_2,$$
$$D_1 = -X_1X_2^2Y_1^2 - X_1X_2Y_1^2Y_2 + X_1Y_1^3 + Y_1^2,$$
$$D_2 = -X_1X_2^2Y_1Y_2 - X_2^2Y_1^2Y_2 + X_1X_2Y_1^2 + X_2Y_1^3 - X_2^2Y_1$$
$$\qquad - X_2Y_1Y_2 - X_1X_2 - X_1Y_1 - X_2Y_1 - 1,$$
$$D_1 - X_1D_0 = (X_1^2X_2^2 + X_1X_2)Y_1 + (-X_1X_2^2 - X_1^2 + 1)Y_1^2 + (X_1^2X_2 + X_1)Y_1Y_2,$$
$$D_2 - X_2D_0 = (X_2^3X_1 - X_2 - X_1)Y_1 + (-X_2X_1 - 1),$$
$$D_1 - Y_1D_0 = X_2(Y_2Y_1^3 + Y_1^2) + (Y_2Y_1^2 - Y_1^4 + Y_1^2),$$
$$D_2 - Y_2D_0 = X_2^2(-Y_2Y_1^2 - Y_1) + X_2X_1(Y_2^2Y_1 + Y_1^2 - 1) + X_2(Y_2^2Y_1^2 + Y_1^3 - Y_1)$$
$$\qquad - X_1(Y_2Y_1^2 + Y_1) + (Y_2^2Y_1 - Y_2Y_1^3 - 1),$$

whence[74]

$$\mathfrak{a} = \{1, X_2, X_2^2, X_1, X_1X_2, X_1X_2^2\}, \quad \mathfrak{b} = \{1, Y_1, Y_1Y_2, Y_1^2, Y_1^2Y_2, Y_1^3\},$$

$$K_0 = \{X_2X_1 + 1\} \quad \text{and} \quad H_0 = \{Y_2Y_1^2 + Y_1\}.$$

---

[74] Note that $X_2X_1 + 1 = -(X_2X_1 + 1)f_1 + (X_2^2 + X_1)f_2$.

saturation step: It consists in replacing K_i with $K' := K_i^+ \cap V$, where, for a vector space K, we define the vector space

$$K^+ := \left\{ v_0 + \sum_{i=1}^{n} X_i v_i, \, v_i \in K, 0 \leq i \leq n \right\}$$

and the notation $K^{[p]}$ means p iterations of the operator $\cdot^+$, starting from K, so that $K^{[p]} = (K^{[p-1]})^+$.

Definition 41.13.2 (Mourrain). *A vector space $V \subset \mathcal{P}$ is said to be* connected *to $e \in V$ if, denoting $E := \mathrm{Span}_k\{e\}$, for each $v \in V \setminus E$ there exists $l > 0$ such that $v \in E^{[l]}$, $v = v_0 + \sum_{i=1}^{n} X_i v_i$, with $v_i \in E^{[l-1]} \cap V, 0 \leq i \leq n$.* ◉

Example 41.13.3.   We have

$$K' := \{X_2 X_1 + 1, X_2(X_2 X_1 + 1)\}.$$

◉

quartering step: Set $d := \dim_k(V) = \dim_k(W)$, $D := d - \dim_k(K')$, choose a basis $\{a_1, \ldots, a_D, a_{D+1}, \ldots, a_d\}$ of V such that $\{a_{D+1}, \ldots, a_d\}$ is a basis of K' and set $A := \mathrm{Span}_k\{a_1, \ldots, a_D\}$.

Let $\mathcal{L}(K') = \{\Lambda \in \mathrm{Hom}(\mathcal{P}, k) : \Lambda(k \leftarrow h) = 0 \text{ for each } k \leftarrow h \in K'\}$ and set

$$B := \{\Lambda | D_0 : \Lambda \in \mathcal{L}(K')\} \supset \{\Lambda | D_0 : \Lambda \in \mathcal{L}(\mathsf{I})\},$$

where $| D_0 : \mathrm{Hom}(\mathcal{P}, k) \to \mathcal{P}_Y$ is the morphism which associates with the functional $\Lambda : \mathcal{P} \to k$ the element $\sum_{\tau \in \mathfrak{a}} \sum_{\omega \in \mathfrak{b}} d_{\tau\omega}^{(0)} \Lambda(\tau) \omega \in \mathcal{P}_Y$.

Choose a basis

$$\{b_1, \ldots, b_r, b_{r+1}, \ldots, b_d\}$$

of W such that $\{b_1, \ldots, b_r\}$ is a basis of B; denote $H' := \mathrm{Span}_k\{b_{r+1}, \ldots, b_d\}$.

Remark that, since $K' \subset \mathsf{I}$, $B \supset \mathcal{L}(\mathsf{I})$ is a generating set of A so that H' could be chosen as a subset of I.

On the basis of the decompositions $V = A \otimes K'$ and $W = B' \otimes H'$ the matrices D_i can be decomposed *quarterly*, as

$$\mathsf{D}_0 = \begin{pmatrix} M_0 & 0 \\ H_0 & L_0 \end{pmatrix}, \quad \mathsf{D}_i = \begin{pmatrix} M_i & K_i \\ H_i & L_i \end{pmatrix}, \quad 1 \leq i \leq n.$$

Example 41.13.3 (cont.).   We set $D = 4$ and

$$a_1 = 1, \quad a_2 = X_2, \quad a_3 = X_2^2, \quad a_4 = X_1, \quad a_5 = X_1 X_2 + 1, \quad a_6 = X_1 X_2^2 + X_2,$$

so that $D_0 = a_1 Y_1^3 - a_2 Y_1^2 Y_2 + a_4 Y_1^2 - a_5 Y_1 Y_2 - a_6 Y_1$.

We thus get

$$b_1 = Y_1^3, \quad b_2 = Y_1^2 Y_2, \quad b_3 = Y_1^2, \quad b_4 = 1, \quad b_5 = Y_1 Y_2, \quad b_6 = Y_1 + Y_1^2 Y_2,$$

and obtain

D_0	b_1	b_2	b_3	b_4	b_5	b_6
a_1	1	0	0	0	0	0
a_2	0	-1	0	0	0	0
a_3	0	0	0	0	0	0
a_4	0	0	1	0	0	0
a_5	0	0	0	0	-1	0
a_6	0	1	0	0	0	-1

,

D_1	b_1	b_2	b_3	b_4	b_5	b_6
a_1	0	1	1	0	0	0
a_2	0	0	1	0	0	0
a_3	0	0	0	0	0	0
a_4	1	0	0	0	0	0
a_5	0	-1	0	0	0	0
a_6	0	0	-1	0	0	0

,

D_2	b_1	b_2	b_3	b_4	b_5	b_6
a_1	0	0	-1	0	0	0
a_2	1	1	0	0	0	-1
a_3	0	0	0	0	0	-1
a_4	0	1	0	0	0	-1
a_5	0	0	1	-1	0	0
a_6	0	0	0	0	-1	0

.

column reduction step: By Lemma 41.12.5, if a column of D_0 represents a polynomial $f \in \mathcal{P}$ then the corresponding column of D_i represents $X_i f \bmod \mathsf{I}$, so that we can deduce that the columns of $\begin{pmatrix} K_i \\ L_i \end{pmatrix}$ and actually of K_i give elements in I, allowing us to extend K' and returning a vector space $K_{i+1} : K' \subseteq K_{i+1} \subset \mathsf{I} \cap V$.

Example 41.13.3 (cont.).   The sixth column of D_2 returns $X_2^2 + X_2 + X_1$.

diagonalization step: By construction, the number $r = \dim(B)$ of columns of M_0 is equal to its rank; so there is an $r \times D$ matrix $M_0^\star$ such that $M_0^\star M_0 = \mathrm{Id}_r$.

We thus multiply each D_i by the matrix $P := \begin{pmatrix} \mathrm{Id}_D & 0 \\ -H_0 M_0^\star & \mathrm{Id}_{d-D} \end{pmatrix}$, obtaining the following decompositions:

$$D_0 = \begin{pmatrix} M_0 & 0 \\ 0 & L_0 \end{pmatrix}, \quad D_i = \begin{pmatrix} M_i & K_i \\ H_i' & L_i' \end{pmatrix}, \quad 1 \leq i \leq n.$$

This corresponds to a change of the basis $\mathsf{a} := \{a_1, \ldots, a_d\}$ of V, which becomes $(P^{-1})^T \mathsf{a}$.

Example 41.13.3 (cont.).   We have $M_0^\star = \begin{pmatrix} 1 & 0 & 0 & 0 \\ 0 & -1 & 0 & 0 \\ 0 & 0 & 0 & 1 \end{pmatrix}$ and we have to

multiply by the matrix

$$\left(\begin{array}{cccc|cc} 1 & 0 & 0 & 0 & 0 & 0 \\ 0 & 1 & 0 & 0 & 0 & 0 \\ 0 & 0 & 1 & 0 & 0 & 0 \\ 0 & 0 & 0 & 1 & 0 & 0 \\ \hline 0 & 0 & 0 & 0 & 1 & 0 \\ 0 & 1 & 0 & 0 & 0 & 1 \end{array} \right) ;$$

this returns, with $a_1 = 1, a_2 = -X_1 X_2^2, a_3 = X_2^2, a_4 = X_1, a_5 = X_1 X_2 + 1, a_6 = X_1 X_2^2 + X_2$,

D_0	b_1	b_2	b_3	b_4	b_5	b_6
a_1	1	0	0	0	0	0
a_2	0	-1	0	0	0	0
a_3	0	0	0	0	0	0
a_4	0	0	1	0	0	0
a_5	0	0	0	0	-1	0
a_6	0	0	0	0	0	-1

D_1	b_1	b_2	b_3	b_4	b_5	b_6
a_1	0	1	1	0	0	0
a_2	0	0	1	0	0	0
a_3	0	0	0	0	0	0
a_4	1	0	0	0	0	0
a_5	0	-1	0	0	0	0
a_6	0	0	0	0	0	0

D_2	b_1	b_2	b_3	b_4	b_5	b_6
a_1	0	0	-1	0	0	0
a_2	1	1	0	0	0	-1
a_3	0	0	0	0	0	-1
a_4	0	1	0	0	0	-1
a_5	0	0	1	-1	0	0
a_6	1	1	0	0	-1	-1

row step: We perform on the rows the same construction as we performed on the columns, thus enlarging H_i to $H_{i+1} : H_i \subseteq H_{i+1} \subseteq I \cap W$.

Example 41.13.3 (cont.).   The sixth row of D_2 returns $-Y_2 Y_1 + Y_1^3 - Y_1$.

Lemma 41.13.4 (Mourrain).   *At the end of the algorithm, writing*

$$\overline{K} := K_{i+1} \subset \mathsf{I} \cap V, \quad \overline{H} := H_{i+1} \subset \mathsf{I} \cap W,$$

denoting by $A \subset V$ and $B \subset W$ the vector spaces such that $\overline{K} \otimes A = V, \overline{H} \otimes B = W$ and setting $d := \dim_k(V) = \dim_k(W)$, and $\delta := d - \dim_k(\overline{K})$ it holds that $\overline{K}^+ \cap V = \overline{K}$ and there exist linearly independent polynomials

$$\{a_1, \ldots, a_\delta, a_{\delta+1}, \ldots, a_d\} \subset V, \quad a_i \in \overline{K}, \quad \delta < i \le d,$$

and a basis $\{b_1, \ldots, b_\delta, b_{\delta+1}, \ldots, b_d\}$ of W with $b_i \in \overline{H}, \delta < i \le d$, with respect to which the Dixon matrices D_i have the decompositions

$$\mathsf{D}_0 = \begin{pmatrix} \mathrm{Id}_\delta & 0 \\ 0 & L_0 \end{pmatrix}, \quad \mathsf{D}_i = \begin{pmatrix} \overline{M}_i & 0 \\ 0 & L_i \end{pmatrix}, \quad 1 \le i \le n.$$

Proof.   At the end of the algorithm,[75]

- the saturation step does not increase $\overline{K}$, so that $\overline{K}^+ \cap V = \overline{K}$,
- the quartering step returns the decompositions

$$\mathsf{D}_0 = \begin{pmatrix} M_0 & 0 \\ H_0 & L_0 \end{pmatrix}, \quad \mathsf{D}_i = \begin{pmatrix} M_i & K_i \\ H_i & L_i \end{pmatrix}, \quad 1 \le i \le n,$$

- since the column reduction step does not increase $\overline{K}$, this means that $K_i = 0$ for each i,
- a variation of the diagonalization step in which we left-multiply by

$$\begin{pmatrix} M_0^{-1} & 0 \\ -H_0 M_0^{-1} & \mathrm{Id}_\delta \end{pmatrix}$$

returns the decompositions

$$\mathsf{D}_0 = \begin{pmatrix} \mathrm{Id}_\delta & 0 \\ 0 & L_0 \end{pmatrix}, \quad \mathsf{D}_i = \begin{pmatrix} \overline{M}_i & 0 \\ H_i' & L_i \end{pmatrix}, \quad 1 \le i \le n,$$

with respect to a suitable basis,
- since the row step does not increase $\overline{H}$, this means that $H_i' = 0$ for each i, thus completing the argument. ⊙

Example 41.13.1 (cont.).   We now have

$$K = \{X_1 X_2 + 1, X_1 X_2^2 + X_2, X_2^2 + X_2 + X_1\}$$

and $H = \{Y_1 + Y_1^2 Y_2, -Y_2 Y_1 + Y_1^3 - Y_1\}$, and we thus choose, as bases for V and W respectively,

$$a_1 = 1, \quad a_2 = -X_1 X_2^2, \quad a_3 = X_1, \quad a_4 = X_2^2 + X_2 + X_1,$$
$$a_5 = X_1 X_2 + 1, \quad a_6 = X_1 X_2^2 + X_2$$

---

[75] Meaning, in the last **Repeat**-loop at whose end K_{i+1} and H_{i+1} are not increased.

and

$$b_1 = Y_1^3, \quad b_2 = Y_1^2 Y_2, \quad b_3 = Y_1^2, \quad b_4 = 1,$$
$$b_5 = -Y_2 Y_1 + Y_1^3 - Y_1, \quad b_6 = Y_1 + Y_1^2 Y_2;$$

with respect to these bases we have

D_0	b_1	b_2	b_3	b_4	b_5	b_6
a_1	1	0	0	0	0	0
a_2	0	-1	0	0	0	0
a_3	0	0	1	0	0	0
a_4	0	0	0	0	0	0
a_5	-1	-1	0	0	1	1
a_6	0	0	0	0	0	-1

D_1	b_1	b_2	b_3	b_4	b_5	b_6
a_1	0	1	1	0	0	0
a_2	0	0	1	0	0	0
a_3	1	0	0	0	0	0
a_4	0	0	0	0	0	0
a_5	0	-1	0	0	0	0
a_6	0	0	0	0	0	0

D_2	b_1	b_2	b_3	b_4	b_5	b_6
a_1	0	0	-1	0	0	0
a_2	1	1	0	0	0	0
a_3	0	1	0	0	0	0
a_4	0	0	0	0	0	-1
a_5	0	0	1	-1	0	0
a_6	0	0	0	0	1	1

The next loop will prove that we have already reached the required solution, since neither K nor H is increased, and can also be used to illustrate the claim of the lemma; K is not increased either by the saturation step or by the column reduction step, and we now left-multiply the D_i by

$$\begin{pmatrix} 1 & 0 & 0 & 0 & 0 & 0 \\ 0 & -1 & 0 & 0 & 0 & 0 \\ 0 & 0 & 1 & 0 & 0 & 0 \\ 0 & 0 & 0 & 1 & 0 & 0 \\ 1 & -1 & 0 & 0 & 1 & 0 \\ 0 & 0 & 0 & 0 & 0 & 1 \end{pmatrix},$$

obtaining the basis

$$a_1 = -X_1 X_2, \quad a_2 = X_1 X_2^2 - X_1 X_2 - 1,$$
$$a_3 = X_1, \quad a_4 = X_2^2 + X_2 + X_1,$$
$$a_5 = X_1 X_2 + 1, \quad a_6 = X_1 X_2^2 + X_2$$

and the matrices

D_0	b_1	b_2	b_3	b_4	b_5	b_6
a_1	1	0	0	0	0	0
a_2	0	1	0	0	0	0
a_3	0	0	1	0	0	0
a_4	0	0	0	0	0	0
a_5	0	0	0	0	1	1
a_6	0	0	0	0	0	-1

D_1	b_1	b_2	b_3	b_4	b_5	b_6
a_1	0	1	1	0	0	0
a_2	0	0	-1	0	0	0
a_3	1	0	0	0	0	0
a_4	0	0	0	0	0	0
a_5	0	0	0	0	0	0
a_6	0	0	0	0	0	0

D_2	b_1	b_2	b_3	b_4	b_5	b_6
a_1	0	0	-1	0	0	0
a_2	-1	-1	0	0	0	0
a_3	0	1	0	0	0	0
a_4	0	0	0	0	0	-1
a_5	-1	-1	0	-1	0	0
a_6	0	0	0	0	1	1

Claim 41.13.5 (Cardinal–Mourrain). *The δ-square matrices $\overline{M}_i^T$ (respectively, $\overline{M}_i$) are the matrices $M([X_i], \mathbf{a})$ (respectively, $M([Y_i], \mathbf{b})$) corresponding to multiplication by the variables X_i (respectively, Y_i) in the basis $\mathbf{a} := \{a_1, \ldots, a_\delta\}$ (respectively, $\mathbf{b} := \{b_1, \ldots, b_\delta\}$) of* A.

Historical Remark 41.13.6.   A preliminary version of the algorithm, without the saturation step, which conjectured the claim above was proposed by Cardinal in his *thesis* in 1993.

The insertion of the saturation step and the proof of the claim, under the further assumption that V is connected to the element 1 (see Definition 41.13.2), is due to Mourrain in 2003. As regards the saturation step, Mourrain comments[76]

The reason why we need to introduce this saturation step is that if we multiply all the [Dixon polynomials] by an element of the form $1 + fg$, $f \in \mathcal{P}$, $g \in \mathcal{P}_Y$ with f, g conveniently chosen, we could obtain matrices of the form $\begin{pmatrix} D_i & 0 \\ 0 & D_i \end{pmatrix}$. Applying only the [...] steps as described by Cardinal, would not allow us to avoid the duplication of the structure of A. Moreover, if f and g are in I, the polynomials $(1 + fg)D_i$ share the same properties, modulo

[76] Mourrain B., Bezoutian and quotient ring structure, *J. Symb. Comp.* **39** (2005), 397–415.

I, as the [Dixon polynomials] D_i. To handle this problem, we add the saturation step, which will "connect" the two blocks, provided that the vector space V is connected to an element e. *This is the hypothesis that will be made hereafter to prove the main theorem.*

This hypothesis is easy to check in practice, and usually we have $e = 1$. Moreover, it is satisfied when the polynomials f_i are monomials. We do not have a proof that this extends by linearity to any polynomial f_i.

[...]

To simplify the proof, we will assume hereafter that $e = 1$. The proof can be extended to any e, by showing that, in this case, e is invertible in A and by dividing by e.[77]

$\boxed{\odot}$

Example 41.13.1 (cont.).   Gauss-reducing $\{a_1, a_2, a_3\}$ and $\{b_1, b_2, b_3\}$ with respect the basis elements of, respectively, K and H we can set

$$a_1' := a_5 + a_1 = 1, \quad a_2' := a_2 - a_6 + a_5 = -X_2, \quad a_3' := X_1$$

and

$$b_1' := Y_1^3, \quad b_2' := b_2 - b_6 = -Y_1, \quad b_3' = Y_1^2$$

and we then have[78]

$$\overline{M}_1 = \begin{pmatrix} 0 & 1 & 1 \\ 0 & 0 & -1 \\ 1 & 0 & 0 \end{pmatrix} \quad \text{and} \quad \overline{M}_2 = \begin{pmatrix} 0 & 0 & -1 \\ -1 & -1 & 0 \\ 0 & 1 & 0 \end{pmatrix}.$$

$\boxed{\odot}$

41.14 Mourrain: Proving Cardinal's Conjecture

We use the same notation as in previous sections; in particular we have $\mathbf{a} := \{a_1, \ldots, a_\delta\}$, $\mathbf{b} := \{b_1, \ldots, b_\delta\}$, $A = \mathrm{Span}_k(\mathbf{a})$, $B = \mathrm{Span}_k(\mathbf{b})$ and we set $\overline{M}_p := M_0^{-1} M_p := \left(m_{ji}^{(p)} \right)$.

Proposition 41.14.1 (Mourrain).   *It holds that:*

(1) $X_p a_i = \sum_{l=1}^{\delta} m_{li}^{(p)} a_l - \kappa_i^{(p)}, \kappa_i^{(p)} \in \overline{K}, \text{ for each } p, i, 1 \le p \le n, 1 \le i \le \delta;$

(2) $Y_p b_i = \sum_{l=1}^{\delta} m_{il}^{(p)} b_l - \sigma_i^{(p)}, \sigma_i^{(p)} \in \overline{H}, \text{ for each } p, i, 1 \le p \le n, 1 \le i \le \delta;$

---

[77] By assumption we have $\mathsf{A} = \mathrm{Span}_k(\mathbf{a})$ and $a_i = e a_i'$ for some a_i', so that

$$\mathsf{A} \ni 1 = \sum_i c_i a_i = \left(\sum_i c_i a_i' \right) e.$$

[78] The reader can check the result using the deglex Gröbner basis of I induced by $X_1 < X_2$, which is

$$\{X_1 X_2 + 1, X_1^2 - X_2 - 1, X_2^2 + X_2 + X_1\},$$

so that, in particular, $Y_1^3 \equiv Y_1 - 1$ and $Y_1^2 \equiv Y_2 + 1$.

(3) $D(X_p X_q) = X_q D(X_p) + Y_p \Big(D(X_q) - X_q D(1) \Big)$ *for* $p < q$;

(4) $D(X_p X_q) = Y_p D(X_q) + X_q \Big(D(X_p) - Y_p D(1) \Big)$ *for* $p < q$;

(5) $D(X_p X_q) = \sum_{1 \leq i,j,l \leq \delta} m_{li}^{(q)} m_{ij}^{(p)} a_l \otimes b_j + X_q \chi_1^{(p,q)} + Y_p \chi_2^{(p,q)} + \chi_3^{(p,q)}$ *for* $p < q$ *and suitable elements* $\chi_1^{(p,q)}, \chi_2^{(p,q)}, \chi_3^{(p,q)} \in \overline{K} \otimes \overline{K}$;

(6) $D(X_p X_q) = \sum_{1 \leq i,j,l \leq \delta} m_{li}^{(p)} m_{ij}^{(q)} a_l \otimes b_j + X_p \chi_4^{(p,q)} + Y_q \chi_5^{(p,q)} + \chi_6^{(p,q)}$ *for* $p < q$ *and suitable elements* $\chi_4^{(p,q)}, \chi_5^{(p,q)}, \chi_6^{(p,q)} \in \overline{K} \otimes \overline{K}$.

Proof.

(1) We have

$$D_p = X_p D_0 + \sum_{i=1}^{\delta} \kappa_i^{(p)} b_i + \sum_{l=\delta+1}^{d} \kappa_l^{(p)} b_l, \quad \kappa_i^{(p)}, \kappa_l^{(p)} \in K_0.$$

By identifying the coefficients of each b_i, $1 \leq i \leq \delta$, we have

$$\sum_{l=1}^{\delta} m_{li}^{(p)} a_l = X_p a_i + \kappa_i^{(p)}, \quad \kappa_i^{(p)} \in K_0 \subset \overline{K}.$$

(2) The proof is similar to that of (1).

(3) We have

$$D(X_p X_q) = \begin{vmatrix} f_1(X_1, \ldots, X_n) & \cdots & f_n(X_1, \ldots, X_n) & X_p X_q \\ \delta_1(f_1) & \cdots & \delta_1(f_n) & 0 \\ \vdots & \ddots & \vdots & \vdots \\ \delta_p(f_1) & \cdots & \delta_p(f_n) & X_q \\ \vdots & \ddots & \vdots & \vdots \\ \delta_q(f_1) & \cdots & \delta_q(f_n) & Y_p \\ \vdots & \ddots & \vdots & \vdots \\ \delta_n(f_1) & \cdots & \delta_n(f_n) & 0 \end{vmatrix}$$

$$= X_q \begin{vmatrix} f_1(X_1, \ldots, X_n) & \cdots & f_n(X_1, \ldots, X_n) & X_p \\ \delta_1(f_1) & \cdots & \delta_1(f_n) & 0 \\ \vdots & \ddots & \vdots & \vdots \\ \delta_p(f_1) & \cdots & \delta_p(f_n) & 1 \\ \vdots & \ddots & \vdots & \vdots \\ \vdots & \ddots & \vdots & \vdots \\ \delta_n(f_1) & \cdots & \delta_n(f_n) & 0 \end{vmatrix}$$

$$+ Y_p \begin{vmatrix} f_1(X_1,\ldots,X_n) & \ldots & f_n(X_1,\ldots,X_n) & 0 \\ \delta_1(f_1) & \ldots & \delta_1(f_n) & 0 \\ \vdots & \ddots & \vdots & \vdots \\ \vdots & \ddots & \vdots & \vdots \\ \delta_q(f_1) & \ldots & \delta_q(f_n) & 1 \\ \vdots & \ddots & \vdots & \vdots \\ \delta_n(f_1) & \ldots & \delta_n(f_n) & 0 \end{vmatrix}.$$

(4) As for (3), we make the development

$$D(X_p X_q) = \begin{vmatrix} \delta_1(f_1) & \ldots & \delta_1(f_n) & 0 \\ \vdots & \ddots & \vdots & \vdots \\ \delta_p(f_1) & \ldots & \delta_p(f_n) & X_q \\ \vdots & \ddots & \vdots & \vdots \\ \delta_q(f_1) & \ldots & \delta_q(f_n) & Y_p \\ \vdots & \ddots & \vdots & \vdots \\ \delta_n(f_1) & \ldots & \delta_n(f_n) & 0 \\ f_1(Y_1,\ldots,y_n) & \ldots & f_n(Y_1,\ldots,Y_n) & Y_p Y_q \end{vmatrix}.$$

(5) By (2) we have

$$D(X_p X_q) = X_q D(X_p) + Y_p \big(D(X_q) - X_q D(1) \big)$$

$$= \sum_{1 \le i,j \le \delta} m_{ij}^{(p)} X_q a_i \otimes b_j + X_q \chi_1 + Y_p \sum_{1 \le i \le \delta} \kappa_i^{(q)} \otimes b_i + Y_p \chi_2$$

$$= \sum_{1 \le i,j \le \delta} m_{ij}^{(p)} \left(\sum_{1 \le l \le \delta} m_{li}^{(q)} a_l - \kappa_i^{(q)} \right) \otimes b_j + X_q \chi_1$$

$$+ \sum_{1 \le i \le \delta} \kappa_i^{(q)} \otimes \left(\sum_{1 \le l \le \delta} m_{il}^{(p)} b_l - \sigma_i^{(p)} \right) + Y_p \chi_2$$

$$= \sum_{1 \le i,j,l \le \delta} m_{li}^{(q)} m_{ij}^{(p)} a_l \otimes b_j - \sum_{1 \le i,j \le \delta} m_{ij}^{(p)} \kappa_i^{(q)} \otimes b_j$$

$$+ \sum_{1 \le i,l \le \delta} m_{il}^{(p)} \kappa_i^{(q)} \otimes b_l + X_q \chi_1 + Y_p \chi_2 + \chi_3$$

$$= \sum_{1 \le i,j,l \le \delta} m_{li}^{(q)} m_{ij}^{(p)} a_l \otimes b_j + X_q \chi_1 + Y_p \chi_2 + \chi_3.$$

(6) This is proved similarly to (5). ⊙

Corollary 41.14.2 (Mourrain). *The matrices* $\overline{M}_p = \left(m_{ji}^{(p)} \right)$ *commute.* ⊙

With a slight abuse of notation we will also denote by $\overline{M}_p$ the map

$$\overline{M}_p : A \to A, \quad a_i \mapsto \sum_{l=1}^{\delta} m_{ml}^{(p)} a_l,$$

which corresponds to multiplication by $X_p \pmod{\overline{K}}$.

Since these operations commute, for each $f(X_1, \ldots, X_n) \in \mathcal{P}$ we define

$$f(\overline{M}) := f(\overline{M}_1, \ldots, \overline{M}_n) : A \to A$$

and $N(f) = f(\overline{M})(1)$, so that N is a map $N : \mathcal{P} \to A$.

Proposition 41.14.3 (Mourrain). *If V is connected to the element 1, the ideal* $\mathsf{H} := \mathbb{I}(\overline{K}) \subset \mathcal{P}$ *generated by $\overline{K}$ satisfies $\mathsf{H} = \mathsf{I}$.*

Proof. By construction we have $\mathsf{H} \subset \mathsf{I}$.

The assumption that $V = A \otimes \overline{K}$ is connected to 1 implies the existence of $c_1, \ldots, c_\delta \in k, \kappa \in \overline{K}$, such that $u = \sum_{i=1}^{\delta} c_i a_i = 1 - \kappa$; thus, in particular, u is invertible in A and there exists $\Lambda \in \mathrm{Hom}(\mathcal{P}, k)$ such that $D_0|\Lambda = u$.

By expanding (41.7) along the last column, we have $0 = D(f_i) = f_i(X_1, \ldots, X_n) D_0 + \sum_{j=1}^{n}(-1)^j \delta_j(f_i) E_j$ for suitable $E_j \in \overline{K} \otimes \mathcal{P}_Y$.

Thus, for each i,

$$f_i = f_i \kappa + f_i u = f_i \kappa + f_i D_0|\Lambda = f_i \kappa - \sum_{j=1}^{n}(-1)^j \delta_j(f_i) E_j|\Lambda \in \mathsf{H}.$$

$\boxed{\odot}$

Proposition 41.14.4 (Mourrain). *Assume that V is connected to 1 and $1 \notin \overline{K}$. Then:*

(1) *for each $f \in V$, $f - f(\overline{M})(1) \in \overline{K}$;*
(2) *for each $a \in A$, $N(a) = a$ and $N(\overline{K}) = \{0\}$;*
(3) *$f - f(\overline{M})(1) \in \mathsf{H}$ for each $f \in \mathcal{P}$;*
(4) *$\ker(N) = \mathsf{H}$.*

Proof.

(1) Assume that for each $g \in \mathrm{Span}_k\{1\}^{[l-1]} \cap V$ we have $g - g(\overline{M})(1) \in \overline{K}$, and let $f \in \mathrm{Span}_k\{1\}^{[l]} \cap V$. Since V is connected to 1, we have $f = \sum_{i=1}^{s} X_{l_i} g_i$ with $1 \le l_i \le n$ and $g_i \in \mathrm{Span}_k\{1\}^{[l-1]} \cap V$. Thus

$$f - f(\overline{M})(1) = \sum_{i=1}^{s} X_{l_i}(g_i - g_i(\overline{M})(1)) + \left(X_{l_i} g_i(\overline{M})(1) - \overline{M}_{l_i} g_i(\overline{M})(1) \right).$$

We have $g_i - g_i(\overline{M})(1) \in \overline{K}$ by the induction assumption and $X_{l_i} g_i(\overline{M})(1) - \overline{M}_{l_i} g_i(\overline{M})(1) \in \overline{K}$ by Lemma 41.12.5. Therefore $f - f(\overline{M})(1) \in \overline{K}^+ \cap V = \overline{K}$, the last equality being due to the saturation step. Since the induction hypothesis is true for $f = 1$, the claim follows.

(2) For any polynomial $a \in A$ and any polynomial $\kappa \in \overline{K}$ we have
- $a - a(\overline{M})(1) \in \overline{K} \cap A = \{0\}$ and
- $\kappa(\overline{M})(1) = \kappa - (\kappa - \kappa(\overline{M})(1)) \in \overline{K} \cap A = \{0\}$,

  which implies that, for each $a \in A$, $N(a) = a$ and $N(\overline{K}) = \{0\}$.
(3) Just a few slight adapations allow to use the same argument[79] as used for (1) to prove, again by induction, that $f - f(\overline{M})(1) \in \mathsf{H}$ for each $f \in \mathcal{P}$.
(4) Since $N(\overline{K}) = \{0\}$, $\mathsf{H} \subset \ker(N)$. Conversely, for each $f \in \ker(N)$,

$$f = f - N(f) = f - f(\overline{M})(1) \in \mathsf{H};$$

thus $\ker(N) \subset \mathsf{H}$. $\qquad \boxdot$

Theorem 41.14.5 (Mourrain). *Assume that V is connected to* 1. *The δ-square matrices $\overline{M}_i^T$ (respectively, $\overline{M}_i$) are the matrices $M([X_i], \mathbf{a})$ (respectively, $M([Y_i], \mathbf{b})$) corresponding to multiplication by the variables X_i (respectively Y_i) in the basis $\mathbf{a} := \{a_1, \ldots, a_\delta\}$ (respectively, $\mathbf{b} := \{b_1, \ldots, b_\delta\}$) of* A.

Proof. If $1 \in \overline{K} \subset \mathsf{I}$ then $\mathsf{A} = \{0\}$ and, by the saturation step, $\overline{K} = V$ and the claim is trivial.

Since V is connected to 1, if $1 \notin \overline{K}$ then we may assume 1 to be an element of the basis A. Then, by the proposition above, $\ker(N) = \mathsf{H} = \mathsf{I}$ and $\mathrm{Im}(N) = A$, so that $A \cong \mathcal{P}/\mathsf{I} = \mathsf{A}$. $\qquad \boxdot$

41.15 Mourrain: A Gröbner-free Solver

Remark 41.15.1. If V is connected to 1, the Cardinal–Mourrain Algorithm returns a Gröbner representation

$$\mathbf{a}, \overline{M}_p, 1 \leq p \leq n,$$

of I.

These data are those required by the Auzinger–Stetter Algorithm (Compare Section 40.8) and thus are sufficient to solve zero-dimensional ideals. $\qquad \boxdot$

Remark 41.15.2 (Mourrain). Let us now set $d_i := \deg(f_i)$, $D := \max_i\{d_i\}$ and $d := 1 + \sum_{i=1}^n (d_i - 1)$.

Denote by ν a bound of the size of the matrices D_i which is at most the number of terms of degree bounded by d, that is, by Stirling's formula, $\mathcal{O}(e^n D^n)$.

If V is connected to 1, a Gröbner representation $\mathbf{a}, \overline{M}_p, 1 \leq p \leq n$, of I, where the basis elements $a_i \in \mathbf{a}$ satisfy $\deg(a_i) \leq d$, can be computed in $\mathcal{O}(n\nu^4)$ arithmetical operations.

---

[79] $\mathcal{P}$ is connected to 1; the claim holds for $f = 1$; $g_i - g_i(\overline{M})(1) \in \mathsf{H}$ by induction; $X_{l_i} g_i(\overline{M})(1) - \overline{M}_{l_i} g_i(\overline{M})(1) \in \mathsf{I}$ by Lemma 41.12.5.

In fact, the Cardinal–Mourrain Algorithm requires to perform at most v loops, each performing linear transformations over n matrices of size v. ⊙

The Cardinal–Mourrain Algorithm can also efficiently substitute Buchberger's Algorithm to provide a good complexity procedure for solving the membership test.

Proposition 41.15.3 (Mourrain). *For each $f \in \mathcal{P}$ it is possible to test whether $f \in \mathsf{I}$ in $\mathcal{O}(nv^4 L)$ arithmetical operations, where L denotes the cost of evaluating $g(\overline{M})$ for $g \in \mathcal{F} \cup \{f\}$.*

Moreover, writing

$$\bar{d} := \deg(f) + \sum_{i=1}^{n}(d_i - 1), \quad \bar{D} := \max\{\deg(f_i), \deg(f)\} \quad and \quad \bar{v} = \mathcal{O}(e^n \bar{D}^n),$$

with complexity $\mathcal{O}(n\bar{v}^4)$ it is possible to decide whether $f \in \mathsf{I}$ and, if this is the case, to produce a representation

$$f \cdot u = \sum_{i=1}^{n} f_i g_i : u, \; g_i \in \mathcal{P}, \quad u \equiv 1 \bmod \mathsf{I}, \quad \deg(u) \le d, \quad \deg(g_i) \le \bar{d}.$$

Proof. Adapt Cardinal–Mourrain's Algorithm (Figure 41.3) by substituting the *saturation step* by the inclusion in K_i and H_i of the n polynomials corresponding to the non-zero columns (respectively, rows) of the matrices $f_i(\overline{M})$, $1 \le i \le n$; the effect, even if V is not connected to 1, is that $f_i(\overline{M}) = 0, 1 \le i \le n$.

Thus if on the one hand we denote by σ the map

$$\sigma : \mathcal{P} \to k^{\delta \times \delta}, \quad f \mapsto \sigma(f) = f(\overline{M}),$$

we have $\mathsf{I} \subset \ker(\sigma)$ since, by construction, $f_i(\overline{M}) = 0, 1 \le i \le n$.

On the other hand, denoting by $u \in V \subset \mathcal{P}$ the element such that[80] $u = D_0|\ell \equiv 1 \bmod \mathsf{I}$, we have, for $f \in \ker(\sigma) \subset \mathcal{P}$,

$$f(X_1, \ldots, X_n) = f(X_1, \ldots, X_n) - f(\overline{M})(u) \in \mathsf{I}.$$

Thus we have $\mathsf{I} = \ker(\sigma)$ and $f \in \mathsf{I} \iff f(\overline{M}) = 0$.

The modified algorithm requires us to perform at most v loops, each of which performs linear transformations over n matrices of size v and evaluations of matrices $f_i(\overline{M})$; thus its complexity is $\mathcal{O}(nv^4 L)$.

Let us now modify the Cardinal–Mourrain Algorithm (Figure 41.3) in a different way: namely, we also take into account the matrix $\mathsf{D}(f)$ and apply the modifications performed by the algorithm not only to the D_i but also to $\mathsf{D}(f)$. The effect is that we obtain not only a basis $\mathbf{a}$ of A and with respect to it the matrices $M([X_i], \mathbf{a})$ representing multiplication by the variables but also the matrix $M_f := M([f], \mathbf{a})$ representing multiplication by f. Such an algorithm has complexity $\mathcal{O}(n\bar{v}^4)$, where $\bar{v} = \mathcal{O}(e^n \bar{D}^n)$ and $\bar{D} := \max\{\deg(f_i), \deg(f)\}$.

---

[80] The existence of such a u, if V is connected to 1, is granted by Proposition 41.14.3 but even without this assumption it is a consequence of the fact that $D_0|$ is an isomorphism, $\mathcal{F}$ being a complete intersection.

We thus have $f \in \mathsf{I} \iff M_f = 0$, and if we expand $\mathsf{D}(f)$ along the first row we have

$$D(f) = f(X_1, \ldots, X_n)D_0 - \sum_{i=1}^{n} f_i E_i$$

for suitable $E_i \in \mathcal{P}_\otimes$.

Lemma 41.12.4 implies that $D(f)|\Lambda = 0$ for each $f \in \mathsf{I}$, and $\Lambda \in \mathcal{L}(\mathsf{I})$, so that we have

$$f \cdot u = f(X_1, \ldots, X_n)D_0|\ell = \sum_{i=1}^{n} f_i E_i|\ell,$$

where $g_i := E_i|\ell \in \mathcal{P}$ and $u \in \mathcal{P}$ satisfies $u = D_0|\ell \equiv 1 \bmod \mathsf{I}$, and is thus invertible in I.

The degree bound is obvious by construction. $\quad\boxdot$

42

Lazard II

The introduction of the Gröbner basis to the computer algebra community activated a new interest in some older bases, such as Macaulay's basis (Sections 23.5 and 23.6) and the related algorithms (see Chapter 30), Hironaka's standard bases (Sections 24.5–24.8) and Ritt's *characteristic sets*.

Ritt's results (dated 1932), strongly influenced by Noether's results on the Decomposition Theorem (Sections 27.3, 27.4 and 32.3) were intended to give an algebraic standpoint to differential equations but, as was already usual for Riquier's followers (Delassus, Janet, Gunther), Ritt translated his results into the algebraic variety setting. In this setting he gave an effective decomposition algorithm, which, through the further application of univariate factorization, returned an irredundant *prime* decomposition of a radical ideal.

As the computer algebra community became aware of Buchberger's result, in China Wu Wen-Tsün was applying a weaker (but for his aims sufficient) version of Ritt's algorithm as a tool toward a "mechanization" of theorem-proving in elementary geometry; Wu's version of Ritt's result omits the hard and useless (for his aims) factorization step, thus returning a decomposition of a radical ideal into unmixed ideals.

In the early 1990s, within the **PoSSo** frame, Lazard, a strong expander and developer of the Kronecker–Duval Philosophy, reformulated Ritt's solver. He avoided the required factorization by means of Duval splitting via his Theorem 11.3.2, thus producing a decomposition into radical unmixed ideals, each defined via a *triangular set*, that is, what I called (in Definition 11.4.1) an admissible Duval sequence.

Later, Möller proposed an algorithm which applies only to zero-dimensional ideals, decomposing them into ideals presented through a triangular set; the theory is based on ideas related to the Gianni–Kalkbrener Theorem 39.3.1 and the algorithm is an adaptation of Traverso's Algorithm 29.3.8.

Once a zero-dimensional ideal is represented in terms of a triangular set, Kronecker–Duval Philosophy requires the transformation of this data into a form suitable for the computation **with** arithmetical expressions of its roots; suitable representations of such roots are available in the older literature, for instance Gröbner's *allgemeine* representation and Kronecker's parametrization; as proved by Alonso *et al.* Kronecker's idea is more suitable than Gröbner's since it gives a representation with smaller bit-size complexity.

An efficient algorithm that will obtain, from a triangular set, a Kronecker parametrization or *Rational Univariate Representation* (RUR), via a suitable computation of matrix traces due to Rouillier, crowns Kronecker–Duval Philosophy.

After presenting Ritt's decomposition theory (Sections 42.1 and 42.2) and the related solvers proposed by Ritt and Wu (Section 42.3), I discuss Lazard's reformulation and expansion of triangular sets (Section 42.4 and (42.5)), the related solver (Section 42.6) and the relation between Lazard's triangular sets and Gröbner bases (Section 42.7).

Next I discuss Möller's approach (Section 42.8) and Rouillier's Rational Univariate Representation (Section 42.9), postponing to Chapter 45 a discussion of algorithms that compute **with** arithmetical expressions for roots, given via the *allgemeine* and RUR representation.

42.1 Ritt: Characteristic Sets for Differential Polynomial Ideals

Let k be a *differential*[1] field of characteristic zero.

Once an indeterminate such as Y is introduced, it is implicitly considered as the first element of an infinite sequence of symbols $Y, Y', Y'', \ldots, Y^{(p)}, \ldots$; Y is then a *differential indeterminate*, whose pth *derivative* is $Y^{(p)}$.[2]

If we now consider n differential indeterminates $Y_1, \ldots, Y_n$, we will denote by Y_{ij} the jth derivative of Y_i. We will write

$$k\{Y_1, \ldots, Y_n\} := k[Y_{ij} : 1 \le i \le n, j \in \mathbb{N}]$$

for the polynomial ring in the infinite set of variables $\{Y_{ij} : 1 \le i \le n, j \in \mathbb{N}\}$, whose elements we call *differential polynomials*. For each $A \in k\{Y_1, \ldots, Y_n\}$, its derivative is the differential polynomial obtained by applying the rules

---

[1] That is, a field endowed with an operation (differentiation) $\cdot' : k \to k$ which satisfies, for each $a, b \in k$,

   (1) $(a + b)' = a' + b'$,
   (2) $(ab)' = a'b + ab'$.

The elements $a \in k$ for which $a' = 0$ are called *constants*. Note that, setting

- $b := 0$ in (1) we have $0' = 0$, and
- $b := 1, a \ne 0$ in (2) we have $1' = 0$.

It is then easy to deduce, from the following equalities,

- $(m + 1)' = m' + 1' = m'$,
- $0 = 0' = (m + (-m))' = m' + (-m)' \implies (-m)' = -(m')$,
- $0 = (mm^{-1})' = m(m^{-1})' + m^{-1}m' \implies (m^{-1})' = -m'/m^2$,

satisfied by each $m \ne 0$, that each $a \in \mathbb{Q} \subset k$ is a constant.

[2] The pth derivative of $Y^{(q)}$ is $Y^{(q+p)}$ for each pair of integers p and q.

140   *Lazard II*

(1) $(a + b)' = a' + b'$,
(2) $(ab)' = a'b + ab'$,
(3) $(Y_i^{(q)})' := Y_i^{(q+1)}$ for each i and q.

Example 42.1.1.   If k is the constant field $\mathbb{Q}(X)$ and $A := XY_1^2 + X^2 Y_{21} \in k\{Y_1, Y_2\}$ then $A' := Y_1^2 + 2XY_1 Y_{11} + 2XY_{21} + X^2 Y_{22}$ and

$$A'' := 4Y_1 Y_{11} + 2XY_{11}^2 + 2XY_1 Y_{12} + 2Y_{21} + 4XY_{22} + X^2 Y_{23}.$$

$\odot$

Definition 42.1.2.   *A subset* $I \subset k\{Y_1, \ldots, Y_n\}$ *is called a* differential ideal *if it is an ideal and satisfies*

$$f \in I \implies f' \in I.$$

$\odot$

Remark 42.1.3.   Note that, given any set $\Lambda \subset k\{Y_1, \ldots, Y_n\}$, the (polynomial) ideal generated by Λ does not necessarily coincide with the differential ideal, which can be only defined as, equivalently,

(1) the set $I \subset k\{Y_1, \ldots, Y_n\}$ such that

- $\Lambda \subset I$,
- $G_1, G_2 \in I \implies G_1 + G_2 \in I$,
- $G \in I, A \in k\{Y_1, \ldots, Y_n\} \implies AG \in I$,
- $G \in I \implies G' \in I$; or

(2) the smallest differential ideal containing Λ; or

(3) the intersection of all differential ideals containing Λ. $\odot$

Historical Remark 42.1.4.   In connection with Historical Remark 30.2.6 it is worthwhile comparing Steinitz's notation with that used by Ritt, which, for a set $\Lambda \subset k\{Y_1, \ldots, Y_n\}$, denotes as

(Λ) the polynomial ideal generated by Λ;
[Λ] the differential ideal generated by Λ;[3]
$\{\Lambda\}$ the radical of [Λ], which is in fact a differential ideal.[4]

---

[3] In fact (compare (3) in Remark 42.1.3 above) this is an *intersection* of differential ideals.

[4] The argument consists in proving that, for each $\pi \in \mathbb{N}, \pi \neq 0$, and each differential polynomial $A \in k\{Y_1, \ldots, Y_n\}$, the following hold:

(1) $A^{\pi-1} A' \in [A^\pi]$;
(2) $A^{\pi-\delta}(A')^{2\delta-1} \in [A^\pi] \implies A^{\pi-\delta-1}(A')^{2\delta+1} \in [A^\pi]$, for each $\delta \in \mathbb{N}, 1 \leq \delta < \pi$,
(3) $(A')^{2\pi-1} \in [A^\pi]$,
(4) $A \in \{\Lambda\} \implies A' \in \{\Lambda\}$.

In fact:

(1) $A^{\pi-1} A' = \pi^{-1}(A^\pi)' \in [A^\pi]$;
(2) $B := (\pi - \delta)A^{\pi-\delta-1}(A')^{2\delta} + (2\delta - 1)A^{\pi-\delta}(A')^{2\delta-2} A'' = \left(A^{\pi-\delta}(A')^{2\delta-1}\right)' \in [A^\pi]$, so that

$$(\pi - \delta)A^{\pi-\delta-1}(A')^{2\delta+1} = A'B - (2\delta - 1)A^{\pi-\delta}(A')^{2\delta-1} A'' \in [A^\pi];$$

Personally, I think that this notation is not a remainder of Steinitz's notation but is an elementary direct use of the obvious sequence $(\cdot), [\cdot], \{\cdot\}$. ◎

Definition 42.1.5 (Ritt). *For a differential polynomial $A \in k\{Y_1, \ldots, Y_n\}$:*

- *the* class *of A,* class(A), *is the value $p \le n$ such that*

$$A \in k\{Y_1, \ldots, Y_p\} \setminus k\{Y_1, \ldots, Y_{p-1}\};$$

- *the order of A w.r.t. Y_i is the value j such that[5]*

$$A \in \mathcal{F}[Y_i, Y_{i1}, \ldots, Y_{ij}] \setminus \mathcal{F}[Y_i, Y_{i1}, \ldots, Y_{i\,j-1}],$$

where we have set $\mathcal{F} := k\{Y_1, \ldots, Y_{i-1}, Y_{i+1}, \ldots, Y_n\}$.

If $A \in k$, A is said to be of class 0.
If $A \in k\{Y_1, \ldots, Y_{i-1}, Y_{i+1}, \ldots, Y_n\}$, its order w.r.t. Y_i is 0.
For $A_1, A_2 \in k\{Y_1, \ldots, Y_n\}$, A_1 is said to be of higher rank than A_2 in Y_i if either A_1 is of higher order than A_2 w.r.t. Y_i or A_1 and A_2 have the same order δ but the degree of A_1 in the variable $Y_{i\delta}$ is higher than that of A_2.
If class$(A_1) = p > 0$, A_2 is said to be reduced w.r.t. A_1 if it is of lower rank in Y_p than A_1. ◎

Definition 42.1.6 (Ritt). *Let $A_1, A_2 \in k\{Y_1, \ldots, Y_n\}$ and for $i \in \{1, 2\}$, denote by*

$p_i := $ class(A_i) *the class of A_i,*
δ_i *the order of A_i w.r.t. Y_{p_i},*
d_i *the degree of A_i in the variable $Y_{p_i\delta_i}$.*

The differential polynomial A_1 is said to be of higher rank than A_2 (denoted as $A_1 \succ A_2$) if

$$\begin{cases} p_1 > p_2 & or \\ p_1 = p_2, \delta_1 > \delta_2, & or \\ p_1 = p_2, \delta_1 = \delta_2, & and\ d_1 > d_2. \end{cases}$$

The differential polynomials A_1 and A_2 are said to be of the same rank (denoted as $A_1 \sim A_2$) if $p_1 = p_2, \delta_1 = \delta_2$ and $d_1 = d_2$. ◎

Both $\preceq$ and $\sim$ are equivalences.

We remark also that the only term ordering which is compatible with $\prec$ is the lexicographical order $<$ induced by

$$Y_1 < Y_{12} < \cdots < Y_{1j} < \cdots < Y_2 < Y_{22} < \cdots < Y_n < Y_{n2} < \cdots .$$

(3) since $A^{\pi - \delta}(A')^{2\delta - 1} \in |A^{\pi}|$ for $\delta = 1$ by (1), iteratively (2) implies that

$$A^{\pi - \delta - 1}(A')^{2\delta + 1} \in |A^{\pi}|$$

for $\delta = \pi - 1$, that is $(A')^{2\pi - 1} \in |A^{\pi}|$.
(4) Let π be such that $A^{\pi} \in |\Lambda|$; then, by the previous result, $(A')^{2\pi - 1} \in |A^{\pi}| \subset |\Lambda|$, whence $A' \in \{\Lambda\}$.

[5] We must consider also the case $j = 0$, where, with a slight abuse of notation, we set $Y_{i0} := Y_i$.

Lemma 42.1.7. *Each set of differential polynomials $\mathcal{A}$ contains a member $A \in \mathcal{A}$ such that $A \preceq B$ for each $B \in \mathcal{A}$.*

Proof. If $\mathcal{A} \cap k \neq \emptyset$ then any constant element answers the requirement. Otherwise, denote by

- p the minimal value such that $\mathcal{A} \cap k\{Y_1, \ldots, Y_p\} \neq \emptyset$,
- δ the minimal value such that $\mathcal{B} := \mathcal{A} \cap k\{Y_1, \ldots, Y_{p-1}\}[Y_p, Y_{p1}, \ldots, Y_{p\delta}] \neq \emptyset$

and choose in $\mathcal{B}$ the element of minimal degree in $Y_{p\delta}$.   ◉

Definition 42.1.8 (Ritt). *A finite set $\{A_1, \ldots, A_r\}$ of differential polynomials is called a* chain *if either*

- $r = 1$ *and $A_1 \neq 0$ or*
- $r > 1$, $\mathrm{class}(A_1) = p > 0$ *and, for each $j > i$, $\mathrm{class}(A_j) > \mathrm{class}(A_i)$ and A_j is reduced w.r.t. A_i.*[6]

The chain $\mathcal{A} := \{A_1, \ldots, A_r\}$ is said to be of higher rank *than the chain $\mathcal{B} := \{B_1, \ldots, B_s\}$ (denoted as $\mathcal{A} \succ \mathcal{B}$) if either*

(1) *there is a j, $j \leq \min\{r, s\}$, such that $A_i \sim B_i$ for $i < j$ and $A_j \succ B_j$, or*
(2) *$s > r$ and $A_i \sim B_i$ for $i \leq r$.*

The chains $\mathcal{A} := \{A_1, \ldots, A_r\}$ and $\mathcal{B} := \{B_1, \ldots, B_s\}$ are said to be of the same rank *(denoted as $\mathcal{A} \sim \mathcal{B}$) iff $s = r$ and $A_i \sim B_i$ for each i.*

If $\mathcal{A} := \{A_1, \ldots, A_r\}$ is a chain for which $\mathrm{class}(A_1) = p > 0$, a differential polynomial F is said to be reduced w.r.t. $\mathcal{A}$ *if it is reduced w.r.t. each $A_i \in \mathcal{A}$.*   ◉

Lemma 42.1.9. *Let*

$$\mathcal{A} := \{A_1, \ldots, A_r\}, \quad \mathcal{B} := \{B_1, \ldots, B_s\}, \quad \mathcal{C} := \{C_1, \ldots, C_t\}$$

be three chains. Then

$$\mathcal{A} \succ \mathcal{B}, \mathcal{B} \succ \mathcal{C} \implies \mathcal{A} \succ \mathcal{C}.$$

Proof. There are four cases:

- Both $\mathcal{A} \succ \mathcal{B}$ and $\mathcal{B} \succ \mathcal{C}$ hold by (1) in Definition 42.1.8: denote by j the smallest value such that $B_j \succ C_j$. Either
  - $A_i \sim B_i \sim C_i$ for $i < j$ and $A_j \succeq B_j \succ C_j$, or
  - there is an $h \leq j$ such that $A_i \sim B_i \sim C_i$ for $i < h$ and $A_h \succ B_h \succeq C_h$.
  For each alternative, $\mathcal{A} \succ \mathcal{C}$ by (1).

- $\mathcal{A} \succ \mathcal{B}$ by (2) while $\mathcal{B} \succ \mathcal{C}$ by (1); let j be the smallest value such that $B_j \succ C_j$.
  - If $r < j \leq t$ then $\mathcal{A} \succ \mathcal{C}$ by (2);
  - If $r \geq j$ then $A_i \sim B_i \sim C_i$ for $i < j$ and $A_j \sim B_j \succ C_j$, so that $\mathcal{A} \succ \mathcal{C}$ by (1).

---

[6] Of course, $r \leq n$.

- $\mathcal{A} \succ \mathcal{B}$ by (1) while $\mathcal{B} \succ \mathcal{C}$ by (2); let j be the smallest value such that $A_j \succ B_j$. Then $A_i \sim B_i \sim C_i$ for $i < j$ and $A_j \succ B_j \sim C_j$; therefore $\mathcal{A} \succ \mathcal{C}$ by (1).
- Both $\mathcal{A} \succ \mathcal{B}$ and $\mathcal{B} \succ \mathcal{C}$ hold by (2), so that $r < s < t$, $A_i \sim B_i \sim C_i$ for $i \leq r$ and $\mathcal{A} \succ \mathcal{C}$ by (2).   ◉

Lemma 42.1.10.   *Each set of chains $\mathfrak{A}$ contains a member $\mathcal{A} \in \mathfrak{A}$ such that $\mathcal{A} \preceq \mathcal{B}$ for each $\mathcal{B} \in \mathfrak{A}$.*   ◉

Proof.   We form a subset $\mathfrak{A}_1 \subset \mathfrak{A}$ from the chains $\mathcal{A} := \{A_1, \ldots, A_r\}$ which satisfy $A_1 \preceq B_1$ for each $\mathcal{B} := \{B_1, \ldots, B_s\} \in \mathfrak{A}$.

If each chain $\mathcal{A} \in \mathfrak{A}_1$ satisfies $\#\mathcal{A} = 1$ then this means that each chain in $\mathfrak{A}_1$ satisfies our requirement.

Otherwise, we form a subset $\mathfrak{A}_2 \subset \mathfrak{A}_1$ collecting the chains $\mathcal{A} := \{A_1, \ldots, A_r\}$ which satisfy $A_2 \preceq B_2$ for each $\mathcal{B} := \{B_1, \ldots, B_s\} \in \mathfrak{A}_1$. If each chain $\mathcal{A} \in \mathfrak{A}_2$ satisfies $\#\mathcal{A} = 2$ then each chain in $\mathfrak{A}_2$ serves our purpose.

Otherwise, we repeat the same construction; since each chain has at most n elements, in the worst case $\mathfrak{A}_n$ returns the required chains.   ◉

Definition 42.1.11.   *For any (finite or infinite) set $\mathsf{G} \subset k\{Y_1, \ldots, Y_n\}$, any chain $\mathcal{A} \subset \mathsf{G}$ such that $\mathcal{A} \preceq \mathcal{B}$ for each chain $\mathcal{B} \subset \mathsf{G}$, whose existence is proved in the lemma above, is called a* characteristic set *of G.*   ◉

Note that $(\mathcal{A}) \subseteq (\mathsf{G})$ but equality does not necessarily hold.

Lemma 42.1.12.   *Let $\mathsf{G} \subset k\{Y_1, \ldots, Y_n\}$ and $\mathcal{A} := \{A_1, \ldots, A_r\} \subset \mathsf{G}$ be a chain, where $\mathrm{class}(A_1) > 0$. The following conditions are equivalent:*

(1) $\mathcal{A}$ is a characteristic set of G;
(2) G contains no $G \in k\{Y_1, \ldots, Y_n\} \setminus \{0\}$ which is reduced w.r.t. $\mathcal{A}$.

Proof.   Assume that $\mathcal{A}$ is not a characteristic set and let $\mathcal{B} := \{B_1, \ldots, B_s\}$ be a characteristic set, so that $\mathcal{B} \prec \mathcal{A}$. If $\mathcal{A} \succ \mathcal{B}$ by (1), there is some $B_i, i \leq r$, such that $B_i \prec A_i$, that is reduced by $\mathcal{A}$; if, instead, $\mathcal{A} \succ \mathcal{B}$ by (2), B_{r+1} is reduced by $\mathcal{A}$.

Suppose now that G contains a differential polynomial $G \neq 0$ which is reduced w.r.t. $\mathcal{A}$. If $\mathrm{class}(G) > \mathrm{class}(A_r)$ then the chain $\mathcal{B} := \{A_1, \ldots, A_r, G\}$ is of lower rank than $\mathcal{A}$; otherwise, denoting as j the highest value for which $\mathrm{class}(A_j) \leq \mathrm{class}(G)$, the chain $\mathcal{B} := \{A_1, \ldots, A_j, G, A_{j+1}, \ldots, A_r\}$ is of lower rank than $\mathcal{A}$.   ◉

Lemma 42.1.13.   *Let $\mathsf{G} \subset k\{Y_1, \ldots, Y_n\}$ and let $\mathcal{A} := \{A_1, \ldots, A_r\} \subset \mathsf{G}$ be a characteristic set of G, where $\mathrm{class}(A_1) > 0$. Let $G \notin \mathsf{G}$ be a non-zero differential polynomial which is reduced w.r.t. $\mathcal{A}$, set $\mathsf{G}' := \mathsf{G} \cup \{G\}$ and let $\mathcal{B}$ be a characteristic set of G'. Then $\mathcal{B} \prec \mathcal{A}$.*   ◉

Algorithm 42.1.14.   Let $\mathsf{G} \subset k\{Y_1, \ldots, Y_n\} \setminus \{0\}$ be a finite set. The following algorithm allows the extraction of a characteristic set $\mathcal{A} := \{A_1, \ldots, A_r\} \subset \mathsf{G}$.

Let us begin by picking an element $A_1 \in G$ which is of least (that is, minimal) rank. If $\mathrm{class}(A_1) = 0$, $\mathcal{A} := \{A_1\}$ is the required characteristic set. If $\mathrm{class}(A_1) > 0$ and no element in G is reduced w.r.t. $\{A_1\}$, again $\mathcal{A} := \{A_1\}$ is the required characteristic set.

Otherwise, each element in G which is reduced w.r.t. $\{A_1\}$ is such that $\mathrm{class}(G) > \mathrm{class}(A_1)$; choose as A_2 any such element of least rank.

Again, either G contains no other element which is reduced w.r.t. $\mathcal{A} := \{A_1, A_2\}$, which is the required characteristic set, or one can choose as A_3 any element which is reduced w.r.t. $\mathcal{A} := \{A_1, A_2\}$ and of least rank among all possible choices.

Inductively repeating the same constructions, a characteristic set $\mathcal{A} := \{A_1, \ldots, A_r\}$ is obtained in a finite number of steps. $\quad\boxed{\odot}$

Definition 42.1.15. *For a differential polynomial $G \in k\{Y_1, \ldots, Y_n\}$ of class $p > 0$ and of order m in Y_p, the* separant *of G is the differential polynomial $\frac{\partial G}{\partial Y_{pm}}$ and its* initial *is the cofficient of the highest power of Y_{pm} in G.*

More precisely, expressing G as a univariate polynomial, that is, setting

$$G := \sum_{i=0}^{d} c_i Y_{pm}^i \in k\{Y_1, \ldots, Y_{p-1}\}[Y_p, Y_{p1}, \ldots, Y_{p\,m-1}][Y_{pm}],$$

with $c_d \neq 0$, its separant is $\frac{\partial G}{\partial Y_{pm}} = \sum_{i=1}^{d} i c_i Y_{pm}^{i-1}$ and its initial is $c_d \in k\{Y_1, \ldots, Y_{p-1}\}[Y_p, Y_{p1}, \ldots, Y_{p\,m-1}]$. $\quad\boxed{\odot}$

Let $\mathcal{A} := \{A_1, \ldots, A_r\}$ be a chain and, for each i, let us denote as S_i and I_i the separant and initial of A_i.

If a differential polynomial G is not reduced w.r.t. $\mathcal{A}$, let us denote by

j the greatest value such that G is not reduced w.r.t. A_j,
p the class of A_j,
m the order of A_j in Y_p;
$h \geq m$ the order of G in Y_p.

We can therefore associate with each differential polynomial G which is not reduced w.r.t. $\mathcal{A}$ a pair

$$\Phi(G) := (j, h), \quad 1 \leq j \leq r, \quad h \in \mathbb{N},$$

and assume that the set of such pairs is well ordered by an ordering $<$ defined by

$$(j, h) > (j', h') \iff \text{either } j > j' \text{ or } j = j' \text{ and } h > h'.$$

Let us consider the possible cases.

(1) If $h > m$, write $l := h - m$ and note that $A_j^{(l)}$ is of order h in Y_p, linear in Y_{ph} and has S_j as initial. The division algorithm performed on

$$k\{Y_1, \ldots, Y_{p-1}\}[Y_p, Y_{p1}, \ldots, Y_{p\,m-1}][Y_{pm}]$$

(see Volume I, p. 12) allows us to compute a value $v \in \mathbb{N}$ and differential polynomials C, D satisfying

$$S_j^v G = C A_j^{(l)} + D.$$

We remark that D is

uniquely determined, if v is chosen as small as possible,

of order less than h in Y_p,

of rank not higher then G in Y_a, $p < a \leq n$,[7]

and so is reduced w.r.t. A_i, $i > j$.

Thus $\Phi(D) := (j', h') < (j, h) = \Phi(G)$, since either $h' < h$ or $h' = m$, and D is reduced by A_i for each $i > j$, so that $j' \leq j$.

(2) If $h = m$ then both G and A_j can be considered as univariate polynomials in

$$k\{Y_1, \ldots, Y_{p-1}, Y_{p+1}, \ldots, Y_n\}[Y_p, Y_{p1}, \ldots, Y_{p\,m-1}][Y_{pm}],$$

where the division algorithm allows us to compute a value $v \in \mathbb{N}$ and differential polynomials C, D satisfying

$$I_j^v G = C A_j + D;$$

we remark that D is uniquely determined, if v is chosen as small as possible and reduced w.r.t. both A_j and each A_i, $i > j$.

Thus $\Phi(D) := (j', h') < (j, h) = \Phi(G)$, since $j' < j$.

Since $<$ is Noetherian, in a finite number of applications of this algorithm we can compute a sequence of differential polynomials $G := D_0, D_1, \ldots, D_t$, where D_t is reduced w.r.t. $\mathcal{A}$, which satisfy the relations

$$P_i^{v_i} D_i = C_i B_i + D_{i+1},$$

with

$v_i \in \mathbb{N}$,
$P_i \in \{S_j, I_j, 1 \leq j \leq r\}$,
$C_i \in k\{Y_1, \ldots, Y_n\}$,
$B_i \in \{A_j^{(h)}, 1 \leq j \leq r, h \in \mathbb{N}\}$.

As a consequence we have

Theorem 42.1.16 (Ritt). *Let $\mathcal{A} := \{A_1, \ldots, A_r\}$ be a chain and, for each i, let us denote as S_i and I_i the separant and initial of A_i. For each differential polynomial G it is possible to compute*

---

[7] In fact, since S_j is free of Y_a, we need only treat the case in which G depends on Y_a and its derivates. Denoting by g the order of G in Y_a, clearly the order of D in Y_a does not exceed g and, in the case where its value is exactly g, the assumption that its degree δ in Y_{ag} is higher than that of G implies that some term of C is divisible by Y_{ag}^δ, so that some term of $CA_j^{(l)}$ is divisible by $Y_{ph} Y_{pg}^\delta$; this is a contradiction since no such term occurs either in $S_j^v G$, since G has no term divisible by Y_{pg}^δ, or in D, which has no term divisible by Y_{ph}.

values $v_i \in \mathbb{N}$
and $w_i \in \mathbb{N}$,
polynomials $C_{jh} \in k\{Y_1, \ldots, Y_n\}$
and a polynomial $R \in k\{Y_1, \ldots, Y_n\}$,

such that

(1) *R is reduced w.r.t. $\mathcal{A}$*
(2) *and is uniquely determined by the values v_i and w_i,*
(3) *the set $\{(j, h) : C_{jh} \neq 0\}$ is finite,*
(4) *$S_1^{v_1} \cdots S_r^{v_r} I_1^{w_1} \cdots I_r^{w_r} G = R + \sum_{j,h} C_{jh} A_j^{(h)}$.*

Moreover, the values v_j and w_j can be assumed to be the minimal values satisfying a relation of this kind. ◙

Definition 42.1.17 (Ritt). *The unique polynomial R determined by the minimal values v_i and w_i satisfying Theorem 42.1.16(4) is called the* remainder *of G w.r.t. $\mathcal{A}$.* ◙

Theorem 42.1.18 (Ritt). *Let*

$\mathsf{I} \subset k\{Y_1, \ldots, Y_n\}$ be a differential ideal,
$\mathcal{A} := \{A_1, \ldots, A_r\}$ be a characteristic set of I,
S_i and I_i, $1 \leq i \leq r$, the separant and initial of A_i,
$X := \prod_{i=1}^{r} S_i I_i$.

For each $G \in k\{Y_1, \ldots, Y_n\}$, denoting as R the remainder of G w.r.t. $\mathcal{A}$, we have

$$G \in \mathsf{I} \implies R = 0 \implies G \in \mathsf{I} : X^{\infty}.$$

If moreover I is prime then $G \in \mathsf{I} \iff R = 0$ and $(\mathcal{A}) = \mathsf{I}$.

Proof. If $G \in \mathsf{I}$ then $R \in \mathsf{I}$; therefore, being reduced w.r.t. $\mathcal{A}$, R is necessarily 0 by Lemma 42.1.12.

Conversely, if $R = 0$ then, with the notation of Theorem 42.1.16, we have

$$S_1^{v_1} \cdots S_r^{v_r} I_1^{w_1} \cdots I_r^{w_r} G = R + \sum_{j,h} C_{jh} A_j^{(h)} \in \mathsf{I}$$

and $G \in \mathsf{I} : X^{\infty}$.

We remark now that the S_i and I_i, being reduced w.r.t. $\mathcal{A}$, are not members of I (Lemma 42.1.12). Therefore, if I is prime and $R = 0$,

$$S_1^{v_1} \cdots S_r^{v_r} I_1^{w_1} \cdots I_r^{w_r} G = R + \sum_{j,h} C_{jh} A_j^{(h)} \in \mathsf{I} \implies G \in \mathsf{I}.$$

◙

Remark 42.1.19. For any prime differential ideal $\mathsf{I} \subset k\{Y_1, \ldots, Y_n\}$, one can pick a maximal set of variables $\{V_1, \ldots, V_d\} \subset \{Y_1, \ldots, Y_n\}$ such that

- $\mathsf{I} \cap k\{V_1, \ldots, V_d\} = 0$,
- for each $Z \in \{Y_1, \ldots, Y_n\} \setminus \{V_1, \ldots, V_d\}$ there is a non-zero differential polynomial $G_Z \in \mathsf{I} \cap k\{V_1, \ldots, V_d, Z\}$.

Up to a relabeling of the variables we have the identification

$$k\{Y_1, \ldots, Y_n\} \cong k\{V_1, \ldots, V_d, Z_1, \ldots, Z_r\},$$

so that

- $\mathsf{I} \cap k\{V_1, \ldots, V_d\} = 0$,
- for each $i, 1 \leq i \leq r$, there is a non-zero differential polynomial in $\mathsf{I} \cap k\{V_1, \ldots, V_d, Z_i\}$.

If for each $i, 1 \leq i \leq r$, we pick any element $A_i \in \mathsf{I} \cap k\{V_1, \ldots, V_d, Z_i\}$ then the set $\{A_1, \ldots, A_r\}$ is naturally a chain, since each A_i is reduced w.r.t. $\{A_1, \ldots, A_{i-1}\}$.

Analogously, if we pick any element $A_1 \in \mathsf{I} \cap k\{V_1, \ldots, V_d, Z_1\}$ and recursively, for $i, 1 < i \leq r$, any element $A_i \in \mathsf{I} \cap k\{V_1, \ldots, V_d, Z_1, \ldots, Z_i\}$ which is reduced w.r.t. $\{A_1, \ldots, A_{i-1}\}$ and of least rank among all possible choices,[8] then $\{A_1, \ldots, A_r\}$ is a characteristic set of I.

We will call $\{V_1, \ldots, V_d\}$ a *maximal set of independent indeterminates*, following Weispfenning (cf. Definition 27.11.4), or a *parametric set of indeterminates*, following Ritt, or the *set of the transcendental variables* of I, following Lazard.　　⊙

42.2 Ritt: Characteristic Sets for Polynomial Ideals

Let us now restrict ourselves to an (algebraic) field k of characteristic zero, without requiring the existence of a differential structure and the polynomial ring $\mathcal{P} := k[X_1, \ldots, X_n]$, and reinterpret the theory developed in the previous section, assuming that k is a differential field in which all derivatives are zero and $\mathcal{P}$ is a subring of $k\{X_1, \ldots, X_n\}$:

$$k[X_1, \ldots, X_n] \subset k\{X_1, \ldots, X_n\}.$$

Then for any set $\Lambda \subset k[X_1, \ldots, X_n]$ we restrict ourselves to considering the polynomial ideal

$$\mathbb{I}(\Lambda) := [\Lambda] \cap k[X_1, \ldots, X_n] = (\Lambda) \cap k[X_1, \ldots, X_n].$$

In this context the same notation and results as those introduced in the previous section are available. In particular:

---

[8] This means that we must pick a reduced polynomial

$$A_i \in k\{V_1, \ldots, V_d, Z_1, \ldots, Z_{i-1}\}[Z_i, Z_{i1}, \ldots, Z_{i\,q-1}][Z_{iq}]$$

of minimal degree in Z_{iq} among all possible choices, where q, the order of A_i in Z_i, is the minimal value for which

$$\mathsf{I} \cap k\{V_1, \ldots, V_d, Z_1, \ldots, Z_{i-1}\}[Z_i, Z_{i1}, \ldots, Z_{i\,q-1}][Z_{iq}] \neq \emptyset.$$

- For a polynomial $A \in k[X_1, \ldots, X_n]$ the *class* of A, class(A), is the value $p \leq n$ such that

$$A \in k[X_1, \ldots, X_p] \setminus k[X_1, \ldots, X_{p-1}],$$

the polynomials of class 0 being the elements in k.

- For any two polynomials $A_1, A_2 \in k[X_1, \ldots, X_n]$, A_1 is said to be of *higher rank than A_2 in X_i* if the degree of A_1 in the variable X_i is higher than that of A_2.

- If class$(A_1) = p > 0$, A_2 is said to be *reduced w.r.t.* A_1 if it is of lower degree in X_p than A_1.

- For any two polynomials $A_1, A_2 \in k[X_1, \ldots, X_n]$, denoting, for $i \in \{1, 2\}$, p_i as the class of A_i and d_i as the degree of A_i in the variable X_{p_i}, we have

$$A_1 \succ A_2 \iff \begin{cases} p_1 > p_2 & \text{or} \\ p_1 = p_2 & \text{and } d_1 > d_2 \end{cases}$$

and

$$A_1 \sim A_2 \iff p_1 = p_2, d_1 = d_2.$$

- A finite set

$$\mathcal{A} := \{A_1, \ldots, A_r\} \subset k[X_1, \ldots, X_n]$$

is called a *chain* (or *ascending set*, or *reduced triangular set*) if either $r = 1$ and $A_1 \neq 0$ or

class$(A_1) = p > 0$,
for each $j > i$, class$(A_j) >$ class(A_i) and
A_j is reduced w.r.t A_i.

- Given any (finite or infinite) set $\mathsf{G} \subset k[X_1, \ldots, X_n]$, in the chain

$$\mathcal{A} := \{A_1, \ldots, A_r\} \subset \mathsf{G}$$

produced by Algorithm 42.1.14, each element A_i is not only reduced w.r.t. $\{A_1, \ldots, A_{i-1}\}$ and of minimal rank among all possible choices, but necessarily its class is higher than that of A_{i-1}.

- For two chains in $k[X_1, \ldots, X_n]$, we have

$$\{A_1, \ldots, A_r\} := \mathcal{A} \succ \mathcal{B} := \{B_1, \ldots, B_s\}$$

if either
(1) there is a j, $j \leq \min\{r, s\}$, such that $A_i \sim B_i$ for $i < j$ and $A_j \succ B_j$, or
(2) $s > r$, and $A_i \sim B_i$ for $i \leq r$,
and

$$\mathcal{A} \sim \mathcal{B} \iff s = r \quad \text{and} \quad A_i \sim B_i \text{ for each } i.$$

- If $\mathcal{A} := \{A_1, \ldots, A_r\} \subset k[X_1, \ldots, X_n]$ is a chain for which class$(A_1) = p > 0$, a polynomial $F \in k[X_1, \ldots, X_n]$ is *reduced w.r.t.* $\mathcal{A}$ if it is reduced w.r.t. each $A_i \in \mathcal{A}$.

- For any set $G \subset k[X_1, \ldots, X_n]$, any chain $\mathcal{A} \subset G$ such that $\mathcal{A} \preceq \mathcal{B}$ for each chain $\mathcal{B} \subset G$ is called a *characteristic set* (or *basic set*) of G.

- For a polynomial

$$G := \sum_{i=0}^{d} c_i X_p^i \in k[X_1, \ldots, X_{p-1}][X_p], \quad c_i \in k[X_1, \ldots, X_{p-1}], \quad c_d \neq 0,$$

of class $p > 0$, its *initial* is its leading polynomial $c_d = \mathrm{Lp}(G)$.

- For a chain $\mathcal{A} := \{A_1, \ldots, A_r\}$ and a polynomial G not reduced w.r.t. $\mathcal{A}$, denoting, for each i, I_i as the initial of A_i, let j be the greatest value such that G is not reduced w.r.t. A_j and let p be the class of A_j; then the division algorithm in

$$k[X_1, \ldots, X_{p-1}, X_{p+1}, \ldots, X_n][X_p]$$

allows the computation of a value $v \in \mathbb{N}$ and polynomials C, D satisfying

$$I_j^v G = C A_j + D,$$

where D is uniquely determined, if v is chosen as small as possible, and is reduced w.r.t. both A_j and each $A_i, i > j$.

- Then (Theorem 42.1.16) for each such G it is possible to compute
  values $w_i \in \mathbb{N}$,
  polynomials $C_j \in k[X_1, \ldots, X_n]$,
  a polynomial $R \in k[X_1, \ldots, X_n]$
such that R is reduced w.r.t. $\mathcal{A}$, and is uniquely determined by the values w_i, and it holds that

$$I_1^{w_1} \cdots I_r^{w_r} G = R + \sum_j C_j A_j.$$

Moreover, the values v_i can be assumed to be the minimal values satisfying a relation of this kind.

- The *remainder of G w.r.t.* $\mathcal{A}$ is the unique polynomial R determined by the minimal values v_i in the formula above.

- For any prime ideal $\mathsf{I} \subset k[X_1, \ldots, X_n]$, we can relabel the variables so that

$$k[X_1, \ldots, X_n] \cong k[V_1, \ldots, V_d, Z_1, \ldots, Z_r]$$

and $\{V_1, \ldots, V_d\}$ is a *maximal set of independent indeterminates* for I, $d := \dim(\mathsf{I})$.

Then each characteristic set of I consists (Remark 42.1.19) of r polynomials $A_i \in \mathsf{I} \cap k[V_1, \ldots, V_d, Z_1, \ldots, Z_i]$ which are reduced w.r.t. $\{A_1, \ldots, A_{i-1}\}$ and of minimal degree in Z_i among all possible choices.

In this context it is worthwhile to record an old-fashioned proof of the following well-known result:

Proposition 42.2.1. *Let* $\mathsf{I} \subset k[X_1, \ldots, X_n]$ *be a prime ideal,* $\dim(\mathsf{I}) := d$, $\{V_1, \ldots, V_d\}$ *be a maximal set of independent indeterminates for* I *and* $K \in k[X_1, \ldots, X_n]$ *a polynomial not contained in* I.

Then the ideal $\mathsf{I}' := \mathsf{I} + \mathbb{I}(K)$ *is such that* $\mathsf{I}' \cap k[V_1, \ldots, V_d] \neq \{0\}$.

Proof (Ritt).   Using the same notation as above, this is Ritt's argument:[9]

We start with the observation that the polynomials in I which involve no Z_i with $i > j$, where $1 \leq j < n - d$, constitute a prime polynomial ideal; we describe this prime ideal[10] by I_j.

I' contains the remainder of K with respect to $\{A_1, \ldots, A_r\}$. Of all nonzero polynomials in I' which are reduced with respect to $\{A_1, \ldots, A_r\}$, let B be the one which is of lowest rank. We say that B is free of the Z.

Suppose that this is not so, and let B be of class $d + p$ with $p > 0$. The initial C of B is not in I. There is a relation

$$C^m A_p = DB + E$$

where E, if not zero, is of lower degree than B in Z_p. We say that E is in I. Let this be false. If $p > 1$, the remainder of E with respect to $A_1, \ldots, A_{p-1}$ is a non zero polynomial contained in I', which is reduced with respect to $\{A_1, \ldots, A_r\}$ and of lower rank than B. If $p = 1$, a similar statement can be made of E itself. Thus[11] E is in I, so that DB is in I. Then[12] D is in I. D is of positive degree[13] in Z_p. As the initial of DB is that of $C^m A_p$, the initial I of D is not in I. If we had $p = 1$, D would be a nonzero polynomial in I which is reduced with respect to $\{A_1, \ldots, A_r\}$; this is because D is of lower degree in Z_p than A_p. Thus $p > 1$. The remainder of D with respect to $A_1, \ldots, A_{p-1}$ is zero.[14] Thus JD, with J some product of powers of the initials of $A_1, \ldots, A_{p-1}$, is linear.[15] in $A_1, \ldots, A_{p-1}$. If we write JD as a polynomial in Z_p its coefficients will be in I_{p-1}. Thus JI is in I_{p-1}. This is false because neither J nor I is in I_{p-1}.

Thus B is free of the Z and our statement is proved.   ◉

Remark 42.2.2.   Since any polynomial $P = P_0$ can be uniquely expressed as

$$P = P_0 = \mathrm{Lp}(P_0) X_{j_0}^{\delta_0} + R_0,$$

where $j_0 = \mathrm{class}(P_0)$, $P_1 := \mathrm{Lp}(P) \in k[X_1, \ldots, X_{j_0-1}]$ and $\deg_{j_0}(R_0) < \delta_0 = \deg_{j_0}(P_0)$, we can define recursively

values $j_i := \mathrm{class}(P_i) < j_{i-1}$ and $\delta_i \in \mathbb{N}$,
polynomials $R_i \in k[X_1, \ldots, X_{j_i}]$ and

---

[9] Ritt J.F., *Differential Algebra*, A.M.S. Colloquium Publications **33** (1950), p. 84.
          I have just adapted the notation.
[10] That is, $\mathsf{I}_j := \mathsf{I} \cap k[V_1, \ldots, V_d, Z_1, \ldots, Z_j]$.
[11] We remark that, by assumption, B is a non-zero polynomial contained in I', which is reduced with respect to $\{A_1, \ldots, A_r\}$ and of lower rank. So we have just reached a contradiction.
[12] $B \in k[V_1, \ldots, V_d] \implies B \notin \mathsf{I}$.
[13] Necessarily the degree in Z_p of B is lower than that of A_p.
[14] $D \in \mathsf{I}$.
[15] That is, $JD \in (A_1, \ldots, A_{p-1}) = \mathbb{I}(A_1, \ldots, A_{p-1})$.

$$P_{i+1} := \mathrm{Lp}_i(P) := \mathrm{Lp}(\mathrm{Lp}_{i-1}(P)) = \mathrm{Lp}(P_i) \in k[X_1, \ldots, X_{j_i-1}] \text{ such that}$$

$$P_i = \mathrm{Lp}_i(P)X_{j_i}^{\delta_i} + R_i, \quad \deg_{j_i}(R_i) < \delta_i = \deg_{j_i}(P_i)$$

until i reaches a value ι for which $P_{\iota+1} = \mathrm{Lp}_\iota(P) \in k$.

Then we have

$$P = \mathrm{Lp}_\iota(P)X_{j_\iota}^{\delta_\iota} X_{j_{\iota-1}}^{\delta_{\iota-1}} \cdots X_{j_1}^{\delta_1} X_{j_0}^{\delta_0} + R_\iota \prod_{i=0}^{\iota-1} X_{j_i}^{\delta_i} + \sum_{h=0}^{\iota-1} R_h \prod_{i=0}^{h-1} X_{j_i}^{\delta_i}.$$

Recalling that the only term ordering which is compatible with $\prec$ is the lexicographical order $<$ induced by $X_1 < X_2 < \cdots < X_n$, we necessarily have

$$\mathrm{lc}_<(P) = \mathrm{Lp}_\iota(P), \quad \mathbf{T}_<(P) = X_{j_\iota}^{\delta_\iota} X_{j_{\iota-1}}^{\delta_{\iota-1}} \cdots X_{j_1}^{\delta_1} X_{j_0}^{\delta_0}.$$

⊡

If we consider a chain

$$\mathcal{A} := \{A_1, \ldots, A_r\} \subset k[X_1, \ldots, X_n] \cong k[V_1, \ldots, V_d, Z_1, \ldots, Z_r],$$

each A_i being of class $d + i$, we can find conditions for $\mathcal{A}$ to be a characteristic set of a prime ideal.

Let us denote

l the ideal generated by $\mathcal{A}$ in $k(V_1, \ldots, V_d)[Z_1, \ldots, Z_r]$,
$\mathsf{l}_j := \mathsf{l} \cap k(V_1, \ldots, V_d)[Z_1, \ldots, Z_j]$,
$L_j := k(V_1, \ldots, V_d)[Z_1, \ldots, Z_j]/\mathsf{l}_j$,
$\pi_j : k(V_1, \ldots, V_d)[Z_1, \ldots, Z_r] \to L_j[Z_{j+1}, \ldots, Z_r]$,
$I := \prod_i I_i$, each I_i denoting the initial of A_i.

Then we have:

Proposition 42.2.3. *With the present notation, $\mathcal{A}$ is a characteristic set of the prime ideal* $\mathsf{l} : I^\infty$ *iff the following conditions hold:*

(1) *if $r = 1$, A_1 is irreducible in $k(V_1, \ldots, V_d)[Z_1]$;*
(2) *if $r > 1$, $\{A_1, \ldots, A_{r-1}\}$ is a characteristic set of the prime ideal l_{r-1} and $\pi_{r-1}(A_r)$ is irreducible in $L_{r-1}[Z_r]$.*

Proof. The proof is given by a reformulation of Theorems 34.1.2 and 34.3.2. ⊡

Algorithm 42.2.4 (Ritt). Let $\mathsf{G} \subset k[X_1, \ldots, X_n]$ be a finite set generating an ideal l.

The following algorithm allows the computation of a characteristic set $\mathcal{A} := \{A_1, \ldots, A_r\}$ of l and a polynomial I such that, denoting by L the ideal generated by $\mathcal{A}$, we have

$$\mathsf{L} \subset \mathsf{l} \subset \mathsf{L} : I^\infty =: \mathsf{H};$$

if, moreover, l is prime then $\mathsf{l} = \mathsf{L}$.

One begins by extracting from G a characteristic set $\mathcal{A}$ using the method described in Algorithm 42.1.14, so that $(\mathcal{A}) \subseteq I$.[16] Next one computes the remainders w.r.t. $\mathcal{A}$ of each member in G and then includes the non-zero remainders in G, producing a larger set G', which, however, satisfies $\mathbb{I}(G') = \mathbb{I}(G) = I$.

If not all such remainders are zero then (Lemma 42.1.13) we have that a characteristic set $\mathcal{A}'$ of G' satisfies $\mathcal{A}' \prec \mathcal{A}$.

Thus the same procedure can be repeated until one arrives at a set G^* which satisfies $\mathbb{I}(G^*) = \mathbb{I}(G) = I$ and a characteristic set $\mathcal{A}^* := \{A_1, \ldots, A_r\}$ extracted from G^*, for which either

- A_1 is of class zero, so that $I = (1)$, or
- all the remainders w.r.t. $\mathcal{A}^*$ of each member in G^* are zero. Then (Theorem 42.1.18), writing $I := \prod_{i=1}^{r} I_i$, the product of all initials I_i of the A_i, we have

$$\mathbb{I}(\mathcal{A}^*) \subset I \subset \mathbb{I}(\mathcal{A}^*) : I^\infty;$$

if, moreover, I is prime then $\mathbb{I}(\mathcal{A}^*) = I$.

Historical Remark 42.2.5.   To put this into better historical perspective, I think it is worthwhile to quote the words of Ritt,[17] referring also to his older book:[18]

The form in which the results of differential algebra are being presented has thus been deeply influenced by the teachings of Emmy Noether, a prime mover of our period, who, in continuing Julius König's development of Kronecker's ideas, brought mathematicians to know algebra as it was never known before.

In this connection, I should like to say something concerning basis theorems. The basis theorem [...] will be seen to play, in the present theory, the role held by Hilbert's theorem in the theories of polynomial ideals and of algebraic manifolds. When I began to work on algebraic differential equations, early in 1930, van der Waerden's excellent *Modern Algebra* had not yet appeared. However, Emmy Noether's work of the twenties was available, and there was nothing to prevent one from learning in her papers the value of basis theorems in decomposition problems. Actually, I became acquainted with the basis theorem principle in the writings of Jules Drach on logical integration.[19]

42.3 Ritt's and Wu's Solvers

Algorithm 42.3.1 (Ritt).   Ritt applied Proposition 42.2.3, Algorithm 42.2.4 and Theorem 42.1.18 as tools for proposing a solver,[20] which, given a finite set $G \subset k[X_1, \ldots, X_n]$ generating an ideal I, applies Algorithm 42.2.4 in order to

---

[16] We can of course assume that A_1 is of positive class; otherwise $I = (1)$ and we are through.

[17] From the preface of his book *Differential Algebra*, A.M.S. Colloquium Publications **33** (1950), p. iv.

[18] Ritt J.F., *Differential Equations from the Algebraic Standpoint*, A.M.S. Colloquium Publications **14** (1932).

[19] The reference is to Drach, J., Essai sur la théorie général de l'integration et sur la classification des Transcendentes, *Ann. Éc. Norm. 3ᵉ série* **15** (1898), 245–384.

[20] Ritt J.F., *Differential Algebra*, A.M.S. Colloquium Publications **33** (1950), pp. 95–98.

extract a characteristic set $\mathcal{A}^*$ satisfying the properties stated by Theorem 42.1.18 and tests whether it satisfies the properties of Proposition 42.2.3.[21]

If the answer is positive then $\mathcal{A}^*$ is a characteristic set of the prime ideal $\mathsf{H} := \mathbb{I}(\mathcal{A}^*) : I^\infty$ and, in this case,[22]

$$\mathcal{Z}(\mathsf{I}) = \mathcal{Z}(\mathsf{H}) \cup \mathcal{Z}(\mathsf{I} + \mathbb{I}(I_1)) \cup \cdots \cup \mathcal{Z}(\mathsf{I} + \mathbb{I}(I_r));$$

the same algorithm is to be recursively applied to each set $\mathsf{G} \cup \{I_i\}$.

If, instead, the answer is negative then we have found (using the notation of Proposition 42.2.3) some factorization[23]

$$\pi_{r-1}(A_r) = c \prod_{i=1}^{s} \pi_{r-1}(g_i)^{e_i}$$

in L_{r-1}, where $g_i \in k[V_1, \ldots, V_d][Z_1, \ldots, Z_{r-1}][Z_r]$ and $c \in k(V_1, \ldots, V_d)$ is a unit; therefore

$$\mathcal{Z}(\mathsf{I}) = \mathcal{Z}(\mathsf{I} + \mathbb{I}(g_1)) \cup \cdots \cup \mathcal{Z}(\mathsf{I} + \mathbb{I}(g_s)).$$

The result corresponds to an irredundant prime decomposition of I. ⊙

Historical Remark 42.3.2. In other words Ritt, in 1950, is applying the notion of "solving" discussed in Section 34.5. In 1978, Wu Wen-Tsün relaxed this notion of "solving", preserving the structure of the corresponding basis (an *ascending set*), which has a shape similar to that of a *Primbasis* (Definition 34.3.3) or an admissible sequence (Definitions 8.2.2 and 11.4.2) but which requires less strong properties (essentially just those implied by Theorem 42.1.18 and Algorithm 42.2.4).

In fact, in his research toward a theorem-proving algorithm in differential (and elementary) geometry, as a tool for testing whether $f(\alpha) = 0$ for each root α of I, where $f \in k\{X_1, \ldots, X_n\}$ is a given differential polynomial and $\mathsf{I} := \mathbb{I}(\mathsf{G}) \subset k\{X_1, \ldots, X_n\}$ a given differential ideal,[24] Wu applied Ritt's theory directly, reducing the problem to a test whether

(1) the remainder of f w.r.t. $\mathcal{A}^*$ is zero and,
(2) by recursive application of the same algorithm, $f(\alpha) = 0$ for each root α of $\mathsf{I} + \{I_i\}, i \leq r$.

---

[21] Essentially Ritt proposes the primality test discussed in Section 35.4:

- the rôle used there by the Gröbner basis G' is taken here by the characteristic set $\mathcal{A}^*$;
- the test $\mathsf{I} : I^\infty = \mathsf{I}$ is not required here, since, in this setting,

$$\mathsf{I} : I^\infty = \left(\mathbb{I}(\mathcal{A}^*) : I^\infty\right) : I^\infty$$
$$= \mathbb{I}(\mathcal{A}^*) : I^\infty$$
$$= \mathsf{I}.$$

[22] It is worthwhile to note that this formula, which is today improperly attributed to Wu Wen-Tsün, is Equation (32) in Ritt J.F., *op. cit.*, p. 98.

[23] As in the case of the prime decomposition algorithm discussed in Section 35.2.

[24] If $K \supset k$ is a differential field extension of k, $\alpha := (a_1, \ldots, a_n) \in K^n$ and $f \in k\{X_1, \ldots, X_n\}$ is a differential polynomial, the evaluation $f(\alpha)$ is by definition the value $\Phi_\alpha(f)$, where $\Phi_\alpha : k\{X_1, \ldots, X_n\} \to K$ is the morphism defined by $\Phi_\alpha(X_{ij}) = a_i^{(j)}$ for each i, j.

In fact, for each root α of I,

(1) if $\prod_{i=1}^{r} I_i(\alpha) \neq 0$ then, by Theorem 42.1.16, $f(\alpha) = 0$ iff the remainder of f w.r.t. $\mathcal{A}$ is zero;

(2) if $I_i(\alpha) = 0$ then α is a root of $I + \mathbb{I}(I_i)$.

For Wu's application, there was therefore no need to factorize the characteristic set $\mathbf{A}^*$ in order to deduce an irredundant prime decomposition of I; all one needed to do was to decompose $\sqrt{I}$ as an intersection of (unmixed) ideals generated by a characteristic set, in order to apply the results implied by Theorem 42.1.18 and Algorithm 42.2.4. Thus the problem was reduced to the zero-testing of "normal forms" of each element in G w.r.t. each such characteristic set.

In the algebraic setting $k[X_1, \ldots, X_n]$, such a characteristic set is a set

$$\mathcal{A} := \{A_1, \ldots, A_r\} \subset k[X_1, \ldots, X_n] \cong k[V_1, \ldots, V_d][Z_1, \ldots, Z_r],$$

where (compare Definition 22.0.1), for each i,

- the degree of A_i in X_j is less than that of A_j, for each $j < i$, and
- $A_i \in k[V_1, \ldots, V_d][Z_1, \ldots, Z_{i-1}][Z_i]$.

That is, it is exactly what we have informally called a *weak admissible sequence*.[25]

◉

Algorithm 42.3.3 (Wu).   In this approach, Algorithm 42.3.1 is simplified as follows. Given a finite set $G \subset k[X_1, \ldots, X_n]$ generating an ideal I, one

- applies Algorithm 42.2.4 in order to extract a basic set $\mathcal{A}^* := \{A_1, \ldots, A_r\}$,
- returns $\mathcal{A}^*$ and the polynomial $I := \prod_{i=1}^{r} I_i$, where I_i denotes the initial of A_i,
- and applies the same algorithm to each set $G \cup \{I_i\}$, $1 \leq i \leq r$. ◉

The *rationale* of this approach is based on the following:

Lemma 42.3.4.   *Let*

$\mathcal{A} := \{A_1, \ldots, A_r\} \subset k[X_1, \ldots, X_n]$ *be a characteristic set,*
L *be the ideal generated by $\mathcal{A}$,*
I_j *be the initial of A_j, for each j,*
$I := \prod_{i=1}^{r} I_i$,
$\mathrm{Rem}(\mathcal{A})$ *be the set of all polynomials whose remainder w.r.t. $\mathcal{A}$ is 0,*[26]
$\mathrm{Sat}(\mathcal{A}) := L : I^\infty$,
$\mathfrak{Z}(\mathcal{A}) := \mathcal{Z}(\mathcal{A}) \setminus \mathcal{Z}(\mathbb{I}(I)) = \{\alpha \in k^n : A_1(\alpha) = \cdots = A_r(\alpha) = 0 \neq I(\alpha)\}$.

Then $\mathcal{Z}(\mathrm{Sat}(\mathcal{A})) = \overline{\mathfrak{Z}(\mathcal{A})}$. ◉

---

[25] Throughout this chapter, I will preserve the language used in the previous volumes; so I will substitute the neologisms of *admissible Ritt* and *Lazard sequence* for the terminology introduced by Lazard and his students, which is in any case reported between parentheses.

[26] Not necessarily an ideal!

Proof. We have[27]

$$\mathcal{Z}(\mathrm{Sat}(\mathcal{A})) = \mathcal{Z}(\sqrt{(\mathcal{A}) : I}) = \overline{\mathcal{Z}(\sqrt{(\mathcal{A})} \setminus \mathcal{Z}(\{I\})} = \overline{\Im(\mathcal{A})}.$$

$\odot$

Corollary 42.3.5 (Ritt). *With the same notation and assumptions as in Lemma 42.3.4, let* $\mathsf{G} \subset k[X_1, \ldots, X_n]$ *be a finite set generating an ideal* I *such that* $\mathcal{A} \subset \mathsf{I}$ *and* $\mathsf{G} \subset \mathrm{Rem}(\mathcal{A})$. *Then:*

(1) $\mathsf{L} \subset \mathsf{I} \subseteq \mathrm{Rem}(\mathcal{A}) \subseteq \mathrm{Sat}(\mathcal{A})$;
(2) *if* I *is prime then* $\mathsf{L} = \mathsf{I} = \mathrm{Rem}(\mathcal{A}) = \mathrm{Sat}(\mathcal{A})$;
(3) $\mathcal{Z}(\mathrm{Sat}(\mathcal{A})) = \overline{\Im(\mathcal{A})}$;
(4) $\overline{\Im(\mathcal{A})} \subseteq \mathcal{Z}(\mathsf{I}) \subseteq \mathcal{Z}(\mathsf{L})$;
(5) $\mathcal{Z}(\mathsf{I}) = \Im(\mathcal{A}) \cup \mathcal{Z}(\mathsf{I} + \mathbb{I}(I_1)) \cup \cdots \cup \mathcal{Z}(\mathsf{I} + \mathbb{I}(I_r))$.

Proof.

(1) and (2) are just a reformulation of Theorem 42.1.18 and (3) is Lemma 42.3.4.
(4) $\overline{\Im(\mathcal{A})} = \mathcal{Z}(\mathrm{Sat}(\mathcal{A})) \subseteq \mathcal{Z}(\mathsf{I}) \subseteq \mathcal{Z}(\mathsf{L})$.
(5) For each $\alpha \in \mathcal{Z}(\mathsf{I})$,

- $\prod_{i=1}^{r} I_i(\alpha) \neq 0 \iff \alpha \in \Im(\mathcal{A})$,
- for some $i \leq r$, we have $I_i(\alpha) = 0 \iff \alpha \in \mathcal{Z}(\mathsf{I} + \{I_i\})$.

$\odot$

Corollary 42.3.6 (Wu). *With the same notation and assumptions as in Lemma 42.3.4 and Corollary 42.3.5, for any* $g \in k[X_1, \ldots, X_n]$, $g \in \mathsf{I}$ *iff*

the remainder of g *w.r.t.* $\mathcal{A}$ *is 0, and*
$g \in \mathsf{I} + \mathbb{I}(I_i)$ *for each* i, $1 \leq i \leq r$.

$\odot$

Example 42.3.7. Let

$$\mathsf{G} := \{X_1^6 - X_1^4, (X_1^4 - 2X_1^2)X_3, (X_1^2 - 1)X_2X_4 + X_3\} \subset k[X_1, X_2, X_3, X_4].$$

---

[27] The formula

$$\mathcal{Z}\left(\sqrt{(\mathcal{A}) : I}\right) = \overline{\mathcal{Z}\left(\sqrt{(\mathcal{A})}\right) \setminus \mathcal{Z}(\{I\})}$$

is a specialization of $\mathcal{Z}(\mathsf{I} : f) = \overline{\mathcal{Z}(\mathsf{I}) \setminus \mathcal{Z}(\{f\})}$ – where I is a radical polynomial ideal and f a polynomial in $k[X_1, \ldots, X_n]$, – whose proof is the following. For $g \in \mathsf{I} : f$ and $\alpha \in \mathcal{Z}(\mathsf{I}) \setminus \mathcal{Z}(\{f\})$ we have $0 = fg(\alpha) = f(\alpha)g(\alpha)$ and $f(\alpha) \neq 0$, so that $g(\alpha) = 0$. Therefore $\mathcal{Z}(\mathsf{I}) \setminus \mathcal{Z}(\{f\}) \subset \mathcal{Z}(\mathsf{I} : f)$.

Conversely, assume that g satisfies $g(\alpha) = 0$ for each $\alpha \in \mathcal{Z}(\mathsf{I}) \setminus \mathcal{Z}(\{f\})$, in other words, for each $\alpha \in \mathcal{Z}(\mathsf{I})$, $f(\alpha) \neq 0 \implies g(\alpha) = 0$; thus $fg(\alpha) = f(\alpha)g(\alpha) = 0$ for each $\alpha \in \mathcal{Z}(\mathsf{I})$. As a consequence $fg \in \sqrt{\mathsf{I}} = \mathsf{I}$ and $g \in \mathsf{I} : f$.

Thus $\{g \in k[X_1, \ldots, X_n] : g(\alpha) = 0, \alpha \in \mathcal{Z}(\mathsf{I}) \setminus \mathcal{Z}(\{f\})\} \subset (\mathsf{I} : f)$ and $\mathcal{Z}(\mathsf{I} : f) \subset \mathcal{Z}(\mathsf{I}) \setminus \mathcal{Z}(\{f\})$.

Since G is a characteristic set of the ideal it generates, according to Ritt's solver (Algorithm 42.3.1) we test the irreducibility of $X_1^6 - X_1^4$, discovering the decomposition

$$\mathcal{Z}(\mathsf{G}) = \mathcal{Z}(\mathsf{G} + \{X_1 - 1\}) \cup \mathcal{Z}(\mathsf{G} + \{X_1 + 1\}) \cup \mathcal{Z}(\mathsf{G} + \{X_1\});$$

the characteristic set of $\mathcal{Z}(\mathsf{G} + \{X_1 - 1\})$ being $\{X_1 - 1, X_3\}$, which is prime, we have found a root $(1, 0) \in k(X_2, X_4)^2$. In the same way $\mathcal{Z}(\mathsf{G} + \{X_1 + 1\})$ gives the root $(-1, 0) \in k(X_2, X_4)^2$; the characteristic set $\mathcal{A} := \{X_1, X_2 X_4 - X_3\}$ of $\mathcal{Z}(\mathsf{G} + \{X_1\})$ returns a root $(0, X_3/X_2) \in k(X_2, X_3)^2$ of the prime $(\mathcal{A}) = (\mathcal{A}) : X_2^\infty$ and the decomposition

$$\mathcal{Z}(\mathsf{G}) = \mathcal{Z}(\mathsf{G} + \{X_1 - 1\}) \cup \mathcal{Z}(\mathsf{G} + \{X_1 + 1\}) \cup \mathcal{Z}((\mathcal{A}) : X_2^\infty) \cup \mathcal{Z}(\mathcal{A} + \{X_2\}).$$

The characteristic set $\mathcal{B} := \{X_1, X_2, X_3\}$ of $\mathcal{A} + \{X_2\})$ is prime and gives the root $(0, 0, 0) \in k(X_4)^3$. So, in conclusion, we have found the (redundant) prime decomposition

$$(\mathsf{G}) = (X_1 - 1, X_3) \cap (X_1 + 1, X_3) \cap (X_1, X_2 X_4 - X_3) \cap (X_1, X_2, X_3)$$

and the manifold decomposition

$$\mathcal{Z}(\mathsf{G}) = \{(1, a, 0, b) : a, b \in k\} \cup \{(-1, a, 0, b) : a, b \in k\}$$
$$\cup \{(0, a, b, b/a) : a, b \in k, a \neq 0\} \cup \{(0, 0, 0, b) : b \in k\}.$$

Wu's solver (Algorithm 42.3.3) instead returns the root decompositions

$$\mathcal{Z}(\mathsf{G}) = 3(\mathsf{G}) \cup \mathcal{Z}(\mathsf{G} + \{X_1^4 - 2X_1^2\}) \cup \mathcal{Z}(\mathsf{G} + \{X_1^2 - 1\}),$$
$$\mathcal{Z}(\mathsf{G} + \{X_1^4 - 2X_1^2\}) = 3(\mathcal{C}_1) \cup \mathcal{Z}(\mathcal{C}_1 + \{X_2\}),$$
$$\mathcal{Z}(\mathcal{C}_1 + \{X_2\}) = 3(\mathcal{C}_3),$$
$$\mathcal{Z}(\mathsf{G} + \{X_1^2 - 1\}) = 3(\mathcal{C}_2),$$

where

$$\mathcal{C}_1 = \{X_1^2, X_2 X_4 - X_3\},$$
$$\mathcal{C}_2 = \{X_1^2 - 1, X_3\},$$
$$\mathcal{C}_3 = \{X_1^2, X_2, X_3\},$$

and

$$3(\mathsf{G}) = \mathcal{Z}(\mathsf{G}) \setminus \mathcal{Z}(\{(X_1^4 - 2X_1^2)(X_1^2 - 1)\}) = \varnothing,$$
$$3(\mathcal{C}_1) = \mathcal{Z}(\mathcal{C}_1) \setminus \mathcal{Z}(\{X_2\}) \qquad = \{(0, a, b, b/a) : a, b \in k, a \neq 0\},$$
$$3(\mathcal{C}_2) = \mathcal{Z}(\mathcal{C}_2) \qquad = \{(1, a, 0, b), (-1, a, 0, b) : a, b \in k\}.$$
$$3(\mathcal{C}_3) = \mathcal{Z}(\mathcal{C}_3) \qquad = \{(0, 0, 0, b) : b \in k\}.$$

42.4　Lazard: Triangular Sets

Preserving the same notation as in Sections 42.2 and 42.3, computational considerations suggested to relax as follows the notion of reduction.

Definition 42.4.1 (Moreno–Maza). *For any two polynomials $A_1, A_2 \in k[X_1, \ldots, X_n]$, where A_1 is of class $p > 0$, A_2 is said to be* initially reduced *w.r.t. A_1 if its initial is of lower degree in X_p than that of A_1.* ◉

Definition 42.4.2 (Aubry et al.).　*A finite non-empty set*

$$\{A_1, \ldots, A_r\} \subset k[X_1, \ldots, X_n]$$

is called

> *a* triangular set *if each A_i is of positive class and there are no two elements having the same class,*
> *an* initially reduced triangular set *if*
> - *A_1 is of positive class,*
> - *for each $j > i$, A_j is of higher class than A_i and*
> - *A_j is initially reduced w.r.t A_i,*
>
> *a* fine triangular set *if*
> - *A_1 is of positive class,*
> - *for each $j > i$, A_j is of higher class than A_i,*
> - *for each j, the remainder of I_j w.r.t. $\{A_1, \ldots, A_{j-1}\}$ is non-zero.* ◉

Remark 42.4.3 (Lazard).　In connection with this generalization, it is easy to realize that all the results stated by Ritt for chains and characteristic sets hold *verbatim* for any triangular set, in a way not dissimilar to that for Buchberger's reduction, where the absence of inter-reduction of the basis does not affect the result on normal-form zero-testing.

This also justifies the relaxation of the notion of reduction.

In particular the notion of a *fine triangular set* allows one to avoid inter-reduction while preserving the degree in X_p of an element of class p and so the rank relation between members of triangular sets. ◉

Definition 42.4.4 (Aubry et al.).　*Let $G \subset k[X_1, \ldots, X_n] \setminus \{0\}$ be a finite set generating the ideal I. A finite non-empty triangular set*

$$\mathcal{A} := \{A_1, \ldots, A_r\} \subset k[X_1, \ldots, X_n]$$

is called:

> *an* admissible Ritt sequence *(or* Ritt characteristic set*) of G if $\mathcal{A} \prec \mathcal{B}$ for each fine triangular set $\mathcal{B} \subset G$;*
> *a* strong admissible Ritt sequence *(or* Wu characteristic set*) of G if there exists a finite set $G^* \subset I$ such that $\mathbb{I}(G^*) = I$ and $G^* \subseteq \mathrm{Rem}(\mathcal{A})$.* ◉

Historical Remark 42.4.5.   Clearly, the notion of the Ritt characteristic set is an elementary adaptation to triangular sets of the notion of the characteristic set (Definition 42.1.11) and that of the Wu characteristic set is a precise description of the particular triangular sets which are produced by Algorithm 42.2.4; both ideas, therefore are explicitly present in Ritt's theory and none is related to Wu's results.

Indeed, Wu's results are not related to the notion and properties of characteristic sets; as discussed in Historical Remark 42.3.2, his relevant contributions consist in using this theory not as a solving tool but as a tool for testing membership (Corollary 42.3.6), and, in this context, to relax the irrelevant requirement of primality.

It is Lazard[28] who, within the Kronecker–Duval philosophical frame discussed in Volume I and of which he was one of the main advocates, suggested the reconsideration of Ritt's solver in Wu's relaxed context and introduced the notion of "triangular sets", thus greatly improving the old notion of "solving" as used by Kronecker, Macaulay, Gröbner and Ritt and presented in Section 34.5.   ⊙

In this setting, let us now consider

- a triangular set $\mathcal{A} := \{A_1, \ldots, A_r\} \subset k[X_1, \ldots, X_n]$ generating the ideal L,
- I_j, the initial of A_j, for each j,
- $I := \prod_{i=1}^{r} I_i$,
- $\mathrm{Rem}(\mathcal{A})$, the set of all polynomials whose remainder w.r.t. $\mathcal{A}$ is 0,
- $\mathrm{Sat}(\mathcal{A}) := \mathsf{L} : I^{\infty}$,
- $\mathfrak{Z}(\mathcal{A}) := \mathcal{Z}(\mathcal{A}) \setminus \mathcal{Z}(\{I\}) = \{\alpha \in \mathsf{k}^n : A_1(\alpha) = \cdots = A_r(\alpha) = 0 \neq I(\alpha)\}$,
- $\mathsf{G} \subset k[X_1, \ldots, X_n]$, a finite set generating an ideal I such that $\mathcal{A} \subset \mathsf{I}$ and $\mathsf{G} \subset \mathrm{Rem}(\mathcal{A})$.

Corollary 42.4.6.   *With the present notation, if $\mathcal{A}$ is a strong admissible Ritt sequence of G then:*

(1) $\mathsf{L} \subset \mathsf{I} \subseteq \mathrm{Rem}(\mathcal{A}) \subseteq \mathrm{Sat}(\mathcal{A})$;
(2) *if I is prime then* $\mathsf{L} = \mathsf{I} = \mathrm{Rem}(\mathcal{A}) = \mathrm{Sat}(\mathcal{A})$;
(3) $\mathcal{Z}(\mathrm{Sat}(\mathcal{A})) = \overline{\mathfrak{Z}(\mathcal{A})}$;
(4) $\overline{\mathfrak{Z}(\mathcal{A})} \subseteq \mathcal{Z}(\mathsf{I}) \subseteq \mathcal{Z}(\mathsf{L})$;
(5) $\mathcal{Z}(\mathsf{I}) = \mathfrak{Z}(\mathcal{A}) \cup \mathcal{Z}(\mathsf{I} + \mathbb{I}(I_1)) \cup \cdots \cup \mathcal{Z}(\mathsf{I} + \mathbb{I}(I_r))$;
(6) *for each $g \in k[X_1, \ldots, X_n]$,*

$$g \in \mathsf{I} \iff g \in \mathrm{Rem}(\mathcal{A}) \cap (\mathsf{I} + \mathbb{I}(I_1)) \cap \cdots \cap (\mathsf{I} + \mathbb{I}(I_r)).$$

⊙

---

[28] In

Lazard D., Solving zero-dimensional algebraic systems, *J. Symb. Comp.* **15** (1992), 117–132.

Lazard D., A new method for solving algebraic systems of positive dimension, *Disc. Appl. Math.* **33** (1991), 147–160.

Lazard D. Systems of algebraic equations (algorithms and complexity), *Symp. Math.* **34** (1993), 84–105.

Proposition 42.4.7. *If, with the present notation, $\mathcal{A}$ is fine then the following conditions are equivalent:*

(1) *$\mathcal{A}$ is an admissible Ritt sequence of* G*;*
(2) $\mathsf{I} \subset \mathrm{Rem}(\mathcal{A})$.

Moreover this implies that

(3) *$\mathcal{A}$ is a strong admissible Ritt sequence of* G*.*

Proof. This is just a reformulation of Lemma 42.1.12. 

42.5 Admissible Lazard Sequence

Lazard reconsidered the notion of triangular sets in the same framework as that used for admissible sequences (Definition 8.2.2) and admissible Duval sequences (Section 11.4).

Let us begin by explicitly interpreting the field k as a quotient field $L_0 := k$ of some domain R_0, that is, we assume that we are given a domain R_0 and a multiplicative system $S_0 \subset R_0$ such that[29]

$$k =: L_0 := \left\{ \frac{a}{b} : a \in R_0, b \in S_0 \right\}.$$

Then let us consider a triangular set[30]

$$\mathcal{A} := \{f_1, \ldots, f_r\} \subset R_0[X_1, \ldots, X_n],$$

where, by definition, we can wlog assume that

$$0 < \mathrm{class}(f_1) < \cdots < \mathrm{class}(f_i) < \mathrm{class}(f_{i+1}) < \cdots < \mathrm{class}(f_r).$$

We also set $d_i := \deg_j(f_i)$, where $j := \mathrm{class}(f_i)$.

We can now partition the variables into "parameters" (or "independent indeterminates") and the rest:

Definition 42.5.1 (Lazard). *A variable X_j is called*

algebraic *for $\mathcal{A}$ if $j = \mathrm{class}(f_i)$ for some $f_i \in \mathcal{A}$ which is said to* introduce X_j*;*[31]
transcendental *for $\mathcal{A}$ if $j \neq \mathrm{class}(f_i)$ for each $f_i \in \mathcal{A}$.*

---

[29] We can for instance take $R_0 := L_0 := k$ and $S_0 := \{1\}$, but for $k := \mathbb{Q}$ we could consider instead $R_0 := \mathbb{Z}$, $S_0 := \mathbb{N} \setminus \{0\}$.

    The reason why we choose to represent the field elements as explicit fractions with denominators in a restricted chosen multiplicative system S_0 is in order to force uniqueness; see the next note.

[30] Each element f_i is chosen with coefficients not in L_0 but in R_0; as a consequence among all possible associated polynomials we can restrict our choice to those such that their leading coefficient is in S_0. If we take $R_0 := L_0 := k$ and $S_0 := \{1\}$ this simply means we require that each f_i is monic.

[31] This concept is present in Ritt J.F., *op. cit.*, p. 34. See footnote 17.

As usual we relabel the variables as

$$k[X_1, \ldots, X_n] \cong k[V_1, \ldots, V_d, Z_1, \ldots, Z_r],$$

so that $\{V_1, \ldots, V_d\}$ (respectively, $\{Z_1, \ldots, Z_r\}$) is the set of transcendental (respectively, algebraic) variables for $\mathcal{A}$.

Then we can recursively define, for each j:

- i, the value such that $\mathcal{A} \cap R_0[X_1, \ldots, X_j] = \{f_1, \ldots, f_i\}$ or, equivalently, the maximal value for which $\mathrm{class}(f_i) \leq j$,
- δ, the value such that $\{V_1, \ldots, V_\delta\} = \{X_1, \ldots, X_j\} \cap \{V_1, \ldots, V_d\}$,
- $R_j := R_0[X_1, \ldots, X_j]/(f_1, \ldots, f_i)$,
- $S_j := \{a \in R_j : a \text{ is not a zero-divisor}\}$;
- L_j, the quotient ring $L_j := \{a/b : a \in R_j, b \in S_j\}$;
- $\pi_j := R_0[X_1, \ldots, X_j] \to R_j$ and $\pi_j := L_0[X_1, \ldots, X_j] \to L_j$, the canonical projections.

In particular, for each j:

- if $X_j = Z_i$ is algebraic, so that it is introduced by f_i and $j = \mathrm{class}(f_i)$, we have
  - $R_j = R_{j-1}[X_j]/\pi_{j-1}(f_i) \cong R_0[X_1, \ldots, X_j]/(f_1, \ldots, f_i)$,
  - $S_j = S_{j-1} = S_0[V_1, \ldots, V_\delta]$,
  - $L_j = L_{j-1}[X_j]/\pi_{j-1}(f_i) \cong L_0(V_1, \ldots, V_\delta)[Z_1, \ldots, Z_i]/(f_1, \ldots, f_i)$;
- if $X_j = V_\delta$ is instead transcendental, we have
  - $R_j = R_0[X_1, \ldots, X_j]/(f_1, \ldots, f_i) \cong R_{j-1}[X_j]$,
  - $S_j = S_{j-1}[X_j] = S_0[V_1, \ldots, V_\delta]$,
  - $L_j = L_{j-1}(X_j) \cong L_0(V_1, \ldots, V_\delta)[Z_1, \ldots, Z_i]/(f_1, \ldots, f_i)$.

Definition 42.5.2.   *We say that*

$$\mathcal{A} := \{f_1, \ldots, f_r\} \subset R_0[X_1, \ldots, X_n]$$

is an admissible Lazard sequence *if, for each h, $1 \leq h \leq r$, setting $j := \mathrm{class}(f_h)$, it holds that:*

(i) (triangular) $\mathrm{class}(f_h) > 0$ *and* $\mathrm{class}(f_h) > \mathrm{class}(f_i)$ *for each $i < h$;*[32]

(ii) (reduced)   *the degree of f_h in the algebraic variable $X_j = Z_i$ is strictly less than that of f_i, that is,* $\deg_j(f_h) < d_i$ *for each $i < h$;*

(iii) (normalized) $\mathbf{T}_<(f_h) \in k[V_1, \ldots, V_d][Z_h]$ *(compare Remark 42.2.2);*

(iv) (R_0-normalized) $\mathrm{lc}(f_h) \in S_0$;

(v) (squarefree) $\mathrm{Res}(\pi_{j-1}(f_h), \pi_{j-1}(f'_h)) \in L_{j-1}$ *is invertible;*[33]

(vi) (primitive) $\mathrm{Cont}(\pi_{j-1}(f_h)) = 1$ *in* $L_{j-1}[X_j]$.

---

[32] As a consequence $\mathcal{A}$ is a triangular set; we therefore use the same notation as above, denoting as $(L_0, \ldots, L_n)$ and $(\pi_1, \ldots, \pi_n)$ the corresponding fields and projections.

[33] As a consequence $\pi_{j-1}(f_h)$ is squarefree.

The sequence $\mathcal{A}$ is called a weak admissible Lazard sequence *(or a regular set) whose associated map is $(\pi_1, \ldots, \pi_n)$ and whose associated tower of simple extensions is $(L_0, \ldots, L_n)$ if it satisfies only conditions* (i)–(iv). ◎

Definition 42.5.3 (Lazard). *Let*

> $\mathcal{A} := \{A_1, \ldots, A_r\} \subset k[X_1, \ldots, X_n]$ *be an admissible Lazard sequence,*
> L *be the ideal generated by $\mathcal{A}$,*
> I_j *be the initial of A_j, for each j,*
> $I := \prod_{i=1}^{r} I_i.$

Then the set

$$3(\mathcal{A}) := \mathcal{Z}(\mathcal{A}) \setminus \mathcal{Z}(\{I\}) = \left\{\alpha \in k^n : A_1(\alpha) = \cdots = A_r(\alpha) = 0 \neq I(\alpha)\right\}$$

is called the quasi-component *associated with $\mathcal{A}$, and the ideal*

$$\mathrm{Sat}(\mathcal{A}) := \mathsf{L} : I^\infty := \mathsf{H}$$

is called the quasi-prime ideal *associated with $\mathcal{A}$.* ◎

Remark 42.5.4 (Lazard). Condition (iii) requires (compare Remark 42.2.2) that, for each h and i, denoting $j := \mathrm{class}(\mathrm{Lp}_i(f_h))$ we have $X_j \in \{V_1, \ldots, V_d\}$. If this is not the case and $X_j = Z_\iota$ for some ι then, either

- $\gcd(\mathrm{Lp}_i(f_h), f_\iota) \neq 1$ and we can find a partial factorization of f_ι, or
- $\gcd(\mathrm{Lp}_i(f_h), f_\iota) = 1 = s\,\mathrm{Lp}_i(f_h) + tf_\iota$ for suitable polynomials $s, t \in k[X_1, \ldots, X_j]$, so that $F_h := sf_h$ is such that $\mathrm{class}(\mathrm{Lp}_i(F_h)) < \mathrm{class}(\mathrm{Lp}_i(f_h))$. ◎

Remark 42.5.5. If either

- the ideal $\mathbb{I}(\mathcal{A}) \subset k[X_1, \ldots, X_n]$ is zero-dimensional and we choose $R_0 := L_0 := k$, $S_0 := \{1\}$, or
- we restrict our considerations to its extension $\mathbb{I}(\mathcal{A})k(V_1, \ldots, V_d)[Z_1, \ldots, Z_r]$ and we choose $R_0 := L_0 := k(V_1, \ldots, V_d)$, $S_0 := \{1\}$

then in both cases all variables are algebraic and a set $\mathcal{A} := \{f_1, \ldots, f_r\}$, where each f_i is chosen monic, is an admissible Lazard sequence iff it is an admissible Duval sequence.

In fact, if $\mathcal{A}$ is an admissible Lazard sequence then (v) allows us to deduce that, since $\mathrm{Res}(f_1, f_1') \in L_0$ (Proposition 6.6.4), f_1 is squarefree in $L_0[X_1]$ and L_1 is a Duval field (Definition 11.4.2) and, inductively, that

- L_{j-1} is a direct sum of fields $L_{j-1} = \oplus_\iota L_{j-1\kappa}$ so that, denoting by $\pi_{j-1\kappa} : L_{j-1} \to L_{j-1\kappa}$ the canonical projection,
- condition (v), $\mathrm{Res}(\pi_{j-1}(f_j), \pi_{j-1}(f_j)') \in L_{j-1}$, implies that

$$\mathrm{Res}(\pi_{j-1\kappa}\pi_{j-1}(f_j), \pi_{j-1\kappa}\pi_{j-1}(f_j)') \in L_{j-1\kappa}$$

- and that each $\pi_{j-1\,\kappa}\pi_{j-1}(f_h)$ is squarefree in $L_{j-1\,\kappa}[Z_j]$,
- so that L_j is a Duval field.

Conversely, (i) and (ii) are satisfied by any admissible sequence and (v) by an admissible Duval sequence, (iii) and (iv) are equivalent to the requirement that the f_i are monic and (vi) is trivially satisfied since we are assuming that $S_0 := \{1\}$. ⊙

Remark 42.5.6 (Lazard). The relation to admissible Ritt sequences is thus expounded by Lazard:[34]

The notion of [an admissible Lazard sequence] is stronger than the notion of characteristic set in the Ritt–Wu Wen-Tsün method: Characteristic sets are only subject to conditions (i) and (ii); but we will see that this is not sufficient; in particular a characteristic set may correspond to an empty "component" of the zero-set. This is avoided by condition (iii). The conditions (iv) to (vi) are needed in order to obtain the uniqueness of the triangular set associated with a quasi-component.

and[35]

Wu Wen-Tsün's algorithm, like Buchberger's one, depends on many choices; moreover, the result of Wu Wen-Tsün's algorithm is not uniquely determined. [...] Thus there is a need for a more canonical algorithm, that is an algorithm in which the result (or even better the intermediate results) is more intrinsic, that is, depends on the algebraic structure of the input and not on the algorithm itself. This definition of "intrinsic" is rather imprecise; it may be better understood by considering the example of an algorithm which is intrinsic, namely the subresultant algorithm, which has the property that the coefficients of the successive remainders may be defined as subdeterminants of [the] Sylvester matrix.
[...]
For getting a canonical result, [Lazard] strengthened the definition of a triangular set by asking that the polynomials in it are squarefree, primitive and monic in some technical sense. With these conditions, the set of the solutions of an algebraic system is uniquely decomposed in so called *quasi-components* which are themselves in one to one correspondence with these strengthened triangular systems. ⊙

Historical Remark 42.5.7. The existence in Wu's solver of empty components

$$3(\mathcal{A}) := \mathcal{Z}(\mathcal{A}) \setminus \mathcal{Z}(\mathbb{I}(I)) = \big\{\alpha \in \mathsf{k}^n : A_1(\alpha) = \cdots = A_r(\alpha) = 0 \neq I(\alpha)\big\}$$

is illustrated by Example 42.3.7 where, for the characteristic set

$$\mathsf{G} := \{f_1, f_2, f_3\} := \{X_1^6 - X_1^4, (X_1^4 - 2X_1^2)X_3, (X_1^2 - 1)X_2X_4 + X_3\},$$

we have

$$I = \prod_{i=1}^{3} I_i = (X_1^4 - 2X_1^2)(X_1^2 - 1) = \sqrt{f_1}(X_1^3 - 2X_1),$$

so that $3(\mathsf{G}) = \emptyset$.

[34] Lazard D., A new method for solving algebraic systems of posisitive dimension, *Disc. Appl. Math.* **33** (1991), p.151.

[35] Lazard D. Systems of algebraic equations (algorithms and complexity), in *Symposia Mathematica* **34** (1993), pp. 84–105, Cambridge University Press.

This of course cannot happen in the old-fashioned Ritt solver, where each "solution" is the characteristic set of a prime. As I have already mentioned in Historical Remark 42.4.5, Lazard's contribution consists in relaxing the notion of "characteristic sets" in order to avoid factorization while preserving both the general structure and the relevant properties.

In order to reach this result, it is clearly sufficient to impose condition (iii), which, as we observed in Remark 42.5.4, can be forced by just applying the Extended Euclidean Algorithm to $\mathrm{Lp}_i(f_h)$ and f_t in all cases in which $X_j = Z_t$ for $j := \mathrm{class}(\mathrm{Lp}_i(f_h))$.

The comparison with the notion of "solving" discussed in Section 34.5 is striking: what Lazard did was simply to substitute *regular sets* for admissible sequences, *quasi-primes* for primes and *quasi-components* for irreducible varieties!   ◉

Theorem 42.5.8 (Aubry *et al.*). *Let*

$$\mathcal{A} := \{f_1, \ldots, f_r\} \subset R_0[X_1, \ldots, X_n]$$

be a weak admissible Lazard sequence whose associated map is $(\pi_1, \ldots, \pi_n)$ and whose associated tower of simple extensions is $(L_0, \ldots, L_n)$. Then, for each $p \in R_0[X_1, \ldots, X_n]$, the following conditions are equivalent:

(1) $\pi_n(p) = 0,$
(2) $p \in \mathrm{Rem}(\mathcal{A}),$
(3) $p \in \mathrm{Sat}(\mathcal{A}).$

Proof. The proof is by induction on n. We begin by remarking that if we call π_0 the identity on R_0 and set $\mathcal{A} := \emptyset$ then the statements become

(1) $p = \pi_0(p) = 0,$
(2) $p \in \mathrm{Rem}(\emptyset),$
(3) $p \in \mathrm{Sat}(\emptyset) = \{0\},$

which are obviously equivalent.

So, we can assume that $n > 0$ and that the theorem holds for $R_0[X_1, \ldots, X_{n-1}]$, and we denote $\mathcal{S} := \mathrm{Sat}(\mathcal{A}) \subset R_0[X_1, \ldots, X_{n-1}]$.

If X_n is transcendental, for each $p = \sum_{i=0}^{d} a_i X_n^i \in R_0[X_1, \ldots, X_{n-1}][X_n]$ the result follows by induction, since we have

(1) $\pi_n(p) = 0 \iff \pi_{n-1}(a_i) = 0$ for each i,
(2) $p \in \mathrm{Rem}(\mathcal{A}) \iff a_i \in \mathrm{Rem}(\mathcal{A})$ for each i,
(3) $p \in \mathrm{Sat}(\mathcal{A}) \iff a_i \in \mathcal{S}$ for each i.

If, instead, $X_n = Z_i$ is algebraic, denote by r the remainder of p w.r.t. f_i and express it as $r = \sum_{l=0}^{d} a_l X_n^l \in R_0[X_1, \ldots, X_{n-1}][X_n]$. Then the result is a consequence of the following claims:

(1) $\pi_n(p) = 0 \iff \pi_n(r) = 0;$
(2) $\pi_n(r) = 0 \iff \pi_{n-1}(r) = 0;$

(3) $\pi_{n-1}(r) = 0 \iff \pi_{n-1}(a_l) = 0$ for each l;
(4) $r \in \mathrm{Rem}(\mathcal{A}) \iff a_l \in \mathrm{Rem}(\mathcal{A})$ for each l;
(5) $\pi_{n-1}(r) = 0 \iff r \in \mathrm{Rem}(\mathcal{A})$;
(6) $\pi_n(p) = 0 \iff p \in \mathrm{Rem}(\mathcal{A})$;
(7) $\pi_n(p) = 0 \iff p \in \mathrm{Sat}(\mathcal{A})$.

The proofs of the above are as follows.

(1) By definition $\pi_n(\mathrm{Lp}(f_i)) = \pi_{n-1}(\mathrm{Lp}(f_i)) \in L_n$ is a unit, while $\pi_n(f_i) = 0$; therefore, from $r = \mathrm{Lp}(f_i)^w p + q f_i$ we have

$$\pi_n(r) = \pi_n(\mathrm{Lp}(f_i))^w \pi_n(p) + \pi_n(q)\pi_n(f_i) = \pi_n(\mathrm{Lp}(f_i))^w \pi_n(p),$$

whence the claim.
(2) Since $\pi_n(r) = 0 \iff \pi_{n-1}(r) \in (\pi_{n-1}(f_i))$, the claim follows because $\deg_n(r) < \deg_n(f_i)$ and $\pi_{n-1}(\mathrm{Lp}(f_i)) \in L_n$ is a unit.
(3) Obvious.
(4) Obvious.
(5) This holds by the inductive assumption.
(6) This holds by the list of implications and by the fact that the remainders of r and p are the same.
(7) Since $\mathrm{Rem}(\mathcal{A}) \subset \mathrm{Sat}(\mathcal{A})$, it is sufficient to prove that $p \in \mathrm{Sat}(\mathcal{A}) \implies \pi_n(p) = 0$.

We have $I^m p \in \mathbb{I}(\mathcal{A})$, where m is a suitable integer and I is the product of all initials of the elements in $\mathcal{A}$.

Clearly $\pi_n(I)$ is a unit, so that $\pi_n(I^m p) = 0$ implies $\pi_n(p) = 0$.

$\square$

Theorem 42.5.9. *Let* $\mathcal{A} := \{A_1, \ldots, A_r\} \subset k[X_1, \ldots, X_n]$ *be a triangular set generating the ideal* L.
Then the following conditions are equivalent:

(1) *$\mathcal{A}$ is a weak admissible Lazard sequence;*
(2) *$\mathcal{A}$ is an admissible Ritt sequence of* $\mathrm{Sat}(\mathcal{A})$;
(3) *$\mathrm{Sat}(\mathcal{A}) = \mathrm{Rem}(\mathcal{A})$.*

Proof.
(1) $\implies$ (3) is Theorem 42.5.8.
(2) $\iff$ (3) is a consequence of Lemma 42.1.12.
(3) $\implies$ (1) Assume that (3) holds while (1) does not. Inductively we can assume that $\mathcal{B} := \mathcal{A} \cap R_0[X_1, \ldots, X_{n-1}]$ is a weak admissible Lazard sequence whose associated map is $(\pi_1, \ldots, \pi_{n-1})$ and whose associated tower of simple extensions is $(L_0, \ldots, L_{n-1})$.

Clearly $X_n = Z_r$ is algebraic and, since we are assuming that (1) does not hold, the initial I_r of A_r must be such that $\pi_{n-1}(I_r)$ is a zero divisor in L_{n-1}, so there is a $p \in k[X_1, \ldots, X_{n-1}]$ such that $\pi_{n-1}(I_r p) = 0$ and $\pi_{n-1}(p) \neq 0$.

Therefore $I_r p \in \mathrm{Sat}(\mathcal{B})$ and the remainder r of p w.r.t. $\mathcal{A}$ is non-zero. Clearly we also have $I_r r \in \mathrm{Sat}(\mathcal{B})$ and $r \in \mathrm{Sat}(\mathcal{A})$. Thus $r \in k[X_1, \ldots, X_{n-1}]$ satisfies $r \in \mathrm{Sat}(\mathcal{A})$ and $r \notin \mathrm{Rem}(\mathcal{A})$, giving the required contradiction. $\boxed{\odot}$

An ideal $[\mathrm{Sat}(\mathcal{A})]$ which is the saturated ideal of a [weak admissible Lazard sequence $\mathcal{A} = \{A_1, \ldots, A_r\}$] is said [to be] *triangularizable*; it is always equi-dimensional of dimension $n - |r|$, where n is the number of variables and $|r|$ the length of $[\mathcal{A}]$. A prime ideal is always triangularizable, and the primes associated to a triangularizable ideal are simply obtained by factoring recursively each $|A_i|$ in the field extensions defined by the factors of $[A_1, \ldots, A_{i-1}]$.

It should be remarked here that [weak admissible Lazard sequences] are a good alternative to Gröbner bases for representing triangularizable and prime ideals in computers: the number of polynomials in a triangular set is always bounded by the number of variables, which is not the case for generating sets of Gröbner bases of prime ideals. Computing a Gröbner basis from a triangular set may be done by any Gröbner base algorithm, and is usually not too difficult. The inverse transformation is very easy for prime ideals.[36]

Actually, the standard way for a complete resolution of a polynomial system consists in the following scheme.
1　Compute a lex Gröbner base, either directly or through a change [of] base ordering. This step checks zero-dimensionality.
2　Deduce from it a set of [weak admissible Lazard sequences].
3　For each of these triangular systems, compute a RUR [Rational Univariate Representation].
4　For each RUR compute a numerical approximation of the solutions together with a bound of the error.[37]

In the sections that follow we will discuss efficient algorithms to compute triangular sets; the first, due to Lazard, returns a weak admissible Lazard sequence and applies also in the non-zero-dimensional case; the second, by Möller, requires zero-dimensionality. In the last section we discuss the notion of an RUR and the related algorithms.

42.6　Lazard's Solver

Let us begin by remarking that, since admissible Lazard sequences and admissible Duval sequences coincide, arithmetical operations in each member L_i of a tower of simple extensions can be performed *à la* Duval.

In particular:

- when $\pi_j(p) \in L_j$, where $p \in k[X_1, \ldots, X_j] \setminus k[X_1, \ldots, X_{j-1}]$ and $X_j = Z_i$ is algebraic, is a *zero-divisor*, it is sufficient to compute, in $L_{j-1}[Z_i]$, $f' := \gcd(\pi_{j-1}(p), \pi_{j-1}(f_i))$ and $f'' := \pi_{j-1}(f_i)/f'$ in order to obtain a Duval splitting

$$L_j \cong (L_{j-1}[Z_i]/f') \oplus (L_{j-1}[Z_i]/f''),$$

[36] Lazard D., Resolution of polynomial systems, in *Proc. ASCM 2000*, World Scientific (2000), pp. 1–8.
[37] Lazard D., On the specification for solvers of polynomial systems, in *Proc. ASCM 2001*, World Scientific (2001), pp. 1–10.

where, denoting as $\pi' : L_j \to L_{j-1}[Z_i]/f'$ and $\pi'' : L_j \to L_{j-1}[Z_i]/f''$ the canonical projections, we have that $\pi'\pi_j(p) = 0$ and $\pi''\pi_j(p)$ is invertible;

- testing the invertibility of $\pi_j(p) \in L_j$ for a polynomial

$$p \in k[X_1, \ldots, X_j] \setminus k[X_1, \ldots, X_{j-1}]$$

consists in testing the invertibility of
- $\pi_{j-1}(\mathrm{Lp}(p)) \in L_{j-1}$ if X_j is transcendental,
- $\mathrm{Res}(\pi_{j-1}(f_h), \pi_{j-1}(f_i)) \in L_{j-1}$ if $X_j = Z_i$ is algebraic;

- computing the inverse of an invertible element $\pi_j(p) \in L_j$, where $p \in k[X_1, \ldots, X_j] \setminus k[X_1, \ldots, X_{j-1}]$ and $X_j = Z_i$ is algebraic, in principle requires one to compute $\gcd(p, f_i)$ in $L_{j-1}[X_j]$, but[38]
two difficulties arise: the first one is that the Euclidean algorithm and its generalizations are defined only for polynomials on integer rings. Fortunately, [Duval's Model] permits us to compute as if the coefficients were in a field if we split when we encounter a zero-divisor.

The second difficulty is to decide which Euclidean algorithm to use: the coefficients being polynomials, an elementary algorithm will generate a swell of coefficients; thus we have to use the subresultant algorithm;[39] but it needs exact quotients which are not well defined in our context.

We suggest the following approach: apply the subresultant algorithm to the input viewed as multivariate polynomials in $[k[X_1, \ldots, X_j]]$; reduce the subresultants, starting from low degrees; the first which does not reduce to zero reduces to a factor of $[f_i]$ viewed as a polynomial [in $L_{j-1}[X_j] = L_{j-1}[Z_i]$];

- Condition (vi) requires one to perform gcd computations over a Duval field $L_{j-1}[X_j] = L_{j-1}[Z_i]$; however, condition (iv) implies that a coefficient of f_i is a member of $k[V_1, \ldots, V_d]$, and so no splitting occurs.

The central procedure of Lazard's Solver is an algorithm **intersect**$(p, \mathcal{A})$, where $p \in k[X_1, \ldots, X_n]$ and $\mathcal{A} \subset k[X_1, \ldots, X_n]$ is an admissible Lazard sequence, and

---

[38] Lazard D., A new method for solving algebraic systems of positive dimension, *Disc. Appl. Math.* **33** (1991), p. 154.

[39] That is, the version of the Euclidean Algorithm proposed by Collins and Brown and briefly discussed in Example 1.6.1 and Historical Remark 1.6.2.

It consists, given two polynomials $P_0, P_1 \in D[X]$, where D is a domain, in producing, by means of a pseudo-division algorithm, a polynomial remainder sequence

$$P_0, P_1, \ldots, P_r = \gcd(P_0, P_1) \in D[X]$$

which satisfies the relations

$$P_{i+2}/c_i = b_i P_i - Q_{i+1} P_{i+1}$$

for suitable $Q_{i+1} \in D[X]$ and $b_i \in D$, and elements $c_i \in D$ that can be predicted. These data also allow us to compute (essentially as in Proposition 1.3.1) polynomials S_i, T_i satisfying Bezout's identities $P_i = P_0 S_i + P_1 T_i$.

In the quoted passage, the proposed approach is to apply the algorithm to $\pi_{j-1}(f_i)$ and $\pi_{j-1}(p)$ in $L_{j-1}[X_j] = L_{j-1}[Z_i]$, where $p \in k[X_1, \ldots, X_j] \setminus k[X_1, \ldots, X_{j-1}]$ and $X_j = Z_i$ is algebraic. The technical problem is that the computation requires zero-testing; the proposal solution requires one to
- compute a PRS $f_i, p, P_2, \ldots, P_r$ of f_i and p in $k[X_1, \ldots, X_{j-1}][X_j]$,
- evaluate $\pi_{j-1}(P_r), \pi_{j-1}(P_{r-1}), \ldots$ until a non-zero element $\pi_{j-1}(P_\rho)$ is produced, which therefore satisfies

$$\pi_{j-1}(P_\rho) = \gcd(\pi_{j-1}(f_i), \pi_{j-1}(p)) \in L_{j-1}[X_j] = L_{j-1}[Z_i].$$

whose output is a finite family $\mathfrak{B} := \{\mathcal{B}_1, \ldots, \mathcal{B}_l\}$ of admissible Lazard sequences which satisfy

$$\mathcal{Z}(\mathbb{I}(p)) \cap \mathfrak{Z}(\mathcal{A}) \subseteq \bigcup_{i=1}^{l} \mathfrak{Z}(\mathcal{B}_i) \subseteq \overline{\mathcal{Z}(\mathbb{I}(p)) \cap \mathfrak{Z}(\mathcal{A})}.$$

Given a finite family $\mathfrak{A} := \{\mathcal{A}_1, \ldots, \mathcal{A}_l\}$ of admissible Lazard sequences we write

$$\mathbf{intersect}(p, \mathfrak{A}) := \bigcup_{i=1}^{l} \mathbf{intersect}(p, \mathcal{A}_i).$$

Finally, given a finite set

$$\mathsf{G} := \{g_1, \ldots, g_m\} \subset k[X_1, \ldots, X_n],$$

the procedure

$$\mathbf{solve}(\mathsf{G}) := \mathbf{intersect}(g_1, \mathbf{intersect}(g_2, \mathbf{intersect}(\cdots \mathbf{intersect}(g_m, \emptyset))))$$

returns, as output, a finite family $\mathfrak{B} := \{\mathcal{B}_1, \ldots, \mathcal{B}_l\}$ of admissible Lazard sequences which satisfy

$$\begin{aligned}
\mathcal{Z}(\mathsf{G}) &= \bigcap_i \mathcal{Z}(\mathbb{I}(g_i)) \\
&= \mathcal{Z}(\mathbb{I}(g_1)) \cap \left(\mathcal{Z}(\mathbb{I}(g_2)) \cap \left(\cdots \left(\mathcal{Z}(\mathbb{I}(g_m)) \cap \mathfrak{Z}(\emptyset) \right) \right) \right) \\
&= \bigcup_{i=1}^{l} \mathfrak{Z}(\mathcal{B}_i).
\end{aligned}$$

Algorithm 42.6.1 (Lazard). The algorithm $\mathbf{intersect}(p, \mathcal{A})$ applies another procedure

$$(r) := \mathbf{normalize}(p, \mathcal{A}),$$

whose input is the polynomial $p \in k[X_1, \ldots, X_n]$ and the admissible Lazard sequence

$$\mathcal{A} := \{f_1, \ldots, f_r\} \subset R_0[X_1, \ldots, X_n] = R_0[V_1, \ldots, V_d][Z_1, \ldots, Z_r]$$

for which we use the same notation as in Section 42.5[40] and which computes two polynomials q and r such that

(1) $\pi_n(qp) = \pi_n(r)$,
(2) $\pi_n(p) = 0 \iff \pi_n(r) = 0$ and
(3) r is reduced, normalized and R_0-normalized.

---

[40] In particular $\{V_1, \ldots, V_d\}$ (respectively, $\{Z_1, \ldots, Z_r\}$) is the set of the transcendental (respectively, algebraic) variables in $\mathcal{A}$.

Here is the procedure:

- set $q := 1$;
- (reduced)   we compute the remainder r of p w.r.t. $\mathcal{A}$;
- (normalized)   while

$$\mathbf{T}_{<}(r) := X_{j_\iota}^{\delta_\iota} X_{j_{\iota-1}}^{\delta_{\iota-1}} \cdots X_{j_1}^{\delta_1} X_{j_0}^{\delta_0} \notin k[V_1, \ldots, V_d], \delta_\iota \neq 0,$$

  then
    - compute (compare Remark 42.5.4) a polynomial[41] $s \in k[X_1, \ldots, X_{j_\iota}]$ for which

$$\mathrm{class}(\mathrm{Lp}_\iota(sr)) < \mathrm{class}(\mathrm{Lp}_\iota(r)),$$

    - compute the remainder r of sr w.r.t. $\mathcal{A}$ and
    - set $q := sq$;
- (R_0-normalized)   if $\mathrm{lc}(r) \notin S_0$ then choose[42] $c \in R_0$ such that $\mathrm{lc}(cr) \in S_0$, and set $r := cr, q := cq$. ◉

Algorithm 42.6.2 (Lazard).   We can now present the procedure

$$\mathfrak{B} := \mathbf{intersect}(p, \mathcal{A}),$$

where $\mathcal{A}$ is the admissible Lazard sequence

$$\mathcal{A} := \{f_1, \ldots, f_r\} \subset R_0[X_1, \ldots, X_n] = R_0[V_1, \ldots, V_d][Z_1, \ldots, Z_r]$$

and we use the same notation as in Section 42.5.[43]

(1) $\mathfrak{B} := \emptyset$, $(r) := \mathbf{normalize}(p, \mathcal{A})$.

(2) If
   - $r = 0$, set $\mathfrak{B} := \{\mathcal{A}\}$ and exit;
   - $r \in k \setminus \{0\}$, set $\mathfrak{B} := \emptyset$ and exit;
   - $r \notin k$, **goto** (3).

(3) Expressing r as

$$r = \mathrm{Lp}(r)X_j^{\delta} + \mathsf{r},$$

   where $j = \mathrm{class}(r)$, $\mathrm{Lp}(r) \in k[X_1, \ldots, X_{j-1}]$, $\deg_j(\mathsf{r}) < \delta = \deg_j(r)$, set

$$\mathfrak{B} := \mathfrak{B} \cup \mathbf{intersect}(\mathsf{r}, \mathbf{intersect}(\mathrm{Lp}(r), \mathcal{A})).$$

(4) Compute $\mathrm{Cont}(r) \in k[X_1, \ldots, X_{j-1}][X_j]$ and set $r := r/\mathrm{Cont}(r)$, thus forcing r to be primitive.

(5) Setting $j = \mathrm{class}(r)$, denote by i, δ the values such that

---

[41] Here a Duval splitting could happen.
[42] If we have $R_0 := L_0 := k$ and $S_0 := \{1\}$, this simply requires us to choose $c := \mathrm{lc}(r)^{-1}$.
[43] In particular $\{V_1, \ldots, V_d\}$ (respectively $\{Z_1, \ldots, Z_r\}$) is the set of the transcendental (respectively, algebraic) variables in $\mathcal{A}$.

- $\mathcal{A} \cap R_0[X_1, \ldots, X_j] = \{f_1, \ldots, f_i\}$,
- $\mathrm{class}(f_i) < j$,
- $\{V_1, \ldots, V_\delta\} = \{X_1, \ldots, X_j\} \cap \{V_1, \ldots, V_d\}$,
- $X_j = V_\delta$ is transcendental,

and set

$$\mathcal{A}_j^- := \{f_1, \ldots, f_i\}, \quad \mathcal{A}_j^+ := \{f_{i+1}, \ldots, f_r\}.$$

(6) Compute $R := \mathrm{Res}(\pi_{j-1}(r), \pi_{j-1}(r')) \in L_{j-1}$.

- If R is invertible, so that $\pi_{j-1}(r)$ is squarefree in $L_{j-1}[X_j]$, **goto** (7).
- If, instead, R is not invertible then
  - compute,[44] using the subresultant algorithm, a factor $r_0 \mid r$ such that, in $L_{j-1}[X_j]$,

$$\pi_{j-1}(r_0) = \gcd(\pi_{j-1}(r), \pi_{j-1}(r')),$$

  - set $r := r/r_0$,
  - **goto** (3).

(7) If $\mathcal{A}_j^+ = \emptyset$, set $\mathfrak{C} := \mathcal{A}_j^- \cup \{r\}$.

(8) If $\mathcal{A}_j^+ \neq \emptyset$,

- compute $\mathfrak{C} := \mathbf{intersect}(f_{i+1}, \mathbf{intersect}(\cdots \mathbf{intersect}(f_r, \mathcal{A}_j^- \cup \{r\})))$,
- set $\mathfrak{C} := \{C \in \mathfrak{C} : \mathbf{normalize}(f_l, C^-) \neq 0 \text{ for each } l, i < l \leq r\}$.[45]

(9) $\mathfrak{B} := \mathfrak{B} \cup \mathbf{intersect}(p, \mathfrak{C})$.   ⊡

Example 42.6.3.   Let us compute $\mathbf{solve}(\mathsf{G})$, where (compare Example 42.3.7)

$$\mathsf{G} := \{f_1, f_2, f_3\} \subset k[X_1, X_2, X_3, X_4]$$

and

$$f_1 := X_1^6 - X_1^4, \quad f_2 := (X_1^4 - 2X_1^2)X_3, \quad f_3 := (X_1^2 - 1)X_2X_4 + X_3.$$

We have

$\mathbf{normalize}(f_3, \emptyset) = f_3$,
$\qquad \mathcal{C}_1 := \{f_3\} = \{(X_1^2 - 1)X_2X_4 + X_3\}, \mathbf{normalize}(f_3, \mathcal{C}_1^-) = f_3$,
$\mathbf{intersect}(f_3, \emptyset) = \{\mathcal{C}_1\} =: \mathfrak{C}_1$,
$\mathbf{normalize}(f_2, \mathcal{C}_1) = f_2$,
$\qquad r_1 := \mathrm{Lp}(f_2) = X_1^4 - 2X_1^2$,
$\qquad \mathbf{normalize}(r_1, \mathcal{C}_1) = r_1$,
$\qquad\qquad r_2 := \sqrt{r_1} = X_1^3 - 2X_1$,
$\qquad\qquad \mathcal{C}_2 := \mathcal{C}_1^- \cup \{r_2\} = \{r_2\} = \{X_1^3 - 2X_1\}$,
$\qquad\qquad \mathbf{normalize}(f_3, \mathcal{C}_2) = X_2X_4 + X_3(X_1^2 - 1) =: r_3$,[46]
$\qquad\qquad r_4 := \mathrm{Lp}(r_3) = X_2, r_5 := r_3 - r_4X_4 = (X_1^2 - 1)X_3$,

---

[44] This can produce a Duval splitting.
[45] Here $C^- = C \cap R_0[X_1, \ldots, X_j]$, $j = \mathrm{class}(r)$.
[46] We have $(X_1^2 - 1)f_3 - X_1X_2X_4r_2 = X_2X_4 + X_3(X_1^2 - 1)$.

$$\mathbf{normalize}(r_4, \mathcal{C}_2^-) = r_4,$$
$$\mathcal{C}_3 := \mathcal{C}_2^- \cup \{r_4\} = \{r_2, r_4\} = \{X_1^3 - 2X_1, X_2\},$$
$$\mathbf{intersect}(r_4, \{\mathcal{C}_2\}) = \{\mathcal{C}_3\},$$
$$\mathbf{normalize}(r_5, \mathcal{C}_3^-) = X_3 =: r_6,[47]$$
$$\mathcal{C}_4 := \mathcal{C}_3^- \cup \{r_6\} = \{r_2, r_4, r_6\} = \{X_1^3 - 2X_1, X_2, X_3\},$$
$$\mathbf{normalize}(r_5, \mathcal{C}_3^-) = r_5,$$
$$\mathbf{intersect}(r_5, \{\mathcal{C}_3\}) = \{\mathcal{C}_4\},$$
$$\mathcal{C}_5 := \mathcal{C}_2 \cup \{r_3\} = \{r_2, r_3\} = \{X_1^3 - 2X_1, X_2X_4 + X_3(X_1^2 - 1)\},$$
$$\mathfrak{C}_2 := \{\mathcal{C}_4, \mathcal{C}_5\},$$
$$\mathbf{intersect}(f_3, \mathcal{C}_2) = \mathfrak{C}_2,$$
$$\mathbf{normalize}(r_1, \mathcal{C}_i^-) \neq 0, i \in \{4, 5\},$$
$$\mathbf{intersect}(r_1, \mathfrak{C}_1) = \mathfrak{C}_2,$$
$$f_2/\mathrm{Cont}(f_2) = X_3 = r_6,$$
$$\mathcal{C}_6 := \mathcal{C}_1^- \cup \{r_6\} = \{r_6\} = \{X_3\},$$
$$\mathbf{normalize}(f_3, \mathcal{C}_6) = (X_1^2 - 1)X_2X_4 =: r_7,$$
$$\mathrm{Lp}(r_7) = (X_1^2 - 1)X_2 =: r_8,$$
$$\mathbf{normalize}(r_8, \mathcal{C}_6) = r_8,$$
$$\mathrm{Lp}(r_8) = (X_1^2 - 1) =: r_9,$$
$$\mathbf{normalize}(r_9, \mathcal{C}_6) = r_9,$$
$$\mathcal{C}_7 := \mathcal{C}_6^- \cup \{r_9\} \cup \mathcal{C}_6^+ = \{r_9, r_6\} = \{X_1^2 - 1, X_3\},$$
$$\mathbf{intersect}(r_9, \{\mathcal{C}_6\}) = \{\mathcal{C}_7\},$$
$$r_8/\mathrm{Cont}(r_8) = X_2 = r_4,$$
$$\mathcal{C}_8 := \mathcal{C}_6 \cup \{r_4\} = \{r_4, r_6\} = \{X_2, X_3\},$$
$$\mathbf{intersect}(r_8, \{\mathcal{C}_6\}) = \{\mathcal{C}_7, \mathcal{C}_8\},$$
$$r_7/\mathrm{Cont}(r_7) = X_4 = r_{10},$$
$$\mathcal{C}_9 := \mathcal{C}_6^- \cup \{r_{10}\} = \{r_6, r_{10}\} = \{X_3, X_4\},$$
$$\mathbf{intersect}(f_3, \{\mathcal{C}_6\}) = \{\mathcal{C}_9\},$$
$$\mathfrak{C}_2 := \{\mathcal{C}_4, \mathcal{C}_5, \mathcal{C}_7, \mathcal{C}_8, \mathcal{C}_9\},$$
$$\mathbf{normalize}(f_2, \mathcal{C}_i^-) \neq 0, i \in \{4, 5, 7, 8, 9\},$$
$$\mathbf{intersect}(f_2, \mathfrak{C}_1) = \mathbf{intersect}(f_2, \mathcal{C}_1) = \mathfrak{C}_2,$$
$$\mathbf{Rem}(f_1, r_2) = 2X_1^2 := r_{11}.$$

By Duval splitting we get

$$r_{12} := X_1,$$
$$\mathbf{normalize}(r_{11}, \{r_{12}\}) = r_{12},$$
$$\mathbf{normalize}(r_4, \{r_{12}\}) = r_4,$$
$$\mathbf{normalize}(r_6, \{r_{12}, r_4\}) = r_6,$$
$$\mathcal{C}_{11} = \{r_{12}, r_4, r_6\} = \{X_1, X_2, X_3\},$$
$$r_{13} := X_1^2 - 2,$$
$$\mathbf{normalize}(r_{11}, \{r_{13}\}) = 4,$$
$$\mathbf{intersect}(f_1, \mathcal{C}_4) = \mathcal{C}_{11},$$
$$\mathbf{Rem}(f_1, r_2) = 2X_1^2 := r_{11}.$$

---

[47] We have $(X_1^2 - 1)r_5 - X_1X_3r_2 = X_3$.

By Duval splitting we get

$$r_{12} := X_1,$$
$$\mathbf{normalize}(r_{11}, \{r_{12}\}) = r_{12},$$
$$\mathbf{normalize}(r_3, \{r_{12}\}) = X_2X_4 - X_3 =: r_{14},$$
$$\mathcal{C}_{12} := \{r_{11}, r_{14}\} = \{X_1, X_2X_4 - X_3\},$$
$$r_{13} := X_1^2 - 2,$$
$$\mathbf{normalize}(r_{11}, \{r_{13}\}) = 4,$$
$$\mathbf{intersect}(f_1, \mathcal{C}_5) = \mathcal{C}_{12},$$
$$\mathbf{normalize}(f_1, \mathcal{C}_7) = 0,$$
$$\mathbf{normalize}(f_1, \mathcal{C}_8) = f_1,$$
$$\sqrt{f_1} = X_1^3 - X_1 =: r_{15},$$
$$\mathbf{intersect}(f_1, \mathcal{C}_8) = \{r_{15}, r_4, r_6\} = \{X_1^3 - X_1, X_2, X_3\} =: \mathcal{C}_{13},$$
$$\mathbf{normalize}(f_1, \mathcal{C}_9) = f_1,$$
$$\sqrt{f_1} = X_1^3 - X_1 =: r_{15},$$
$$\mathbf{intersect}(f_1, \mathcal{C}_9) = \{r_{15}, r_6, r_{10}\} = \{X_1^3 - X_1, X_3, X_4\} =: \mathcal{C}_{14},$$
$$\mathbf{solve}(\mathsf{G}) = \mathbf{intersect}(f_1, \mathfrak{C}_2) = \mathfrak{C}_3 := \{\mathcal{C}_{11}, \mathcal{C}_{12}, \mathcal{C}_7, \mathcal{C}_{13}, \mathcal{C}_{14}\},$$

so that

$$
\begin{aligned}
\mathcal{Z}(\mathcal{C}_{11}) &= \mathcal{Z}(\{X_1, X_2, X_3\}) &&= \{(0, 0, 0, a), a \in k\}, \\
\mathcal{Z}(\mathcal{C}_{12}) &= \mathcal{Z}(\{X_1, X_2X_4 - X_3\}) &&= \{(0, a, b, b/a), a, b \in k, a \neq 0\}, \\
\mathcal{Z}(\mathcal{C}_7) &= \mathcal{Z}(\{X_1^2 - 1, X_3\}) &&= \{(x, a, 0, b), x \in \{1, -1\}, a, b \in k\}, \\
\mathcal{Z}(\mathcal{C}_{13}) &= \mathcal{Z}(\{X_1^3 - X_1, X_2, X_3\}) &&= \{(x, 0, 0, a), x \in \{0, 1, -1\}, a \in k\}, \\
\mathcal{Z}(\mathcal{C}_{14}) &= \mathcal{Z}(\{X_1^3 - X_1, X_3, X_4\}) &&= \{(x, a, 0, 0), x \in \{0, 1, -1\}, a \in k\},
\end{aligned}
$$

where

$$\mathcal{Z}(\mathcal{C}_{13}) \cup \mathcal{Z}(\mathcal{C}_{14}) \subset \mathcal{Z}(\mathcal{C}_{11}) \cup \mathcal{Z}(\mathcal{C}_{12}) \cup \mathcal{Z}(\mathcal{C}_7).$$

$\boxdot$

Algorithm 42.6.4 (Lazard).   For removing redundant components, Lazard proposed an algorithm **inclusion?**(T, U), which is performed on the set of quasi-components ordered by increasing dimension and in which each quasi-component T is compared with each component U of higher dimension to test whether $T \subset U$.

The procedure consists in checking whether **normalize**$(f, T) = 0$ for each $f \in U$, the answer being positive iff all tests are successful. $\boxdot$

Of course, the tests produce a Duval-splitting in T.

Example 42.6.5.   For instance, with the present example, the test

$$\mathbf{inclusion?}(\mathcal{Z}(\mathcal{C}_i), \mathcal{Z}(\mathcal{C}_{12})), \quad i \in \{13, 14\},$$

returns the splittings

$$\mathcal{Z}(\mathcal{C}_{13}) = \mathcal{Z}(\mathcal{C}_7) \cup \mathcal{Z}(\mathcal{C}'_{13}), \quad \mathcal{C}'_{13} = \mathcal{Z}(\{X_1, X_2, X_3\}) = \{(0, 0, 0, a), a \in k\},$$
$$\mathcal{Z}(\mathcal{C}_{14}) = \mathcal{Z}(\mathcal{C}_7) \cup \mathcal{Z}(\mathcal{C}'_{14}), \quad \mathcal{C}'_{14} = \mathcal{Z}(\{X_1, X_2, X_3\}) = \{(0, a, 0, 0), a \in k\},$$

the answer being positive for the components $\mathcal{Z}(\mathcal{C}'_i)$. $\boxdot$

42.7 Ritt Bases and Gröbner Bases

Let k be a field of characteristic zero,

$$\mathcal{P} := k[X_1, \ldots, X_n], \quad \mathcal{T} := \{X_1^{a_1} \cdots X_n^{a_n} : (a_1, \ldots, a_n) \in \mathbb{N}^n\},$$

and $<$ be the lexicographical ordering on $\mathcal{T}$ induced by $X_1 < \cdots < X_n$.

Let $\mathsf{I} \subset \mathcal{P}$ and let $G := \{g_1, \ldots, g_s\}$ be the reduced Gröbner basis of I, ordered in such a way that

$$\mathbf{T}(g_1) < \mathbf{T}(g_2) < \cdots < \mathbf{T}(g_{s-1}) < \mathbf{T}(g_s),$$

and denote, for each i, $1 \le i \le n$, $G_i := G \cap k[X_1, \ldots, X_i]$.

Adapting Definition 42.5.1, we say that a variable X_i is

algebraic for I if there is a $g \in G_i \setminus G_{i-1}$,
transcendental for I if $G_i = G_{i-1}$.

As usual we relabel the variables as follows:

$$k[X_1, \ldots, X_n] \cong k[V_1, \ldots, V_d, Z_1, \ldots, Z_r],$$

so that $\{V_1, \ldots, V_d\}$ (respectively, $\{Z_1, \ldots, Z_r\}$) is the set of the transcendental (respectively, algebraic) variables for G.

With each reduced lex Gröbner basis G we associate a set $\mathcal{M}(G)$ inductively (on the rank $r := r(\mathbb{I}(G))$ of the ideal generated, by G as follows:

- if $r = 1$ then set $\mathcal{M}(G) := \{g_1\}$;
- if $r > 1$, denoting by i the value for which $Z_r = X_i$, so that $G_i \setminus G_{i-1} \ne \emptyset$ and (as we will prove below) $\mathcal{M}(G_{i-1})$ is a triangular set, then we set
  - $\mathcal{M}(G) := \mathcal{M}(G_{i-1})$ if $G_i \subset \mathrm{Rem}(\mathcal{M}(G_{i-1}))$,
  - if, instead, $G_i \not\subset \mathrm{Rem}(\mathcal{M}(G_{i-1})$ then we set $\mathcal{M}(G) := \mathcal{M}(G_{i-1}) \cup \{g_j\}$, where j is the minimal value for which the remainder of g_i w.r.t. $\mathcal{M}(G_{i-1})$ is non-zero.

Definition 42.7.1 (Aubry *et al.*).   *The set $\mathcal{M}(G)$ defined above is called the* median set *of I, where I is the ideal generated by the reduced lex Gröbner basis G.*   ◉

Example 42.7.2.   For

$$G := \{X_1 X_2, X_2 X_3, X_3 X_4\} \in k[X_1, X_2, X_3, X_4] \cong k[V_1][Z_1, Z_2, Z_3]$$

we set

$$\mathcal{M}(G_2) = \{X_1 X_2\},$$
$$\mathcal{M}(G_3) = \mathcal{M}(G_2), \text{ since } X_2 X_3 \in \mathrm{Rem}(\mathcal{M}(G_2)),$$
$$\mathcal{M}(G) = \mathcal{M}(G_2) \cup \{X_3 X_4\} = \{X_1 X_2, X_3 X_4\}.$$

◉

Proposition 42.7.3 (Aubry *et al.*).   *Let $G \subset \mathcal{P}$ be a reduced lex Gröbner basis generating an ideal I and let $\mathcal{M}(G)$ be its median set. Then:*

(1) $\mathcal{M}(G)$ *is a non-empty triangular set;*
(2) $\mathcal{M}(G) \subseteq \mathsf{I} \subseteq \mathrm{Rem}(\mathcal{M}(G))$;
(3) $\mathcal{M}(G)$ *is a fine triangular set;*
(4) $\mathcal{M}(G)$ *is an admissible Ritt sequence;*
(5) $\mathcal{M}(G)$ *is initially reduced.*

Proof.

(1) Obvious.
(2) The only non-trivial result is the inclusion $\mathsf{I} \subseteq \mathrm{Rem}(\mathcal{M}(G))$. Assume that there is an $f \in \mathsf{I}$ for which $f \notin \mathrm{Rem}(\mathcal{M}(G))$ and denote by r its remainder w.r.t. $\mathcal{M}(G)$, remarking that $r \in \mathsf{I}$.

   Therefore there is a $g \in G$ such that $\mathbf{T}(g) \mid \mathbf{T}(r)$. Since r (and so also $\mathbf{T}(r)$) is reduced, the same is true for $\mathbf{T}(g)$. Let $i := \mathrm{class}(g)$ and $\mathcal{A} := \{h \in \mathcal{M}(G) : \mathrm{class}(h) < i\}$. Remark that either

   - $G_i \subset \mathrm{Rem}(\mathcal{A})$ and $\mathcal{A} = \mathcal{M}(G_i)$, or
   - $G_i \not\subset \mathrm{Rem}(\mathcal{A})$ and there is an $h \in G_i$ such that $\{h\} = \mathcal{M}(G_i) \setminus \mathcal{A}$,

   so that there are three cases:
   (a) $g \in \mathrm{Rem}(\mathcal{A})$ or[48]
   (b) $g = h \in \mathcal{M}(G) \setminus \mathcal{A}$ or
   (c) $g \neq h \implies \mathbf{T}(g) > \mathbf{T}(h)$.[49]
   However, all these cases reduce to a contradiction:
   (a) contradicts the assumption that $\mathbf{T}(g)$ is reduced;
   (b) ditto;
   (c) cannot hold for the same reason; in fact, $\mathrm{class}(h) = i = \mathrm{class}(g)$ and $\mathbf{T}(g) > \mathbf{T}(h)$ imply that $\deg_i(\mathbf{T}(g)) \geq \deg_i(\mathbf{T}(h)) = \deg_i(h)$, which in turn implies that $\mathbf{T}(g)$ is reduced by h.
(3) Assume that for some $g \in \mathcal{M}(G)$ the remainder of $\mathrm{Lp}(g)$ w.r.t.

   $$\mathcal{A} := \{h \in \mathcal{M}(G) : \mathbf{T}(h) < \mathbf{T}(g)\} = \{h \in \mathcal{M}(G) : \mathrm{class}(h) < \mathrm{class}(g)\}$$

   is zero, so that, writing $j := \mathrm{class}(g)$ and $g' \in k[X_1, \ldots, X_j]$ the polynomial such that $g = \mathrm{Lp}(g)X_j^{\deg_j(g)} + g'$, $\deg_j(g') < \deg_j(g)$, the remainder r of g and that of g' w.r.t. $\mathcal{A}$ are the same. In particular, $r \in \mathsf{I} \subseteq \mathrm{Rem}(\mathcal{M}(G))$, whence $r = 0$ and $g \in \mathrm{Rem}(\mathcal{A})$, contradicting the construction of $\mathcal{M}(G)$.
(4) This follows from (2) and Proposition 42.4.7.
(5) If this does not hold then on the one hand there is a smallest (w.r.t. $j := \mathrm{class}(g)$) element $g \in G$ which is not initially reduced w.r.t.

   $$\mathcal{A} := \{h \in \mathcal{M}(G) : \mathbf{T}(h) < \mathbf{T}(g)\} = \{h \in \mathcal{M}(G) : \mathrm{class}(h) < j\} =: \{A_1, \ldots, A_\rho\}.$$

---

[48] $g \notin \mathrm{Rem}(\mathcal{A})$, so that $G_i \not\subset \mathrm{Rem}(\mathcal{A})$ and $\mathcal{M}(G_i) \setminus \mathcal{A} = \{h\}$.
[49] We assume $\mathrm{lc}(f) = 1$ for each $f \in G$, so that $\mathbf{T}(g) = \mathbf{T}(h) \implies g = h$.

On the other hand the remainder of $\mathrm{Lp}(g)$ is non-zero w.r.t. $\mathcal{A}$. Then Theorem 42.1.18 implies that, for suitable integers w_i and denoting as I_i the initial of A_i,

$$I_1^{w_1} \cdots I_r^{w_r} g = I_1^{w_1} \cdots I_r^{w_r} \mathrm{Lp}(g) X_j^{\deg_j(g)} + I_1^{w_1} \cdots I_r^{w_r} g'$$

reduces w.r.t. $\mathcal{A}$ to a polynomial $t := R X_j^{\deg_j(g)} + g'' \in \mathsf{l}$, for which the following hold:

- t is reduced w.r.t. $\mathcal{A}$,
- $\mathrm{class}(t) = j, \deg_j(t) = \deg_j(g)$,
- $\mathbf{T}(t) = \mathbf{T}(R) X_j^{\deg_j(g)} < \mathbf{T}(\mathrm{Lp}(g)) X_j^{\deg_j(g)} = \mathbf{T}(g)$.

Then, necessarily, $\mathbf{T}(t)$ is divided by $\mathbf{T}(h)$ for some $h \in G_j$ for which $\mathrm{Lp}(h) \in \mathrm{Rem}(\mathcal{A})$; therefore $\mathbf{T}(h) \in \mathrm{Rem}(\mathcal{A})$ and $\mathbf{T}(t) \in \mathrm{Rem}(\mathcal{A})$, contradicting the assumption that t is reduced. $\boxed{\odot}$

It is possible to recover Ritt's Corollary 42.3.5 as follows: for each algebraic variable $Z_i = X_j$, denote by A_i the smallest[50] polynomial in $G_j \setminus G_{j-1}$ and denote $\mathcal{A}(G) := \{A_1, \ldots, A_r\}$.

With this notation we have:

Theorem 42.7.4. *If G generates a prime ideal l then:*

(1) $\mathcal{M}(G) = \mathcal{A}(G)$;
(2) $\mathsf{l} = \mathrm{Rem}(\mathcal{M}(G)) = \mathrm{Sat}(\mathcal{M}(G))$;
(3) $\overline{3(\mathcal{M}(G))} = \mathcal{Z}(\mathsf{l})$;
(4) $3(\mathcal{M}(G)) \neq \emptyset$.

Proof.

(1) It is sufficient to show that for each j the remainder of the initial $\mathrm{Lp}(A_j)$ w.r.t. $\mathcal{B} := \{A_1, \ldots, A_{j-1}\}$ is non-zero.

   If it were zero, we would get a contradiction from $\mathrm{Lp}(A_j) \in \mathsf{l}$ but this is impossible since A_j is a member of a reduced Gröbner basis.
(2) Since $\mathsf{l} \subset \mathrm{Rem}(\mathcal{M}(G))$ by Proposition 42.7.3 above, the result follows from Corollary 42.3.5.
(3) This follows by Corollary 42.3.5(4).
(4) $\mathcal{Z}(\mathsf{l}) \neq \emptyset$. $\boxed{\odot}$

Example 42.7.5. If G generates just a radical ideal l, it could happen that $3(\mathcal{M}(G)) = \emptyset$.

Set

$$F := \{X_1^2 - 2, X_2^2 - 2, (X_1 - X_2)X_3, (X_1 + X_2)X_4\} \in k[X_1, X_2, X_3, X_4].$$

---

[50] Recall that the elements in G are enumerated in such a way that $\mathbf{T}(g_i) < \mathbf{T}(g_{i+1})$, so the "smallest" polynomial in $G_j \setminus G_{j-1}$ is also the polynomial $g \in G_j \setminus G_{j-1}$ having the $<$-minimal value $\mathbf{T}(g)$.

The Gröbner basis is $G := F \cup \{X_3 X_4\}$, and $\mathcal{M}(G) = F$. Clearly, we have for the product of the initials

$$(X_1 - X_2)(X_1 + X_2) = X_1^2 - X_2^2 \in (X_1^2 - 2, X_2^2 - 2).$$

Thus $\mathfrak{Z}(\mathcal{M}(G)) = \emptyset$. ⊚

42.8 Möller's Zero-dimensional Solver

Lemma 42.8.1. *Let* $\mathsf{J} \subset \mathcal{Q}$ *be a zero-dimensional ideal and let* $h \in \mathcal{Q}$. *It holds that*

$$\mathcal{Z}(\mathsf{J} : h^\infty) = \{\alpha \in \mathcal{Z}(\mathsf{J}) : h(\alpha) \neq 0\}.$$

Proof. Let us consider the irredundant primary decomposition $\mathsf{J} = \bigcup_{i=1}^{r} \mathfrak{q}_i$, where, for each i, $\mathfrak{p}_i$ denotes the associate maximal $\mathfrak{p}_i = \sqrt{\mathfrak{q}_i}$. We have (Theorem 26.3.2(19))

$$\mathsf{J} : h^\infty = \bigcup_{i=1}^{r} \mathfrak{q}_i : h^\infty$$

and (Corollary 27.2.12)

$$\mathfrak{q}_i : h^\infty = \begin{cases} \mathcal{Q} & \text{iff } h \in \mathfrak{p}_i, \\ \mathfrak{q}_i & \text{iff } h \notin \mathfrak{p}_i, \end{cases}$$

whence the claim follows easily. ⊚

Proposition 42.8.2 (Möller). *Let* $\mathsf{J} \subset \mathcal{Q}$ *be a zero-dimensional ideal and let* $H := \{h_1, \dots, h_t\} \subset \mathcal{Q}$ *be a set of polynomials such that,*

$$\mathcal{Z}(H) \subset \mathcal{Z}(\mathsf{J}).$$

Setting $\mathsf{J}_t := \mathsf{J}$ *and* $\mathsf{J}_i := \mathsf{J} + \mathbb{I}(h_{i+1}, \dots, h_t)$, $1 \leq i < t$, *it holds that*

$$\mathcal{Z}(\mathsf{J}) = \mathcal{Z}(H) \bigsqcup \bigsqcup_{i=1}^{t} \mathcal{Z}(\mathsf{J}_i : h_i^\infty).$$

Proof. Clearly

$$\mathcal{Z}(\mathsf{J}) \setminus \mathcal{Z}(H) = \{\alpha \in \mathcal{Z}(\mathsf{J}) : \text{there exists } i, i \leq t, h_i(\alpha) \neq 0\}$$

$$= \bigsqcup_{i=1}^{t} \{\alpha \in \mathcal{Z}(\mathsf{J}) : h_t(\alpha) = \cdots = h_{i+1}(\alpha) = 0 \neq h_i(\alpha)\}$$

$$= \bigsqcup_{i=1}^{t} \mathcal{Z}(\mathsf{J}_i : h_i^\infty),$$

the last equality following from Lemma 42.8.1. ⊚

The intended application of Proposition 42.8.2 requires efficient algorithms in order to compute, given a zero-dimensional ideal $\mathsf{J} \subset \mathcal{Q}$ and a polynomial $h \in \mathcal{Q} \setminus \{0\}$, both $\mathsf{J} + (h)$ and $\mathsf{J} : h^\infty$; different techniques are discussed in Sections 26.3–26.7. In connection, Möller anticipated some version of Caboara–Traverso ideas (Section 26.6); in particular he stated:

Lemma 42.8.3 (Möller). *Let $\mathfrak{a} = \mathbb{I}(g_1, \ldots, g_s) \subset \mathcal{Q}$ and $h \in \mathcal{Q} \setminus \{0\}$. Then*

$$\mathsf{M} := \{(u, v) \in \mathcal{Q}^2 : u - hv \in \mathfrak{a}\}$$

is a module with basis $F := \{(g_i, 0), 1 \le i \le s\} \cup \{(h, 1)\}$.

Moreover, we fix any term ordering $\prec$ on $\mathcal{W}$ and denote

- *$\{e_1, e_2\}$ the canonical basis of $\mathcal{Q}^2$,*
- *$\mathcal{W}^{(2)} = \{\tau e_i, : \tau \in \mathcal{W}, i \in \{1, 2\}\}$,*
- *$\prec_2$ the $\prec$-compatible term ordering on $\mathcal{W}^{(2)}$ defined by*

$$\tau e_i \prec_2 \tau' e_j \iff \begin{cases} i > j & or \\ i = j & and\ \tau \prec \tau', \end{cases}$$

- *G the Gröbner basis of M w.r.t. $\prec_2$,*
- *$G_0 := \{b \in \mathcal{Q} : (0, b) \in G\}$,*
- *$G_1 := \{a \in \mathcal{Q} : (a, b) \in G\}$.*

Then $(\mathfrak{a} : h) = \mathbb{I}(G_0)$ and $\mathfrak{a} + (h) = \mathbb{I}(G_1)$.

Proof. Obviously $F \subset \mathsf{M}$. For each $(u, v) \in \mathsf{M}$ there are $f_i \in \mathcal{Q}$ such that $u - hv = \sum_i f_i g_i$, so that

$$(u, v) = \sum_i f_i (g_i, 0) + v(h, 1);$$

this proves that $\mathsf{M} = \mathbb{I}(F)$.

The relation $\mathbb{I}(G_1) = \mathbb{I}(g_1, \ldots, g_s, h) = \mathfrak{a} + (h)$ is obvious.

The other claim is a direct consequence of the trivial equivalence

$$(0, v) \in \mathsf{M} \iff -hv \in \mathfrak{a} \iff v \in (\mathfrak{a} : h).$$

$\boxdot$

In order to deduce $\mathfrak{a} : h^\infty$, Möller proposed to apply the same algorithm iteratively in order to deduce $\mathfrak{a} : h^i$ iteratively; the result is then obtained at stabilization.[51]

Remark 42.8.4. When, as has been assumed, J is zero-dimensional, Möller also proposed to apply Traverso's Algorithm 29.3.8 in order to deduce $\mathsf{J} + (h)$ and a suitable variation of the FGLM algorithm in order to compute $(\mathsf{J} : h)$.

Namely, denoting $\{\tau_1, \ldots, \tau_u\}$ as $\mathbf{N}(\mathsf{J})$ and, for each $f \in \mathcal{Q}$,

---

[51] Compare the discussion after Lemma 26.3.9.

$$\mathbf{Rep}(f, \mathbf{N}(\mathsf{J})) := (\gamma(f, \tau_1, \mathbf{N}(\mathsf{J})), \ldots, \gamma(f, \tau_u, \mathbf{N}(\mathsf{J})))$$

as its *Gröbner description*, in order to obtain $(\mathsf{J} : h)$ Möller's Algorithm is applied to the functionals

$$\ell_i : \mathcal{Q} \to k : f \mapsto \gamma(fh, \tau_i, \mathbf{N}(\mathsf{J}))$$

in the same way as that in which the FGLM Algorithm is obtained by applying Möller's Algorithm to the functionals

$$\ell_i : \mathcal{Q} \to k : f \mapsto \gamma(f, \tau_i, \mathbf{N}(\mathsf{J})).$$

⊙

Now let $\mathsf{J} \subset \mathcal{Q}$ be a *zero-dimensional* ideal and let $G := \{g_1, \ldots, g_s\}$, $\mathrm{lc}(g_i) = 1$, be its Gröbner basis w.r.t. the lex ordering induced by $Z_1 < Z_2 < \cdots < Z_r$, ordered so that $\mathbf{T}(g_1) < \mathbf{T}(g_2) < \cdots < \mathbf{T}(g_s)$.

The assumption that the ideal is zero-dimensional trivially implies that $g_s \in K[Z_1, \ldots, Z_r] \setminus K[Z_1, \ldots, Z_{r-1}]$ and that $\deg_r(g_i) := d_i < d_s := \deg_r(g_s)$ for each $i < s$.

As a consequence (compare Kalkbrener's Theorem 26.2.6),[52] $\{\mathrm{Lp}(g_i), 1 \le i < s\}$ is a Gröbner basis w.r.t. $<$; moreover, since J is zero-dimensional we also have $\mathrm{Lp}(g_s) \in k$.

Theorem 42.8.5 (Möller). *With the present notation we have*

$$\mathbb{I}(g_1, \ldots, g_{s-1}) : g_s = \mathbb{I}(\mathrm{Lp}(g_1), \ldots \mathrm{Lp}(g_{s-1})).$$

Proof. If $h \in \mathsf{J}$ and $\mathbf{T}(h) < \mathbf{T}(g_s)$ then there is a $j < s$ such that $\mathbf{T}(g_j) \mid \mathbf{T}(h)$ and there are $c \in k \setminus \{0\}, \tau \in \mathcal{W}$ such that $h' := h - c\tau g_j$ satisfies $h' \in \mathsf{J}$ and $\mathbf{T}(h') < \mathbf{T}(h) < \mathbf{T}(g_s)$.

Thus, for each $h \in \mathsf{J}$ for which $\mathbf{T}(h) < \mathbf{T}(g_s)$, it holds that $h \in \mathbb{I}(g_1, \ldots, g_{s-1})$.

For each i, $1 \le i < s$, set $h_i := \mathrm{Lp}(g_i)g_s - Z_r^{d_s - d_i} g_i$; since

$$\mathbf{T}(\mathrm{Lp}(g_i))\mathbf{T}(g_s) = \mathbf{T}(\mathrm{Lp}(g_i)Z_r^{d_s} = \mathbf{T}(\mathrm{Lp}(g_i)Z_r^{d_i})Z_r^{d_s - d_i} = \mathbf{T}(g_i)Z_r^{d_s - d_i}$$

we have $\mathbf{T}(h_i) < \mathbf{T}(g_s)$ and, since $h_i \in \mathsf{J}$, we have $h_i \in \mathbb{I}(g_1, \ldots, g_{s-1})$, whence $\mathrm{Lp}(g_i)g_s \in \mathbb{I}(g_1, \ldots, g_{s-1})$ and $\mathrm{Lp}(g_i) \in \mathbb{I}(g_1, \ldots, g_{s-1}) : g_s$. We have thus proven the inclusion $\mathbb{I}(\mathrm{Lp}(g_1), \ldots, \mathrm{Lp}(g_{s-1})) \subseteq \mathbb{I}(g_1, \ldots, g_{s-1}) : g_s$.

Conversely, let us consider a polynomial $g \in \mathbb{I}(g_1, \ldots, g_{s-1}) : g_s, g \ne 0$.

Since $gg_s \in \mathbb{I}(g_1, \ldots, g_{s-1})$, there is an $i < s$ such that $\mathbf{T}(g_i) \mid \mathbf{T}(gg_s) = \mathbf{T}(g)Z_r^{d_s}$ and there are $c \in k \setminus \{0\}, \tau \in \mathcal{W} \cap k[Z_1, \ldots, Z_{r-1}]$ such that

$$\mathbf{T}(gg_s) = c\tau Z_r^{d_s - d_i} \mathbf{T}(g_i) = c\tau \mathbf{T}(\mathrm{Lp}(g_i))Z_r^{d_s} = c\tau \mathbf{T}(\mathrm{Lp}(g_i)g_s).$$

---

[52] Apparently Kalkbrener's Theorem 26.2.6 and the weaker Möller's Theorem 42.8.5, see below, are independent.

Writing $g' := g - c\tau \, \mathrm{Lp}(g_i)$ and remarking that[53]

$$g' \in (\mathbb{I}(g_1, \ldots, g_{s-1}) : g_s) + \mathbb{I}(\mathrm{Lp}(g_1), \ldots \mathrm{Lp}(g_{s-1})) \subseteq \mathbb{I}(g_1, \ldots, g_{s-1}) : g_s$$

we have that either

- $\mathbf{T}(g') < \mathbf{T}(g)$ and $g - g' \in \mathbb{I}(\mathrm{Lp}(g_1), \ldots \mathrm{Lp}(g_{s-1}))$, or
- $g' = 0$ and $g \in \mathbb{I}(\mathrm{Lp}(g_1), \ldots \mathrm{Lp}(g_{s-1}))$.

Thus, by $<$-induction we deduce that

$$\mathbb{I}(g_1, \ldots, g_{s-1}) : g_s \subseteq \mathbb{I}(\mathrm{Lp}(g_1), \ldots \mathrm{Lp}(g_{s-1})).$$

$\boxdot$

Corollary 42.8.6 (Möller). *With the present notation and setting $H := \{\mathrm{Lp}(g_i), 1 \le i < s\} \cup \{g_s\}$ we have*

$$\mathcal{Z}(H) \subset \mathcal{Z}(\mathsf{J}).$$

$\boxdot$

Algorithm 42.8.7 (Möller).   With the present notation, the algorithm described in Figure 42.1 produces a triangular-set decomposition of the zero-dimensional ideal J.

In fact $\mathfrak{T}$ is obtained, according to Proposition 42.8.2, by the disjoint union of

- $\mathfrak{T}'$, which gives the triangular-set decomposition of $\mathbb{I}(H)$, and
- $\mathfrak{T}_i$, which gives the triangular-set decomposition of $\mathsf{J}_i : \mathrm{Lp}(g_i)^\infty$, for $i, l < i < s$,[54] where l is the value for which $\{g_1, \ldots, g_l\} = H \cap K[Z_1, \ldots, Z_{r-1}]$.

The correctness of the **While**-loop is a direct consequence of the ordering of the basis elements[55] and the fact that

$$\mathrm{Lp}(g_{s-i}) \in \mathsf{J} \implies \mathbb{I}(G_{s-i}) : \mathrm{Lp}(g_{s-i})^\infty = \mathcal{Q} \iff \mathcal{Z}\left(\mathbb{I}(G_{s-i}) : \mathrm{Lp}(g_{s-i})^\infty\right) = \emptyset.$$

$\boxdot$

---

[53] Recall that we have just proved the inclusion

$$\mathbb{I}(\mathrm{Lp}(g_1), \ldots \mathrm{Lp}(g_{s-1})) \subseteq \mathbb{I}(g_1, \ldots, g_{s-1}) : g_s.$$

[54] We remark that Proposition 42.8.2(6) apparently requires one to compute, before the **While**-loop,
a reduced Gröbner basis G'_s of $\mathbb{I}(g_1, \ldots, g_s) : g_s^\infty$,
a reduced Gröbner basis G_{s-1} of $\mathbb{I}(g_1, \ldots, g_s) + \mathbb{I}(g_s)$,
the triangular set decomposition of $\mathbb{I}(G'_s)$,
but this computation is trivial and returns $G'_s = \{1\}$ and $G_{s-1} = \{g_1, \ldots, g_s\}$.

[55] For each $l \ge i$ we have

$$\mathrm{Lp}(g_{s-i}) \in \mathsf{J} \iff g_{s-i} \in k[Z_1, \ldots, Z_{r-1}]$$
$$\implies g_{s-l} \in k[Z_1, \ldots, Z_{r-1}]$$
$$\iff \mathrm{Lp}(g_{s-l}) \in \mathsf{J}.$$

$\mathfrak{T} := \mathbf{Solve}(\mathsf{J})$
where
 $\mathsf{J} \subset \mathcal{Q}$ is a zero-dimensional ideal,
 $<$ is the lex ordering induced by $Z_1 < \cdots < Z_r$,
 $\{g_1, \ldots, g_s\}$, $\mathrm{lc}(g_i)=1$, $\mathbf{T}(g_1) < \cdots < \mathbf{T}(g_s)$, is the reduced Gröbner basis of J w.r.t. $<$,
 $\mathfrak{T} = \{\mathsf{t}_1, \ldots, \mathsf{t}_v\}$ is a finite set of triangular sets such that

$$\mathcal{Z}(\mathsf{J}) = \bigsqcup_{j=1}^{v} \mathcal{Z}(\mathbb{I}(\mathsf{t}_j)).$$

Let G be the reduced Gröbner basis w.r.t. $<$ of $\mathbb{I}(\mathrm{Lp}(g_1), \ldots, \mathrm{Lp}(g_{s-1}))$;
$\mathfrak{U} := \mathbf{Solve}\big(\mathbb{I}(\mathrm{Lp}(g_1), \ldots, \mathrm{Lp}(g_{s-1}))\big)$,
$\mathfrak{T}' := \{\mathsf{t} \cup \{\mathrm{NF}(g_s, \mathsf{t})\} : \mathsf{t} \in \mathfrak{U}\}$,
$i = 1$, $G_{s-1} := \{g_1, \ldots, g_s\}$.
While $\mathrm{Lp}(g_{s-i}) \notin \mathsf{J}$ **do**
 Compute a reduced Gröbner basis G'_{s-i} of $\mathbb{I}(G_{s-i}) : \mathrm{Lp}(g_{s-i})^{\infty}$,
 Compute a reduced Gröbner basis G_{s-i-1} of $\mathbb{I}(G_{s-i}) + \mathbb{I}\big(\mathrm{Lp}(g_{s-i})\big)$,
 $\mathfrak{T}_i := \mathbf{Solve}\big(\mathbb{I}(G'_{s-i})\big)$,
 $\mathfrak{T} := \mathfrak{T}' \cup \mathfrak{T}_i$,
 $i := i + 1$.

Figure 42.1. Möller's Algorithm

Example 42.8.8.   To illustrate the algorithm let us consider Example 39.2.3 where, writing

$$H_1 := \{Z_1^2 - Z_1, g_3, g_4\},$$
$$\cup\, \{2Z_1 Z_3 - 2Z_3 + 3Z_2^2 + 6Z_1 Z_2 - 9Z_2 - 2Z_1 + 2\}$$
$$\cup\, \{2Z_2 Z_3 - 2Z_3 + 3Z_2^2 - 4Z_1 Z_2 - 5Z_2 + 4Z_1 + 2\},$$
$$h_1 := 2Z_3^2 - 8Z_3 + 15Z_2^2 + 30Z_1 Z_2 - 45Z_2 + 6,$$
$$H_2 := \{Z_1^2 - Z_1, Z_1 Z_2\},$$
$$h_2 := Z_2^2 - 2Z_2,$$
$$h_3 := 2Z_3 + 3Z_2 - 4Z_1 - 2,$$
$$H_3 := \{Z_1^2 - Z_1, Z_1 Z_2\},$$
$$h_4 := Z_2^2 - Z_2,$$
$$h_5 := Z_3 + 3Z_2 - 2Z_1 - 1,$$

we obtain

$$\mathsf{J}_1 := \{g_1, \ldots, g_8\},$$
$$\mathsf{J}_2 := \mathbb{I}(\{\mathrm{Lp}(g_i), 1 \le i \le 7\}) = \mathbb{I}(Z_1 - 2, Z_2),$$
$$\mathsf{t}_1 := \{Z_1 - 2, Z_2\},$$
$$\mathbf{Solve}(\mathsf{J}_2) := \{\mathsf{t}_1\},$$
$$\mathrm{NF}(g_8, \mathsf{t}_1) = Z_3^3 - 3Z_3^2 + 2Z_3,$$

$$t_1 := \{Z_1 - 2, Z_2, Z_3^3 - 3Z_3^2 + 2Z_3\},$$
$$J_3 := J_1 : (Z_1 - 2) = \mathbb{I}(H_1 \cup \{h_1\}),$$
$$J_5 := \mathbb{I}(\{\mathrm{Lp}(h) : h \in H_1\}) = \mathbb{I}(Z_1 - 1, Z_2 - 1),$$
$$t_2 := \mathbb{I}(\{Z_1 - 1, Z_2 - 1\}),$$
$$\mathbf{Solve}(J_5) := \{t_2\},$$
$$\mathrm{NF}(h_1, t_2) = Z_3^2 - 4Z_3 + 3,$$
$$t_2 := \{Z_1 - 1, Z_2 - 1, Z_3^2 - 4Z_3 + 3\},$$
$$J_6 := J_3 : (Z_2 - 1) = \mathbb{I}(H_2 \cup \{h_2, h_3\}),$$
$$J_7 := \mathbb{I}(\{\mathrm{Lp}(h) : h \in H_2\}) = \mathbb{I}(Z_1, Z_1^2 - Z_1) = \mathbb{I}(Z_1),$$
$$t_3 := \{Z_1, h_2, h_3\} = \{Z_1, Z_2^2 - 2Z_2, 2Z_3 + 3Z_2 - 2\},$$
$$J_8 := J_6 : Z_1 = \mathbb{I}(Z_1 - 1, Z_2),$$
$$t_4 := \{Z_1 - 1, Z_2, h_3\} = \{Z_1 - 1, Z_2, 2Z_3 - 6\},$$
$$\mathbf{Solve}(J_6) := \{t_3, t_4\},$$
$$J_9 := (J_3 + \mathbb{I}(Z_2 - 1)) : (Z_1 - 1) := \mathbb{I}(Z_1, Z_2 - 1, Z_3 + 2),$$
$$t_5 := \{Z_1, Z_2 - 1, Z_3 + 2\},$$
$$\mathbf{Solve}(J_9) := \{t_5\},$$
$$\mathbf{Solve}(J_3) := \{t_i, 2 \le i \le 5\},$$
$$J_4 := (J_3 + \mathbb{I}(Z_2 - 1)) : (Z_1 + Z_2 - 2) = \mathcal{Q},$$
$$\mathbf{Solve}(J_4) := \emptyset,$$
$$\mathbf{Solve}(J_1) := \{t_i, 1 \le i \le 5\},$$

and

$$\mathcal{Z}(J) = \{b_j : 1 \le j \le 9\} = \bigsqcup_{i=1}^{5} \mathcal{Z}(t_i),$$

with

$$\mathcal{Z}(t_1) = \{b_3, b_8, b_9\}, \; \mathcal{Z}(t_2) = \{b_6, b_7\}, \; \mathcal{Z}(t_3) = \{b_1, b_4\}, \; \mathcal{Z}(t_4) = \{b_5\}, \; \mathcal{Z}(t_5) = \{b_2\}.$$

$$\boxed{\odot}$$

42.9 Rouillier: Rational Univariate Representation

Let us assume we are given a zero-dimensional ideal $J \subset \mathcal{Q}$ via the Gröbner representation

$$\mathbf{b} = \{[b_1], \dots, [b_s]\} \subset A = \mathcal{Q}/J, \qquad A_h := \left(a_{ij}^{(h)}\right) = M([Z_h], \mathbf{b}), \qquad 1 \le h \le r,$$

and let us remark that, via a direct application of the Alonso–Raimondo–Traverso Algorithm (Remark 40.8.1), we can reduce, with good complexity, to the case in which

(1) J is radical,
(2) we have a linear form $Y := \sum_h c_h Z_h$ which is a *separating element* of $\mathcal{Z}(J)$.

We recall that

Definition 42.9.1. *A polynomial $f \in Q$ is called a* separating element *of $\mathcal{Z}(\mathsf{J})$ iff, for each $\alpha, \beta \in \mathcal{Z}(\mathsf{J})$, we have $\alpha \neq \beta \implies f(\alpha) \neq f(\beta)$.* ▣

The application of the Alonso–Raimondo–Traverso Algorithm has the further advantage that the separating linear form $Y := \sum_h c_h Z_h$ thus obtained is an *allgemeine* coordinate for J, so that (Corollary 34.3.4) the Alonso–Raimondo–Traverso Algorithm returns a triangular set

$$(g_0(Y), Z_1 - g_1(Y), \ldots, Z_r - g_r(Y) \subset K[Y, Z_1, \ldots, Z_r] \tag{42.1}$$

of the ideal $\mathsf{J}^+ := \mathsf{J} + \left(Y - \sum_h c_h Z_h\right) \subset K[Y, Z_1, \ldots, Z_r]$, where $g_i \in K[Y]$, $\deg(g_i) < \deg(g_0) = \#(\mathcal{Z}(\mathsf{J}))$ and (since J is radical) g_0 is squarefree.

Example 42.9.2. For the radical ideal $\mathsf{J} \subset \mathbb{C}[Z_1, Z_2, Z_3]$ discussed in Examples 39.2.3 and 40.3.3 and the separating element or *allgemeine* coordinate $Y = -3Z_1 + Z_2 + 3Z_3$, we have

$$g_0 = Y^9 + Y^8 - 90Y^7 - 142Y^6 + 2489Y^5$$
$$+ 4689Y^4 - 20880Y^3 - 31428Y^2 + 45360Y,$$
$$389188800g_1 = -8611Y^8 + 29288Y^7 + 697698Y^6$$
$$- 2278040Y^5 - 15347699Y^4 + 56296512Y^3$$
$$+ 44649972Y^2 - 473227920Y + 778377600,$$
$$640640g_2 = 19Y^8 - 108Y^7 - 1426Y^6 + 7808Y^5$$
$$+ 31851Y^4 - 167652Y^3 - 185004Y^2 + 955152Y,$$
$$778377600g_3 = -24917Y^8 + 102316Y^7 + 1972926Y^6$$
$$- 7718320Y^5 - 43595053Y^4 + 180492084Y^3$$
$$+ 164226564Y^2 - 1073833200Y + 1556755200.$$

▣

Proposition 42.9.3 (Alonso–Becker–Roy–Wörmann). *With the current notation, and setting*

$$\mathcal{Z}(\mathsf{J}) := \{\alpha_1, \ldots, \alpha_s\} \subset K^r, \qquad \alpha_i = (a_1^{(i)}, \ldots, a_r^{(i)}), \qquad \beta_i := \sum_h c_h a_h^{(i)},$$

there are polynomials $h_1(Y), \ldots, h_r(Y) \in K[Y]$, $\deg(h_i) < \deg(g_0)$, such that

$$\mathsf{J}^+ = \mathbb{I}\left(g_0(Y), g_0'(Y)Z_1 - h_1(Y), \ldots, g_0'(Y)Z_r - h_r(Y)\right) \subset K[Y, Z_1, \ldots, Z_r]. \tag{42.2}$$

Moreover, for each ι, $1 \leq \iota \leq r$, we have

$$h_\iota(Y) = \sum_{i=1}^s a_\iota^{(i)} \prod_{j \neq i} (Y - \beta_j). \tag{42.3}$$

Proof. Since $g_0(Y)$ is squarefree, $g_0'(Y)$ is invertible in $K[Y]/g_0$; thus the $h_i :=$ **Rem**$(g_0'g_i, g_0)$, satisfy the required property.

Since $g_0'(Y) = \sum_{i=1}^{s} \prod_{j \neq i} (Y - \beta_j)$, for each l, $1 \leq l \leq s$, we have

$$g_0'(\beta_l)a_\iota^{(l)} = a_\iota^{(l)} \sum_{i=1}^{s} \prod_{j \neq i} (\beta_l - \beta_j)$$
$$= a_\iota^{(l)} \prod_{j \neq l} (\beta_l - \beta_j)$$
$$= \sum_{i=1}^{s} a_\iota^{(i)} \prod_{j \neq i} (\beta_l - \beta_j)$$
$$= h_\iota(\beta_l).$$

Remark 42.9.4. Compare Proposition 42.9.3 with Kronecker's result (41.3); the only difference is that here the assumption of the *primality* of the ideal J is relaxed to *radicality*.

Remark 42.9.5 (Alonso–Becker–Roy–Wörmann). Denoting by S the size of the elements $a_{ij}^{(\iota)}$ in the matrices A_ι, clearly (42.3) implies that the coefficients in the Kronecker parametrization (42.2) have size $\mathcal{O}(Ss)$, giving a strong advantage over the $\mathcal{O}(Ss^2)$ size of the coefficients of the *allgemeine* basis (42.1).

Example 42.9.6. In the setting discussed in Example 42.9.2 we have

$$g_0'(Y) = 9Y^8 + 8Y^7 - 630Y^6 - 852Y^5 + 12445Y^4$$
$$+ 18756Y^3 - 62640Y^2 - 62856Y + 45360,$$
$$h_1(Y) = 9Y^8 + 5Y^7 - 638Y^6 - 668Y^5 + 13655Y^4$$
$$+ 15591Y^3 - 92178Y^2 - 76896Y + 90720,$$
$$h_2(Y) = 5Y^8 - 348Y^6 - 62Y^5 + 7155Y^4 + 2790Y^3 - 39852Y^2 - 20088Y,$$
$$h_3(Y) = 7Y^8 + 65Y^7 - 380Y^6 - 3966Y^5 + 3455Y^4$$
$$+ 56421Y^3 - 5562Y^2 - 191160Y + 90720.$$

Corollary 42.9.7 (Alonso–Becker–Roy–Wörmann).
For each $f \in \mathcal{Q}$, there is an $h_f(Y) \in K[Y]$ such that

$$g_0'(Y)f(Z_1, \ldots, Z_r) - h_f(Y) \in \mathsf{J}^+, \deg(h_f) < \deg(g_0).$$

Proof. It is sufficient to set $h_f(Y) :=$ **Rem**$(f(h_1(Y), \ldots, h_r(Y)), g_0(Y))$.

Both Proposition 42.9.3 and Corollary 42.9.7 are existential results; thus we need a computational definition of $h_f(Y)$. To obtain it we consider a new variable S, the extension field $K(S)$, the ideal

$$J^e := JK(S)[Z_1, \ldots, Z_r] \subset K(S)[Z_1, \ldots, Z_r] = \mathcal{Q} \otimes_K K(S),$$

the algebra $\bar{\mathsf{A}} := K(S)[Z_1, \ldots, Z_r]/J^e = \mathsf{A} \otimes_K K(S)$, the element $\bar{f} := Y + Sf \in \bar{\mathsf{A}}$ and the matrix $A_{\bar{f}}$.

Theorem 42.9.8 (Alonso–Becker–Roy–Wörmann). *With the current notation, it holds that:*

(1) *the minimal polynomial $m_{\bar{f}}(T) \in K(S)[T]$ of $A_{\bar{f}}$ is given by*

$$m_{\bar{f}}(T) = \prod_{i=1}^{s} (T - \beta_i - Sf(\alpha_i)) \in K[S, T];$$

(2) *setting $p(S, T) := \partial m_{\bar{f}}/\partial S$, we have*

$$p(0, T) = -\sum_{i=1}^{s} f(\alpha_i) \prod_{\substack{j=1 \\ j \neq i}}^{r} (T - \beta_j);$$

(3) *for each $\alpha \in \mathcal{Z}(\mathsf{J})$ it holds that $f(\alpha) = p(0, \alpha)/g_0'(\alpha)$;*
(4) $h_f(T) = p(0, T).$

Proof. Part (1) is obvious, (2) requires a trivial verification and (4) is a direct consequence of (3), so we have just to prove (3); for each ι, $1 \leq \iota \leq s$, we have

$$p(0, \alpha_\iota) = -\sum_{i=1}^{s} f(\alpha_i) \prod_{\substack{j=1 \\ j \neq i}}^{s} (\beta_\iota - \beta_j)$$

$$= -f(\alpha_\iota) \prod_{\substack{j=1 \\ j \neq \iota}}^{r} (\beta_\iota - \beta_j)$$

$$= -f(\alpha_\iota) g_0'(\alpha_\iota).$$

$\boxed{\odot}$

Remark 42.9.9 (Alonso–Becker–Roy–Wörmann). The computation of the Kronecker parametrization (42.2) does not require one to assume that J is radical; assuming as known $\mathsf{s} = \#\mathcal{Z}(\mathsf{J})$ (an efficient way of computing it is discussed in Corollary 42.9.13 below), and denoting by $\chi(T)$ and $m(T)$ respectively the characteristic and minimal polynomials of A_Y, we have

$$g_0(T) = \sqrt{\chi(T)} = \sqrt{m(T)}.$$

Once a linear form Y is fixed, the minimal polynomial $m(T)$ of A_Y which coincides with the minimal polynomial of the element $Y \in \mathsf{A}$ can be obtained directly by checking the successive powers $[1], [Y], [Y^2], \ldots$ for linear dependency; Y is then a separating element iff $\deg(\sqrt{m}) = \#\mathcal{Z}(\mathsf{J})$.

In order to make this approach effective, all one needs is the availability of a finite set of linear forms which contains at least one separating element. Such a finite set is provided by the Chistov–Grigoriev Corollary 35.6.4. ◎

Remark 42.9.10. Alternatively, once $\#\mathcal{Z}(\mathsf{J})$ is known, a separating linear form Y can be obtained by an easy adaptation of the Alonso–Raimondo Algorithm 35.7.1[56] that avoids the evaluation of the swelling coefficients of the $g_i, i > 0$:

(1) Set $Y := \sum_i a_i Z_i$, $J := 1$.
(2) By linear algebra on the Gröbner descriptions of

$$[1], [Y], [Y^2], \ldots$$

compute the minimal polynomial $m(T) \in K[T]$ such that $m(T) \in \mathsf{J}^+$.
(3) If $d := \deg(\sqrt{m}) < \#\mathcal{Z}(\mathsf{J})$ then set $j := J$ and
    (a) while $j \leq r$ verify whether the elements

$$[(\sqrt{m})'(Y)Z_j], [1], [Y], [Y^2], \ldots, [Y^{d-1}]$$

    are linearly dependent.
    (b) If so, set $j := j + 1$ and go to (3a).
    (c) If instead they are linearly independent, set $J := j$, $Y := Y + cZ_j$, and $a_j := a_j + c$ and go to (2).
(4) When $\deg(\sqrt{m}) = \#\mathcal{Z}(\mathsf{J})$, we have that
    • $\mathsf{J}^+ + (\sqrt{m})$ is radical,
    • $Y := \sum_i a_i Z_i$ is a separating linear form and
    • $\sqrt{m}$ is its minimal polynomial. ◎

Lemma 42.9.11. *Let $f \in \mathcal{Q}$ be a separating element of $\mathcal{Z}(\mathsf{J})$ and let $\mathsf{s} = \#\mathcal{Z}(\mathsf{J})$. Then $\{1, [f], [f^2], \ldots, [f^{\mathsf{s}-1}]\}$ is a K-linearly independent set of A.*

Proof. Assume that $m(T) := \sum_{i=0}^{\mathsf{s}-1} a_i T^i \in k[Y]$ is such that $m(f) \equiv 0 \bmod \mathsf{J}$; since f is a separating element of $\mathcal{Z}(\mathsf{J})$, the polynomial $m(T)$ has s distinct roots $\{f(\alpha), \alpha \in \mathcal{Z}(\mathsf{J})\}$, giving a contradiction. ◎

For every polynomial $h \in \mathcal{Q}$, we can consider the bilinear map

$$\ell_h : \mathsf{A} \times \mathsf{A} \to K, (p, q) \mapsto \mathrm{Tr}(A_{hpq})$$

and the corresponding quadratic form associated with ℓ_h

$$Q_h(x_1, \ldots, x_s) := \sum_{j,l} \gamma_{j,l}^{(h)} x_j x_l,$$

---

[56] Compare also Remark 40.8.1.

which satisfies

$$\ell_h(p, p) = \mathrm{Tr}(A_{hp^2}) = \sum_{j,l} \gamma_{j,l}^{(h)} c_j c_l = Q_h(c_1, \ldots, c_s)$$

for each $p = \sum_{i=1}^{s} c_i[b_i] \in \mathsf{A} = \mathrm{Span}_K\{[b_1], \ldots, [b_s]\}$.

Proposition 42.9.12 (Rouillier). *For a polynomial $h \in \mathcal{Q}$, the quadratic form Q_h has $\#\{\alpha \in \mathcal{Z}(\mathsf{J}) : h(\alpha) \neq 0\}$ as its rank.*

Proof. Let $f \in \mathcal{Q}$ be a separating element of $\mathcal{Z}(\mathsf{J})$. By the above lemma, the set $\{1, [f], [f^2], \ldots, [f^{\mathsf{s}-1}]\}$ is K-linearly independent and thus can be completed to a basis

$$\{1, [f], [f^2], \ldots, [f^{\mathsf{s}-1}], [b_{\mathsf{s}+1}], \ldots, [b_s]\} = \{[b_1], \ldots, [b_s]\}$$

of A. For each $p \in \mathcal{Q}$, let $c_j \in K$ be such that

$$[p] = \sum_{j=0}^{\mathsf{s}-1} c_j[f]^j + \sum_{j=\mathsf{s}+1}^{s} c_j[b_j].$$

Thus, setting $Y_i := \sum_{j=1}^{s} c_j b_j(\alpha_i)$, $1 \leq i \leq \mathsf{s}$, by Corollary 40.5.2 we have

$$Q_h = \mathrm{Tr}(A_{hp^2}) = \sum_{i=1}^{\mathsf{s}} s_i h(\alpha_i) \left(\sum_{j=1}^{s} c_j b_j(\alpha_i) \right)^2 = \sum_{i=1}^{\mathsf{s}} s_i h(\alpha_i) Y_i^2.$$

The matrix

$$\begin{pmatrix} 1 & f(\alpha_1) & f(\alpha_1)^2 & \cdots & f(\alpha_1)^{\mathsf{s}-1} \\ 1 & f(\alpha_2) & f(\alpha_2)^2 & \cdots & f(\alpha_2)^{\mathsf{s}-1} \\ \vdots & \vdots & \vdots & \ddots & \vdots \\ 1 & f(\alpha_{\mathsf{s}}) & f(\alpha_{\mathsf{s}})^2 & \cdots & f(\alpha_{\mathsf{s}})^{\mathsf{s}-1} \end{pmatrix}$$

is invertible – being Vandermonde since $f \in \mathcal{Q}$ separates $\mathcal{Z}(\mathsf{J})$ – and is a submatrix of the matrix associated with the linear forms that define the linear change of variables Y_i.

Thus the Y_i are linearly independent and (compare Theorem 13.5.2) the rank of Q_h is the number of roots of J which are not roots of h. $\boxed{\odot}$

Corollary 42.9.13. *The rank of*

$$Q_1(x_1, \ldots, x_s) := \sum_{j,l} \gamma_{j,l}^{(1)} x_j x_l = \sum_{j,l} \mathrm{Tr}(A_{b_j b_l}) x_j x_l$$

is $\mathsf{s} = \#\mathcal{Z}(\mathsf{J})$. $\boxed{\odot}$

Once a Kronecker parametrization (42.2) is obtained via Theorem 42.9.8, the multiplicity of each root can be obtained in the following way. Denoting by B_f, for each $f \in \mathcal{Q}$, the matrix representing the endomorphism

$$\bar{\Phi} : \mathcal{Q}/\sqrt{\mathsf{J}} \to \mathcal{Q}/\sqrt{\mathsf{J}}, \quad \bar{\Phi}([g]) \mapsto [fg],$$

we have the following:

Lemma 42.9.14 (Alonso–Becker–Roy–Wörmann). *Let Y be a separating linear form of $\mathcal{Z}(\mathsf{J})$. Then:*

(1) *The matrix*
$$\begin{pmatrix} \mathrm{Tr}(B_1) & \cdots & \mathrm{Tr}(B_{Y^{s-1}}) \\ \vdots & \ddots & \vdots \\ \mathrm{Tr}(B_{Y^{s-1}}) & \cdots & \mathrm{Tr}(B_{Y^{2s-2}}) \end{pmatrix}$$
is invertible;

(2) *Let $a_0, \ldots, a_{s-1} \in K$ be the unique solution of the linear system*

$$\begin{pmatrix} \mathrm{Tr}(B_1) & \cdots & \mathrm{Tr}(B_{Y^{s-1}}) \\ \vdots & \ddots & \vdots \\ \mathrm{Tr}(B_{Y^{s-1}}) & \cdots & \mathrm{Tr}(B_{Y^{2s-2}}) \end{pmatrix} \begin{pmatrix} a_0 \\ \vdots \\ a_{s-1} \end{pmatrix} = \begin{pmatrix} \mathrm{Tr}(A_1) \\ \vdots \\ \mathrm{Tr}(A_{Y^{s-1}}) \end{pmatrix}$$

and set $F(Y) := \sum_{l=0}^{s-1} a_l Y^l$; then

$$F(\alpha_i) = s_i = \mathrm{mult}(\alpha_i, \mathsf{J}), \quad \forall \alpha_i \in \mathcal{Z}(\mathsf{J}).$$

Proof.

(1) This holds since the matrix is Hankel.

(2) We have, for each i, $1 \leq i \leq s$,

$$\sum_{i=1}^{s} s_i \beta_i^j = \mathrm{Tr}(A_{Y^j})$$
$$= \sum_{l=0}^{s-1} a_l \, \mathrm{Tr}(B_{Y^{j+l}})$$
$$= \sum_{l=0}^{s-1} a_l \sum_{i=1}^{s} \beta_i^{j+l}$$
$$= \sum_{i=1}^{s} \left(\sum_{l=0}^{s-1} a_l \beta_i^l \right) \beta_i^j$$
$$= \sum_{i=1}^{s} F(\alpha_i) \beta_i^j.$$

Since the matrix $\left(\beta_i^j \right)$ is Vandermonde, it follows that $F(\alpha_i) = s_i$ for each i $\quad \boxed{\odot}$

Up to now, in order to represent the roots of I, we have alternatively

- assumed J to be radical in order to apply Corollary 34.3.4, or
- applied Proposition 42.9.3, using the minimal polynomial $m(Y)$ of $Y \in A$.

In both cases we lose the multiplicity of each root, but we recover these multiplicities via Lemma 42.9.14.

An improvement allows us to remove the requirement that J be radical and to use directly the characteristic polynomial $\chi(T) = \prod_{i=1}^{s}(T - \beta_i)^{s_i}$, thus deducing the multiplicities directly.

Proposition 42.9.15 (Rouiller). *Let $J \subset Q$ be a zero-dimensional ideal, not necessarily radical, and set*

$$\mathcal{Z}(J) := \{\alpha_1, \ldots, \alpha_s\} \subset K^r, \quad \alpha_i = (a_1^{(i)}, \ldots, a_r^{(i)}), \, s_i := \mathrm{mult}(\alpha_i, J).$$

Let $f \in Q$ be a separating element of $\mathcal{Z}(J)$; denote by $\chi := \chi_f := \sum_{i=0}^{s} c_i T^{s-i}$ the characteristic polynomial of A_f and set $\beta_i := f(\alpha_i)$.

For each $h \in Q$ set $\gamma_h(T) := \sum_{i=1}^{s} s_i h(\alpha_i) \prod_{\substack{j=1 \\ j \neq i}}^{s}(T - \beta_j).$

Then:

(1) $\chi = \chi_f = \prod_{i=1}^{s}(T - \beta_i)^{s_i}$;

(2) $\chi'(T)/\chi(T) = \sum_{j \geq 0} \mathrm{Tr}(A_{f^j})/T^{j+1}$;

(3) $(s - i)c_i = \sum_{j=0}^{i} c_{i-j} \, \mathrm{Tr}(A_{f^j})$ *for $i = 0, \ldots, s$*;

(4) $\gamma_h(\beta_l) = s_l h(\alpha_l) \prod_{\substack{j=1 \\ j \neq l}}^{s}(\beta_l - \beta_j)$;

(5) $h(\alpha_l) = \gamma_h(\beta_l)/\gamma_1(\beta_l)$ *for $1 \leq l \leq s$*;

(6) $\gamma_h(T) = \sqrt{\chi(T)} \sum_{j \geq 0} \mathrm{Tr}(A_{hf^j})/T^{j+1}$;

(7) *setting $\sqrt{\chi} := \sum_{i=0}^{s} a_i T^{s-i}$, it holds that*

$$\gamma_h(T) = \sum_{i=0}^{s-1} \sum_{j=0}^{s-i-1} \mathrm{Tr}(A_{hf^j}) a_i T^{s-i-j-1};$$

(8) $\gamma_1(T) = \chi'(T)/\gcd(\chi(T), \chi'(T))$;

(9) $\gcd(\gamma_1(T), \chi(T)) = 1$;

(10) *for each i, $s_i = \gamma_1(\beta_i)/(\sqrt{\chi})'(\beta_i)$*;

(11) *the squarefree decomposition (Definition 4.7.2) $\chi = \prod_l \chi_l^l$ of χ is obtained via*
$$\chi_l := \gcd(\gamma_1(T) - l(\sqrt{\chi})'(T), \sqrt{\chi}).$$

Proof.

(1) See Corollary 40.5.2.

(2) We have

$$\frac{\chi'(T)}{\chi(T)} = \sum_{i=1}^{s} \frac{s_i}{T - \beta_i}$$

$$= \sum_{i=1}^{s} \frac{s_i}{T} \frac{1}{1 - \beta_i/T}$$

$$= \sum_{i=1}^{s} \frac{s_i}{T} \sum_{j \geq 0} \left(\frac{\beta_i}{T}\right)^j$$

$$= \sum_{j \geq 0} \frac{\sum_{i=1}^{s} s_i \beta_i^j}{T^{j+1}}$$

$$= \sum_{j \geq 0} \frac{\mathrm{Tr}(A_{f^j})}{T^{j+1}}.$$

(3) We have

$$\sum_{i=0}^{s} (s - i) c_i T^{s-i-1} = \chi'(T)$$

$$= \chi(T) \sum_{j \geq 0} \frac{\mathrm{Tr}(A_{f^j})}{T^{j+1}}$$

$$= \sum_{l=0}^{s-1} \sum_{j=0}^{s-l-1} c_l \, \mathrm{Tr}(A_{f^j}) T^{s-l-j-1}$$

$$= \sum_{i=0}^{s-1} \sum_{j=0}^{i} c_{i-j} \, \mathrm{Tr}(A_{f^j}) T^{s-i-1}.$$

(4) We have $\gamma_h(\beta_l) = \sum_{i=1}^{s} s_i h(\alpha_i) \prod_{\substack{j=1 \\ j \neq i}} (\beta_l - \beta_j) = s_l h(\alpha_l) \prod_{\substack{j=1 \\ j \neq l}} (\beta_l - \beta_j).$

(5) Obvious.

(6) We have

$$\frac{\gamma_h(T)}{\sqrt{\chi(T)}} = \sum_{i=1}^{s} \frac{s_i h(\alpha_i)}{T - \beta_i} = \sum_{j \geq 0} \frac{\sum_{i=1}^{s} s_i h(\alpha_i) \beta_i^j}{T^{j+1}} = \sum_{j \geq 0} \frac{\mathrm{Tr}(A_{hf^j})}{T^{j+1}}.$$

(7) This is a direct consequence of (6).

(8) Obvious.

(9) Obvious.

(10) We have

$$\frac{\gamma_1(T)}{(\sqrt{\chi})'(T)} = \frac{\sum_{i=1}^{s} s_i \prod_{\substack{j=1 \\ j \neq i}}^{s} (T - \beta_j)}{\sum_{i=1}^{s} \prod_{\substack{j=1 \\ j \neq i}}^{s} (T - \beta_j)}$$

whence the claim.

(11) The claim follows easily from

$$\gamma_1(T) - l(\sqrt{\chi})'(T) = \sum_{i=1}^{s}(s_i - l)\prod_{\substack{j=1\\j\neq i}}^{s}(T - \beta_j)$$

$$= \sum_{\substack{j=1\\s_i\neq l}}^{s}(s_i - l)\prod_{\substack{j=1\\j\neq i}}^{s}(T - \beta_j)$$

$$= \left(\prod_{\substack{j=1\\s_i=l}}^{s}(T - \beta_i)\right)\left(\sum_{\substack{j=1\\s_i\neq l}}^{s}(s_i - l)\prod_{\substack{j=1\\s_i\neq l}}^{s}(T - \beta_j)\right).$$

⊙

Definition 42.9.16 (Rouillier). *Let $J \subset Q$ be a zero-dimensional ideal. A Univariate Representation (χ, Φ) of J is an assignment of polynomials $\chi(T), \gamma_0(T), \gamma_1(T), \ldots, \gamma_r(T) \in K[T]$ that defines the K-isomorphism*

$$\Phi : \{\alpha \in K : \chi(\alpha) = 0\} \to \mathcal{Z}(J) : \alpha \mapsto \left(\frac{\gamma_1(\alpha)}{\gamma_0(\alpha)}, \ldots, \frac{\gamma_r(\alpha)}{\gamma_0(\alpha)}\right),$$

which satisfies $\mathrm{mult}(\Phi(\alpha), J) = \mathrm{mult}(\alpha, \mathbb{I}(\chi))$.

If, moreover, $f \in Q$ is a separating element of $\mathcal{Z}(J)$ and $\chi := \chi_f$ is the characteristic polynomial of A_f, the univariate representation (χ, Φ) is called the Rational Univariate Representation (RUR) of J associated with f. ⊙

Corollary 42.9.17 (Rouillier). *With the assumptions and notations of Proposition 42.9.15 and setting*

$$\Phi : \{f(\alpha), \alpha \in \mathcal{Z}(J)\} \to \mathcal{Z}(J) : \beta = f(\alpha) \mapsto \left(\frac{\gamma_{Z_r}(\beta)}{\gamma_1(\beta)}, \ldots, \frac{\gamma_{Z_1}(\beta)}{\gamma_1(\beta)}\right) = \alpha,$$

(χ_f, Φ) is the Rational Univariate Representation of J associated with f. ⊙

Remark 42.9.18 (Rouillier). For the rational univariate representation

$$(\chi, \Phi), \Phi : \{\alpha \in K : \chi(\alpha) = 0\} \to \mathcal{Z}(J) : \alpha \mapsto \left(\frac{\gamma_1(\alpha)}{\gamma_0(\alpha)}, \ldots, \frac{\gamma_r(\alpha)}{\gamma_0(\alpha)}\right)$$

of J and the factorization $\chi(T) = \prod_i \chi_i(T)$, $\gcd(\chi_i, \chi_j) = 1$, setting

$$\gamma_{ji}(T) := \mathbf{Rem}(\gamma_j(T), \chi_i) \quad \text{for each } i, j$$

and

$$\Phi_i : \{\alpha \in K : \chi_i(\alpha) = 0\} \to \left\{\left(\frac{\gamma_{1i}(\alpha)}{\gamma_{0i}(\alpha)}, \ldots, \frac{\gamma_{ni}(\alpha)}{\gamma_{0i}(\alpha)}\right)\right\}$$

we have that (χ_i, Φ_i) is a Rational Univariate Representation, for each i, and it holds that

$$\mathcal{Z}(\mathsf{J}) = \bigsqcup_i \left\{ \left(\frac{\gamma_{1i}(\alpha)}{\gamma_{0i}(\alpha)}, \dots, \frac{\gamma_{ni}(\alpha)}{\gamma_{0i}(\alpha)} \right) : \alpha \in \mathsf{K}, \chi_i(\alpha) = 0 \right\}.$$

Remark 42.9.19. If J is radical, the representation of $\mathcal{Z}(\mathsf{J})$ given in Proposition 42.9.3 is an RUR of J associated with to $\sum_h c_h Z_h$.

Algorithm 42.9.20 (Rouillier). Given a zero-dimensional ideal $\mathsf{J} \subset \mathcal{Q}$ via a Gröbner representation

$$\mathbf{b} = \{[b_1], \dots, [b_s]\} \subset \mathsf{A} = \mathcal{Q}/\mathsf{J}, \quad A_h := \left(a_{ij}^{(h)} \right) = M([Z_h], \mathbf{b}), \quad 1 \le h \le r,$$

and assuming that matrices A_{b_i} represent the endomorphisms $\Phi_{b_i} : \mathsf{A} \to \mathsf{A}$, an RUR of J associated with a linear form can be computed by the following procedure:

(1) For each $j, l, 1 \le j, l \le s$ compute $\mathrm{Tr}(A_{b_j b_l})$.

(2) Compute $\mathsf{s} := \#\mathcal{Z}(\mathsf{J})$ as the rank of $Q_1(x_1, \dots, x_s) := \sum_{j,l} \mathrm{Tr}(A_{b_j b_l}) x_j x_l$ (Corollary 42.3.6).

(3) Choose (via Corollary 35.6.4) a linear form $Y := \sum_h c_h Z_h$ and compute, via Proposition 42.9.15(3), the characteristic polynomial χ_Y of A_Y. Repeat this until $\deg(\sqrt{\chi_Y}) = \#\mathcal{Z}(\mathsf{J})$, thus granting that Y is a separating element.

Alternatively deduce via Remark 42.9.10 a separating linear form $Y := \sum_h c_h Z_h$ and compute the characteristic polynomial χ_Y of A_Y via Proposition 42.9.15(3).

(4) Compute $\gamma_1(T), \gamma_{Z_1}(T), \dots, \gamma_{Z_r}(T)$ via Proposition 42.9.15(6), (7).

It can be proved that the complexity of the algorithm is $\mathcal{O}(s^3 + rs^2)$ arithmetic operations in K and that, in the case $K = \mathbb{Q}$, the cost is $\mathcal{O}((s^3 + rs^2)\mathsf{M}(ls^2))$ binary arithmetic operations, where l denotes the bit-size of the entries of the matrices A_h and $\mathsf{M}(\ell)$ denotes the cost of multiplying two integers of bit-size ℓ.

Remark 42.9.21 (Dahen). Clearly (compare Proposition 42.9.3) the relation between the elements of the *allgemeine* basis (42.1) and the RUR (42.2) is given by $h_i := \mathbf{Rem}(g_0' g_i, g_0)$.

The better behaviour of Kronecker's parametrization w.r.t. *allgemeine* bases can be generalized to triangular sets.

In fact, given a *triangularizable* zero-dimensional ideal $\mathsf{J} \subset \mathbb{Q}[Z_1, \dots, Z_r]$ via a set $\mathsf{G} := (g_1, \dots, g_m) \subset \mathbb{Q}[Z_1, \dots, Z_r]$, denoting by

$$T := (f_1, \dots, f_r) \subset \mathbb{Q}[Z_1, \dots, Z_r]$$

its triangular set and, for each i, $1 \leq i < r$, denoting by

$$N_{i+1} := NF\left(f_{i+1}\prod_{j=1}^{i}\frac{\partial f_j}{\partial Z_j}, \mathbb{I}(f_1, \ldots, f_i)\right)$$

the normal form of $f_{i+1}\prod_{j=1}^{i}\partial f_j/\partial Z_j$ w.r.t. the Gröbner basis $(f_1, \ldots, f_i)$, both theoretical and practical analyses suggest that the *height*[57] of the triangular set $N := (f_1, N_2, \ldots, N_r)$ of J is better than that of T.[58]

[57] For a set $\mathsf{G} := (g_1, \ldots, g_m) \subset \mathbb{Q}[Z_1, \ldots, Z_r]$, its height is the value $h(\mathsf{G}) := \max(\log(c(g_i, \tau)) : 1 \leq i \leq r, \tau \in \mathcal{W})$.

[58] We have $h(T) = \mathcal{O}(rhd^{2r})$ and $h(N) = \mathcal{O}(rhd^r)$, where $d = \max(\deg(g_i))$ and $h = h(\mathsf{G})$.

43

Lagrange II

Given a (squarefree) polynomial $f \in k[T]$, a natural question is to determine its Galois group (Definition 14.1.3) over k.[1] Using classical techniques for the representation of a group of finite order as a permutation group (Section 43.1) and as a permutation group of the set of roots of a separable polynomial (Section 43.2), Lagrange resolvents (Sections 43.3 and 43.5) and Cauchy modules (Section 43.4), recently the problem has been completely solved (Sections 43.6 and 43.7) by Annick Valibouze and Jean-Marie Arnaudies for polynomials f of degree bounded by 11.

43.1 Representation of Groups as Permutation Groups

Let G be a finite group.[2]

Example 43.1.1.   Throughout Section 43.1, as an example we take as G the alternating group A_4, whose 12 elements are given by

$$
\begin{array}{llllll}
g_1 & = & \mathrm{Id}, & g_2 & = & (1,3,2), & g_3 & = & (1,2),(3,4), \\
g_4 & = & (1,2,3), & g_5 & = & (2,4,3), & g_6 & = & (2,3,4), \\
g_7 & = & (1,4,3), & g_8 & = & (1,4,2), & g_9 & = & (1,3,4), \\
g_{10} & = & (1,3),(2,4), & g_{11} & = & (1,2,4), & g_{12} & = & (1,4),(2,3)
\end{array}
$$

---

[1] Other equally natural questions are

- how to solve f by radicals, in the case where this is possible, and
- the *Galois inverse problem* of determining, given a finite group G, a polynomial $f \in k[T]$ for which G is its Galois group.

These questions are not discussed in this book.

[2] For this theory, see Burnside W., *Theory of Groups of Finite Order*, Cambridge University Press (1911), Chapter XII.

and whose corresponding multiplication table is

	1	2	3	4	5	6	7	8	9	10	11	12
1	1	2	3	4	5	6	7	8	9	10	11	12
2	2	4	7	1	3	8	5	10	12	6	9	11
3	3	6	1	9	8	2	11	5	4	12	7	10
4	4	1	5	2	7	10	3	6	11	8	12	9
5	5	10	4	11	6	1	12	7	2	9	3	8
6	6	9	11	3	1	5	8	12	10	2	4	7
7	7	8	2	12	10	4	9	3	1	11	5	6
8	8	12	9	7	2	3	10	11	6	4	1	5
9	9	3	8	6	11	12	1	2	7	5	10	4
10	10	11	12	5	4	7	6	9	8	1	2	3
11	11	5	6	10	12	9	4	1	3	7	8	2
12	12	7	10	8	9	11	2	4	5	3	6	1

Now impose on the set of the subgroups $H \subset G$ the relation

$$H_1 \sim H_2 \iff \text{ there exists } \tau \in G : H_1 = \tau^{-1} H_2 \tau$$

and write $\mathcal{E} := \{\mathcal{C}_1, \ldots, \mathcal{C}_s\}$, the set of all the conjugacy classes.

We associate with each such class its *degree*,

$$\deg(\mathcal{C}_i) := [G : H], \quad H \in \mathcal{C}_i,$$

and its *weight*,

$$\mathsf{w}(\mathcal{C}_i) := \#G / \deg(\mathcal{C}_i) = \#H, \quad H \in \mathcal{C}_i.$$

We enumerate $\mathcal{E}$ in such a way that[3]

$$\mathsf{w}(\mathcal{C}_1) \leq \mathsf{w}(\mathcal{C}_2) \leq \cdots \leq \mathsf{w}(\mathcal{C}_s),$$

and we impose a partial ordering $\preceq$ on $\mathcal{E}$, setting $\mathcal{C}_i \preceq \mathcal{C}_j$ if the following equivalent conditions hold:

- there are $H \in \mathcal{C}_i$ and $H' \in \mathcal{C}_j$ such that $H \subset H'$;
- for each $H \in \mathcal{C}_i$ there is $H' \in \mathcal{C}_j$ such that $H \subset H'$.

Since in this setting we have $[G : H'] = [G : H][H : H']$, we also have $\mathcal{C}_i \preceq \mathcal{C}_j \implies i \leq j$.

Example 43.1.2. $G = \mathsf{A}_4$ has five conjugacy classes, each consisting of the subgroups of order (respectively) $1, 2, 3, 4, 12$:

$\mathcal{C}_1$ has $\deg(\mathcal{C}_1) = 12$, $\mathsf{w}(\mathcal{C}_1) = 1$, and consists of

$$H_{11} \quad := \quad H_1 \quad := \quad \{\mathrm{Id}\};$$

---

[3] In particular: $\mathcal{C}_1 = \{\{\mathrm{Id}_G\}\}$ and $\mathcal{C}_s = \{G\}$.

C_2 has $\deg(C_2) = 6$, $\mathsf{w}(C_2) = 2$, and consists of

$$
\begin{aligned}
H_{21} &:= H_2 &&:= \{\mathrm{Id}, g_3\}, \\
H_{22} &:= g_4 H_2 g_4^{-1} &&= \{\mathrm{Id}, g_{10}\}, \\
H_{23} : &:= g_2 H_2 g_2^{-1} &&= \{\mathrm{Id}, g_{12}\};
\end{aligned}
$$

C_3 has $\deg(C_3) = 4$, $\mathsf{w}(C_3) = 3$, and consists of

$$
\begin{aligned}
H_{31} &:= H_3 &&:= \{\mathrm{Id}, g_2, g_4\}, \\
H_{32} &:= g_7 H_3 g_7^{-1} &&= \{\mathrm{Id}, g_5, g_6\}, \\
H_{33} &:= g_5 H_3 g_5^{-1} &&= \{\mathrm{Id}, g_7, g_9\}, \\
H_{34} &:= g_6 H_3 g_6^{-1} &&= \{\mathrm{Id}, g_8, g_{11}\};
\end{aligned}
$$

C_4 has $\deg(C_4) = 3$, $\mathsf{w}(C_4) = 4$, and consists of

$$
H_{41} := H_4 := \{\mathrm{Id}, g_3, g_{10}, g_{12}\};
$$

C_5 has $\deg(C_5) = 1$, $\mathsf{w}(C_5) = 12$, and consists of

$$
H_{51} := H_5 := G.
$$

Naturally, we have

$$
C_1 \prec C_2 \prec C_4 \prec C_5, \quad C_1 \prec C_3 \prec C_5.
$$

Now let E be a finite set, $n := \#E$, and denote by S_E the group of the permutations of the set E.

Definition 43.1.3. *Each group morphism* $\Phi : G \to \mathsf{S}_E$ *is called a* representation *of G as a permutation group of degree n.*

This representation is said to be

- faithful *if* $\ker(\Phi) = \{\mathrm{Id}_G\}$,
- transitive *if for each $x, x_0 \in E$ there is a $g \in G : \Phi(g)(x_0) = x$.*

A representation $\Psi : G \to \mathsf{S}_F$ *is called* equivalent *to Φ (denoted as $\Psi \sim \Phi$) if there is a bijection $\Theta : E \to F$ satisfying $\Psi(g) = \Theta \circ \Phi(g) \circ \Theta^{-1}$ for each $g \in G$:*

$$
\begin{array}{ccc}
F & \xrightarrow{\Psi(g)} & F \\
\Theta \uparrow & & \Theta \uparrow \\
E & \xrightarrow{\Phi(g)} & E
\end{array}
$$

We remark that:

(1) $\ker(\Phi) = \bigcap_{x \in E} \{g \in G : \Phi(g)(x) = x\}$;
(2) if Ψ is equivalent to Φ we have
  - $\ker(\Phi) = \ker(\Psi)$,
  - $\{\{g \in G : \Phi(g)(x) = x\} : x \in E\} = \{\{g \in G : \Psi(g)(x) = x\} : x \in F\}$;

(3) if Φ is transitive, the set

$$C_\Phi := \{\{g \in G : \Phi(g)(x) = x\} : x \in E\} \in \mathcal{E}$$

is a conjugacy class, which is called the conjugacy class *associated* with Φ.

Assuming Φ to be transitive, fixing $x_0 \in E$, and denoting

- $H_0 := \{g \in G : \Phi(g)(x_0) = x_0\}$,
- $(G/H_0)_l := \{gH_0 : g \in G\}$, the set of the left classes of H_0,
- Θ the bijection $\Theta : E \to (G/H_0)_l$ defined by $\Theta(x) = \{g \in G : \Phi(g)(x_0) = x\}$,
- $\rho : G \to \mathsf{S}_{(G/H_0)_l}$ the representation $\rho(h)(H') = hH'$, for each $h \in G$ and each $H' \in (G/H_0)_l$;

then

Lemma 43.1.4. *It holds that $\rho \sim \Phi$.* $\quad\boxed{\odot}$

Corollary 43.1.5. *For a finite group G, denote by $\bar{\mathfrak{E}}$ the set of all transitive representations of G as a permutation group and by $\mathfrak{E}$ the set of the equivalence classes of the transitive representations of G as a permutation group: thus $\mathfrak{E} := \bar{\mathfrak{E}}/\sim$.*

We obtain a bijection between $\mathfrak{E}$ and $\mathcal{E}$ (see after Example 43.1.1) by associating with each $\Gamma \in \mathfrak{E}$ the conjugacy class C_Φ associated with each $\Phi : G \to \mathsf{S}_E$ belonging to Γ.

In particular, there are transitive representations of G as a permutation group of degree n only if $n = [G : H]$ for some subgroup $H \subset G$. $\quad\boxed{\odot}$

On the basis of this, we can associate with each $C_i \in \mathcal{E}$ a transitive representation $\rho_i : G \mapsto \mathsf{S}_{d_i}$, $d_i = \#C_i$ over the set of left classes of some $H \in C_i$.

Example 43.1.6. With $G := \mathsf{A}_4$ and $H_2 = \{\mathrm{Id}, g_3\}$ we can consider the left classes of H_2, which are

$$
\begin{array}{lllllll}
L_{21} & := & \{\mathrm{Id}, g_3\}, & L_{22} & := & \{g_2, g_7\}, & L_{23} & := & \{g_4, g_5\}, \\
L_{24} & := & \{g_6, g_{11}\}, & L_{25} & := & \{g_8, g_9\}, & L_{26} & := & \{g_{10}, g_{12}\},
\end{array}
$$

thus giving the representation $\rho_2 : G \to \mathsf{S}_6$, in which

$$
\begin{array}{llll}
\rho_2(g_1) & = & \mathrm{Id}, & \rho_2(g_2) & = & (1,2,3)(4,5,6), \\
\rho_2(g_3) & = & (2,4)(3,5) & \rho_2(g_4) & = & (1,3,2)(4,6,5), \\
\rho_2(g_5) & = & (1,3,4)(2,6,5), & \rho_2(g_6) & = & (1,4,3)(2,5,6), \\
\rho_2(g_7) & = & (1,2,5)(3,6,4), & \rho_2(g_8) & = & (1,5,4)(2,6,3), \\
\rho_2(g_9) & = & (1,5,2)(3,4,6), & \rho_2(g_{10}) & = & (1,6)(2,4), \\
\rho_2(g_{11}) & = & (1,4,5)(2,3,6), & \rho_2(g_{12}) & = & (1,6)(3,5).
\end{array}
$$

In the same way, with $H_3 = \{\mathrm{Id}, g_2, g_4\}$, the left classes of H_3 are

$$
\begin{array}{llllll}
L_{31} & := & \{\mathrm{Id}, g_2, g_4\}, & L_{32} & := & \{g_3, g_6, g_9\}, \\
L_{33} & := & \{g_5, g_{10}, g_{11}\}, & L_{34} & := & \{g_7, g_8, g_{12}\},
\end{array}
$$

thus giving the representation $\rho_3 : G \to \mathsf{S}_4$, in which

$$
\begin{aligned}
\rho_3(g_1) &= \text{Id}, & \rho_3(g_2) &= (2,4,3), \\
\rho_3(g_3) &= (1,2)(3,4), & \rho_3(g_4) &= (2,3,4), \\
\rho_3(g_5) &= (1,3,2), & \rho_3(g_6) &= (1,2,3), \\
\rho_3(g_7) &= (1,4,2), & \rho_3(g_8) &= (1,4,3), \\
\rho_3(g_9) &= (1,2,4), & \rho_3(g_{10}) &= (1,3)(2,4), \\
\rho_3(g_{11}) &= (1,3,4), & \rho_3(g_{12}) &= (1,4)(2,3).
\end{aligned}
$$

Finally, the left classes of $H_4 = \{\text{Id}, g_3, g_{10}, g_{12}\}$ are

$$
\begin{aligned}
L_{41} &:= \{\text{Id}, g_3, g_{10}, g_{12}\}, \\
L_{42} &:= \{g_2, g_6, g_7, g_{11}\}, \\
L_{43} &:= \{g_4, g_5, g_8, g_9\},
\end{aligned}
$$

thus giving the representation $\rho_4 : G \to \mathsf{S}_3$, in which

$$
\begin{aligned}
\rho_4(g_1) &= \rho_4(g_3) = \rho_4(g_{10}) = \rho_4(g_{12}) = \text{Id}, \\
\rho_4(g_2) &= \rho_4(g_6) = \rho_4(g_7) = \rho_4(g_{11}) = (1,2,3), \\
\rho_4(g_4) &= \rho_4(g_5) = \rho_4(g_8) = \rho_4(g_9) = (1,3,2).
\end{aligned}
$$

Let $\Phi : G \to \mathsf{S}_E$ be a (not necessarily transitive) representation of G as a permutation group. Let $C \in \mathcal{E}$ be a conjugacy class and let $H \subset G$ be any member of C; the number

$$
m := \#\{e \in E : \Phi(h)(e) = e : h \in H\}
$$

is clearly independent of the choice of $H \in C$ but does depend on the representation Φ of G as a permutation group and on the conjugacy class C.

Definition 43.1.7. *Such a number m is called the* mark *of C in the representation Φ.*

More generally, if we consider the H-orbits of E and denote as α_i the number of orbits consisting of i elements,[4] such numbers are independent of the choice of H and depend only on Φ and C.

We remark that we have the relation $\#E = n = \sum_i i\alpha_i$.

Example 43.1.8.   Let us choose the representation $\rho_4 : G \to \mathsf{S}_3$ of G as a permutation group and the group $H := H_2$.

Then we have three orbits of cardinality 1, so that $m = 3$.

If instead we choose $\rho_5 : G \to \mathsf{S}_{12}$ for $H := H_2$, we obviously have six orbits, all of cardinality 2.

Now setting $C := C_i \preceq C_j =: C'$, using freely the current notation[5] and writing

---

[4] So that, in particular $m = \alpha_1$.
[5] In particular $H \in C$, $H' \in C'$ and $H \subset H'$.

- $B_{C'} := \{C \in (G/H)_l : C \subset C'\}$ for each $C' \in (G/H')_l$,
- $\mathsf{B} := \{B_{C'} : C' \in (G/H')_l\}$,

we have the partition $(G/H)_l = \sqcup_{B \in \mathsf{B}} B$ in terms of the left-action of G. Thus we have a left-action of G on B which can be identified, via $C' \mapsto B_{C'}$, with that on $(G/H')_l$.

Let us fix, for any conjugacy class $\mathcal{C}_j$, an associated transitive representation $\rho_j :$ $G \to \mathsf{S}_{\mathcal{C}_j}$, $\rho_j(g) : H' \mapsto gH'$.

By the considerations above regarding the action of G on B, the mark of C in the representation ρ_j[6] is independent not only of the choice of $H \in \mathcal{C}_i$ but also of the choice of transitive representation ρ_j. Such a number therefore depends only on the pair $(\mathcal{C}, \mathcal{C}') = (\mathcal{C}_i, \mathcal{C}_j)$ and is called the *incidence number* or *mark* of $(\mathcal{C}_i, \mathcal{C}_j)$. It will be denoted $J(\mathcal{C}_i, \mathcal{C}_j) = m_i^j$. It satisfies

$$
m_i^j = \begin{cases} 0 & i > j, \\ \#G & i = j = 1, \\ 1 & j = s = \#\mathcal{E}, \\ \deg(\mathcal{C}_j) & i = 1. \end{cases}
$$

Example 43.1.9.   The matrix of *incidence numbers* for A_4 is

	1	2	3	4	5
1	12	6	4	3	1
2	0	2	0	3	1
3	0	0	1	0	1
4	0	0	0	3	1
5	0	0	0	0	1

Let $\Phi : G \to \mathsf{S}_E$ be a representation of G as a permutation group and let us consider any orbit $\omega := \{\Phi(g)(x) : g \in G\}$; by restriction we thus obtain a representation $G \to \mathsf{S}_\omega$ which necessarily is equivalent to one of the representations defined by $\mathcal{C}_i$, $1 \le i \le s$. Denoting by a_i, for each i, the number of orbits ω thus equivalent to $\mathcal{C}_i$, we can therefore associate with Φ

[the] symbol $[\sum_{i=1}^s a_i \mathcal{C}_i]$ denoting that the representation $[\Phi]$ is made up of a_1 representations equivalent to $[\mathcal{C}_1]$, a_2 representations equivalent to $[\mathcal{C}_2]$, and so on.[7]

Using modern notation, we associate with Φ an element $\sum_{i=1}^s a_i \mathcal{C}_i$ in the free $\mathbb{Z}$-module $\mathfrak{L}_G = \sum_{i=1}^s \mathbb{Z}\mathcal{C}_i$ with basis $\mathcal{E}$.

In conclusion we have

Proposition 43.1.10.   *The application* $U : \Phi \mapsto \sum_{i=1}^s a_i \mathcal{C}_i$ *defines a bijection between the set of all equivalence classes of the representations of G as a*

---

[6] That is, the number of elements $H' \in \mathcal{C}_j$ which satisfy $\rho_j(h)(H') = H'$ for each $h \in H$.

[7] Burnside W., *op. cit.*, p. 238.

permutation group of degree n and the set of elements $\sum_{i=1}^{s} a_i \mathcal{C}_i \in \mathfrak{L}_G$ such that $a_i \in \mathbb{N}$ for each i and $\sum_{i=1}^{s} a_i \deg(\mathcal{C}_i) = n$.

If moreover we denote, for each j, by μ_j the mark of $\mathcal{C}_j$ in the representation Φ we have the relations

$$\mu_j = \sum_{i=1}^{s} a_i m_i^j.$$

$\boxdot$

In the same way, if we consider the H-orbits of the elements in $\mathcal{C}_j$ and we denote by $a(v)_i^j$ the number of such orbits consisting of v elements[8] then such numbers are independent also of the choice of transitive representation ρ_j, thus depending only on the pair $(\mathcal{C}_i, \mathcal{C}_j)$.

Note that we have the relation

$$\sum_v v a(v)_i^j = \deg(\mathcal{C}_j).$$

Let us consider any sequence $A_i^j := (a(1)_i^j, a(2)_i^j, \ldots, a(v)_i^j, \ldots)$ and let us remark that $a(v)_i^j \neq 0 \implies v \leq \deg(\mathcal{C}_j)$, so that, in particular, there are at most a finite number of values v for which $a(v)_i^j \geq 1$.

The sequences A_i^j can be stored more compactly as

$$B_i^j := [(a_1, v_1), \ldots, (a_r, v_r)] : r \geq 1, \quad 1 \leq v_1 < v_2 < \cdots, \quad a_i \geq 1 \text{ for each } i,$$

where v_l are the values for which $a_l := a(v_l)_i^j \neq 0$.

Definition 43.1.11.   *The* partition array *defined by G is the array, whose rows and columns are indexed by the conjugacy classes of G and whose (i, j)th entry is B_i^j.*

$\boxdot$

Example 43.1.12.   The array (A_i^j) for $\mathbf{A}_4$ is as follows:

	1	2	3	4	5
1	$(12, 0, \ldots)$	$(6, 0, \ldots)$	$(4, 0, \ldots)$	$(3, 0, \ldots)$	$(1, 0, \ldots)$
2	$(0, 6, 0, \ldots)$	$(2, 2, 0, \ldots)$	$(0, 2, 0\ldots)$	$(3, 0, \ldots)$	$(1, 0, \ldots)$
3	$(0, 0, 4, 0, \ldots)$	$(0, 0, 2, 0, \ldots)$	$(1, 0, 1, 0, \ldots)$	$(0, 0, 1, 0, \ldots)$	$(1, 0, \ldots)$
4	$(0, 0, 0, 3, 0, \ldots)$	$(0, 3, 0, \ldots)$	$(0, 0, 0, 1, 0, \ldots)$	$(3, 0, \ldots)$	$(1, 0, \ldots)$
5	$(0, \ldots, 0, 1, 0, \ldots)$	$(0, \ldots, 0, 1, 0, \ldots)$	$(0, 0, 0, 1, 0, \ldots)$	$(0, 0, 1, 0, \ldots)$	$(1, 0, \ldots)$

and the corresponding partition array defined by $\mathbf{A}_4$ is

	1	2	3	4	5
1	$[(12, 1)]$	$[(6, 1)]$	$[(4, 1)]$	$[(3, 1)]$	$[(1, 1)]$
2	$[(6, 2)]$	$[(2, 1), (2, 2)]$	$[(2, 2)]$	$[(3, 1)]$	$[(1, 1)]$
3	$[(4, 3)]$	$[(2, 3)]$	$[(1, 1), (3, 1)]$	$[(1, 3)]$	$[(1, 1)]$
4	$[(3, 4)]$	$[(3, 2)]$	$[(1, 4)]$	$[(3, 1)]$	$[(1, 1)]$
5	$[(1, 12)]$	$[(1, 6)]$	$[(1, 4)]$	$[(1, 3)]$	$[(1, 1)]$

$\boxdot$

---

[8] So that, in particular $m_i^j = a(1)_i^j$.

Proposition 43.1.13. *The rows of the partition array defined by G are all different.*

Proof. For any values $i, j, 1 \leq j < i \leq s$, it is sufficient to show that $B_j^j \neq B_i^j$. Let us fix elements $H_i \in C_i$ and $H_j \in C_j$ and denote by $N_j := \{g \in G : gH_jg^{-1}\}$ the normalizer of H_j in G. Thus the claim follows from

$$a(1)_j^j = [N_j : H_j] \neq 0 = m_i^j = a(1)_i^j.$$

$\textcircled{\odot}$

43.2 Representation as Permutation Group of Roots

Let $f(T) := T^n + a_1 T^{n-1} + \cdots + a_i T^{n-i} + \cdots + a_{n-1}T + a_n \in k[T]$ be a monic separable polynomial of degree n over a field k; denote

k the algebraic closure of k,
$\mathfrak{R}_f := \{\alpha_1, \ldots, \alpha_n\} \subset \mathsf{k}$ the set of the roots of f,
$K_f := k[\alpha_1, \ldots, \alpha_n], k \subset K_f \subset \mathsf{k}$, the splitting field of f,
$G(K_f/k)$ the Galois group of f

so that, in particular,

$$f(T) = T^n + \sum_{i=1}^{n} a_i T^{n-i} = \prod_{j=1}^{n} (T - \alpha_j).$$

We have the natural representation of $G(K_f/k)$ as a permutation group of the roots of f:

$$G(K_f/k) \hookrightarrow \mathsf{S}_{\mathfrak{R}_f} = \mathsf{S}_n, \qquad \sigma \mapsto s_\sigma : \sigma(\alpha_i) = \alpha_{s_\sigma(i)};$$

note that

- the representation is independent of the enumeration of the roots,[9]
- $G(K_f/k)$ is transitive.

Assume now that f has a factorization $f = \prod_{j=1}^{r} p_j$ into irreducible components. The Galois group $G(K_f/k)$ then operates transitively over each $\mathfrak{R}_{p_j}$; denoting, for each j, by C_{i_j} the conjugacy classes of the subgroup $G(K_{p_j}/k) \subset G(K_f/k)$, we have $G(K_f/k) = \sum_j C_{i_j} \in \mathcal{L}_{G(K_f/k)}$.

Conversely, let us assume k to be infinite and let $K, k \subset K \subset \mathsf{k}$ be a finite extension of k, $[K : k] = n$; let us consider a faithful representation $G(K/k) \to \mathsf{S}_n$ of degree n and the corresponding orbits $\omega_1, \ldots, \omega_r$ of $\{1, 2, \ldots, n\}$; if we set $n_i := \#\omega_i$, we can re-enumerate both the orbits and the elements in $\{1, 2, \ldots, n\}$ in such

---

[9] More precisely, let $t \in \mathsf{S}_n$ and define $\beta_i := \alpha_{t(i)}$ for each i; then, for each $\sigma \in G(K_f/k)$, we have

$$\sigma(\beta_i) = \sigma(\alpha_{t(i)}) = \alpha_{s_\sigma t(i)} = \beta_{t^{-1}s_\sigma t(i)} \quad \text{for each } i;$$

so a different enumeration of the roots simply gives an equivalent representation of $G(K_f/k)$.

a way that $\omega_1 = \{1, \ldots, n_1\}, \omega_2 = \{n_1 + 1, \ldots, n_1 + n_2\}, \ldots, \omega_r = \{\sum_{i=1}^{r-1} n_i + 1, \ldots, n\}, n_1 \leq n_2 \leq \cdots \leq n_r$. We can than fix any element $a_i \in \omega_i, n_{i-1} < a_i \leq n_i$, and denote

$$G_i := \{g \in G(K/k) : g(a_i) = a_i\},$$
$$K_i := \{\alpha \in K : g(\alpha) = \alpha, g \in G_i\},$$
$$\xi_i \in K_i, \text{ a primitive element such that } K_i = k[\xi_i],$$
$$P_i \in k[T], \text{ its minimal polynomial,}$$
$$\mathfrak{R}_i := \{\xi_{i1}, \ldots, \xi_{in_i}\}, \text{ the conjugates of } \xi_i.$$

Then, since k is assumed to be infinite, we can assume that the sets $\mathfrak{R}_i$ are disjoint; therefore $f := \prod_i P_i$ is separable. Clearly

$$\{\sigma \in G(K/k) : \sigma(\xi_{ij}) = \xi_{ij} \text{ for each } i, j\} = \{\mathrm{Id}_{G(K/k)}\},$$

so that $K = k\left(\xi_{11}, \ldots, \xi_{ij}, \ldots, \xi_{rn_r}\right)$.

Corollary 43.2.1. *If k is infinite, for each finite extension K, $k \subset K \subset \mathsf{k}$, the Galois field $G(K/k)$ can be faithfully represented as $G(K_f/k)$ for a separable polynomial $f \in k[T]$.* ◉

43.3 Universal Lagrange Resolvent

Let us fix an integer n and let us denote

$$\mathcal{A} := k[X_1, \ldots, X_n],$$
$$\mathcal{F} := k(X_1, \ldots, X_n),$$
$$\sigma_1, \ldots, \sigma_n, \text{ the elementary symmetric functions of } X_1, \ldots, X_n,$$
$$\mathcal{I} \subset \mathcal{A}, \text{ the ideal generated by } \sigma_1, \cdots, \sigma_n,$$
$$\mathcal{S} := k[\sigma_1, \ldots, \sigma_n],$$
$$\mathcal{K} := k(\sigma_1, \ldots, \sigma_n),$$
$$F(T) \in \mathcal{S}[T], \text{ the polynomial}$$

$$F(T) = T^n + \sum_{i=1}^{n}(-1)^i \sigma_i T^{n-i} = \prod_{j=1}^{n}(T - X_j).$$

We have that:

(1) $\mathcal{F} \supset \mathcal{K}$ is an algebraic extension $\mathcal{F} = \mathcal{K}_F$;
(2) $\mathcal{A}$ is the integral closure of $\mathcal{S}$ in $\mathcal{F}$;
(3) the representation of $G(\mathcal{F}/\mathcal{K}) \hookrightarrow \mathsf{S}_{\{X_1,\ldots,X_n\}} = \mathsf{S}_n$ as a permutation group of the roots of F is an isomorphism.

Under this isomorphism, we can therefore associate with each subgroup $H \subset \mathsf{S}_n$ the corresponding invariant field

$$\mathsf{I}(H) := \{\alpha \in \mathcal{F} : h(\alpha) = \alpha, \text{ for each } h \in H\},$$

and so it follows that:

(4) $\mathcal{F} \supset \mathsf{I}(H)$ is an algebraic extension;

(5) $\mathsf{I}(H)$ is a separable extension of $\mathcal{K}$;

(6) $H = G(\mathcal{F}/\mathsf{I}(H))$;

(7) the integral closure of $\mathcal{S}$ in $\mathsf{I}(H)$ is $\mathcal{A}_H := \mathcal{A} \cap \mathsf{I}(H)$;

(8) there are elements $\Psi \in \mathcal{A}_H$ which are $\mathcal{K}$-primitive for $\mathsf{I}(H)$, that is, which satisfy $\mathsf{I}(H) = \mathcal{K}[\Psi]$;

(9) $\mathcal{A}_H$, being integrally closed and noetherian, is a finite $\mathcal{S}$-module;

(10) $\mathsf{I}(H) = \{a/b : a \in \mathcal{A}_H, b \in \mathcal{S}, b \neq 0\} \subset k(\sigma_1, \ldots, \sigma_n)[X_1, \ldots, X_n] = \mathcal{K}[X_1, \ldots, X_n];$[10]

(11) $\mathsf{I}(H)$ is the fraction field of $\mathcal{A}_H$;

(12) the rank of $\mathcal{A}_H$ as a $\mathcal{S}$-module is $\dim_{\mathcal{K}}(\mathsf{I}(H)) = [\mathsf{S}_n : H] := n!/\#H$;

(13) the finite set $\mathcal{B} := \{X_1^{a_1} X_2^{a_2} \cdots X_n^{a_n}, 0 \leq a_i < i\}$, $\#\mathcal{B} = n!$, is both a k-basis of $\mathcal{A}/\mathcal{I}$ and a basis of $\mathcal{A}$ as an $\mathcal{S}$-module.

On the basis of (8) we can introduce the following:

Definition 43.3.1. *Each $\mathcal{K}$-primitive element $\Psi \in \mathcal{A}_H$ for $\mathsf{I}(H)$ is called a* resolvent *of H and is said to be* homogeneous *if it is a homogeneous polynomial in $\mathcal{A}$.*

The minimal polynomial $\mathcal{L}_\Psi \in \mathcal{S}[T]$ of Ψ over $\mathcal{K}$ is called the Lagrange resolvent *of H associated with Ψ.* ⊙

Example 43.3.2. The Vandermonde determinant

$$\Psi := \prod_{i>j}(X_j - X_i) = \begin{vmatrix} 1 & 1 & \cdots & 1 \\ X_1 & X_2 & \cdots & X_n \\ X_1^2 & X_2^2 & \cdots & X_n^2 \\ \vdots & \vdots & \ddots & \vdots \\ X_1^{n-1} & X_2^{n-1} & \cdots & X_n^{n-1} \end{vmatrix}$$

is a $\mathcal{K}$-primitive element for A_4; its minimal polynomial is $\mathcal{L}_\Psi := T^2 - \mathrm{Disc}(F) \in \mathcal{S}[T]$, where $\mathrm{Disc}(F) := \prod_{i>j}(X_j - X_i)^2$ is the discriminant of the polynomial $F(T) \in \mathcal{S}[T]$. ⊙

Lemma 43.3.3. *If $\Psi \in \mathcal{A}$ and $H := \{g \in \mathsf{S}_n : g(\Psi) = \Psi\}$ then Ψ is a resolvent of H.* ⊙

Remark 43.3.4. Each group H possesses homogeneous resolvents provided that k is infinite. In fact, if $\Psi \in \mathcal{A}_H$ is a $\mathcal{K}$-primitive element of degree d, denote by Ψ_t the homogenization of Ψ by $t\sigma_1$, that is,

$$\Psi_t = (t\sigma_1)^d \Psi\left(\frac{X_1}{t\sigma_1}, \ldots, \frac{X_n}{t\sigma_1}\right).$$

---

[10] In fact if $x \in \mathsf{I}(H)$ then there are $\alpha, \beta \in \mathcal{A}, \beta \neq 0$ such that $x = \alpha/\beta$. It is sufficient to define

$$b := \prod_{\sigma \in \mathsf{S}_n} \sigma(\beta) \in \mathcal{S} \setminus \{0\}, \quad a := \alpha \prod_{\sigma \in \mathsf{S}_n \setminus \{\mathrm{Id}\}} \sigma(\beta) \in \mathcal{A}_H$$

in order to obtain $x = a/b$.

Consider now a set of $e := [\mathsf{S}_n : H]$ elements

$$\{\tau_1, \ldots, \tau_e\} \subset \mathsf{S}_n$$

such that $\{\tau_1 H, \ldots, \tau_e H\}$ is the set of all the left classes, and let us set $P_t(T) := \prod_{i=1}^{e}(T - \tau_i(\Psi_t))$ and denote by $\mathcal{D}(t) := \mathrm{Disc}(P_t) \in \mathcal{S}[t]$ its discriminant. Since $P_{1/\sigma_1}(T) = \mathcal{L}_\Psi$, $\mathcal{D}(1/\sigma_1) \neq 0$; thus $\mathcal{D}(t) \neq 0$ and, k being infinite, there is a $\lambda \in k$ such that $\mathcal{D}(\lambda) \neq 0$. Thus Ψ_λ, which is homogeneous of degree d, is $\mathcal{K}$-primitive and is the required homogeneous resolvent. Its corresponding Lagrange resolvent is $\mathcal{L}_{\Psi_\lambda} = P_\lambda(T)$.   ⊙

Theorem 43.3.5 (Lagrange). *Let $H \subset \mathsf{S}_n$ be a subgroup and $\Psi \in \mathcal{A}_H$ a resolvent of H. Denoting by $\Delta_\Psi := \mathrm{Disc}(\mathcal{L}_\Psi)$ the discriminant of the Lagrange resolvent of Ψ, we have*

$$\mathcal{A}_H \subset \left\{ \frac{f(\Psi)}{\Delta_\Psi}, f \in \mathcal{S}[T] \right\}.$$

Proof. Let $e = [\mathsf{S}_n : H]$ and $\{\tau_1, \ldots, \tau_e\} \subset \mathsf{S}_n$ be such that $\{\tau_1 H, \ldots, \tau_e H\}$ is the set of all the left cosets of H, with $\tau_1 = \mathrm{Id}_{\mathsf{S}_n}$. Set $\Psi_i := \tau_i(\Psi)$ for each i, so that

$$\mathcal{L}_\Psi = \prod_{i=1}^{e}(T - \Psi_i) = T^e + \sum_{i=1}^{e}(-1)^i C_i T^{e-i}, \quad C_i \in \mathcal{S}.$$

If we denote by $S_1, \ldots, S_{e-1}$ the elementary symmetric functions on $\Psi_2, \ldots, \Psi_e$ then, since

$$T^{e-1} + \sum_{j=1}^{e-1}(-1)^j S_j T^{e-j-1}$$

$$= \prod_{i=2}^{e}(T - \Psi_i)$$

$$= \frac{\mathcal{L}_\Psi}{T - \Psi_1}$$

$$= T^{e-1} + (\Psi_1 - C_1)T^{e-2} + (\Psi_1^2 - C_1\Psi_1 - C_2)T^{n-i} + \cdots$$

is a polynomial in $\mathcal{S}[\Psi][T]$, we have $S_j \in \mathcal{S}[\Psi]$ for each j.

Considering $g \in \mathcal{A}_H$, set $g_i := \tau_i(g)$ for each i and, for each $m, 0 \leq m < e$, set

$$h_m := \sum_{j=1}^{e} g_j \Psi_j^m = \sum_{j=1}^{e} \tau_j(g\Psi^m) \in \mathcal{S}.$$

The g_i can be solved *à la* Cramer in terms of the h_m: setting

$$\mathcal{D} := \begin{vmatrix} h_0 & 1 & \cdots & 1 \\ h_1 & \Psi_2 & \cdots & \Psi_e \\ \vdots & \vdots & \ddots & \vdots \\ h_{e-1} & \Psi_2^{e-1} & \cdots & \Psi_e^{e-1} \end{vmatrix} \quad \text{and} \quad \delta_\Psi := \begin{vmatrix} 1 & 1 & \cdots & 1 \\ \Psi_1 & \Psi_2 & \cdots & \Psi_e \\ \vdots & \vdots & \ddots & \vdots \\ \Psi_1^{e-1} & \Psi_2^{e-1} & \cdots & \Psi_e^{e-1} \end{vmatrix},$$

so that

$$\delta_\Psi^2 = \prod_{1 \le i < j \le e} (\Psi_j - \Psi_i)^2 = \mathrm{Disc}(\mathcal{L}_\Psi) = \Delta_\Psi,$$

we have $g = \delta_\Psi \mathcal{D} / \Delta_\Psi$.

We can prove our claim if we can show that $\mathcal{E} := \delta_\Psi \mathcal{D} \in \mathcal{S}[\Psi]$: we know that $\mathcal{E}$ can be expressed as $E(\Psi_2, \ldots, \Psi_n)$ where $E \in \mathcal{S}[\Psi][T_2, \ldots, T_n]$ is symmetric in $T_2, \ldots, T_e$; thus the Fundamental Theorem on Symmetric Functions (Theorem 6.2.4) grants the existence of a polynomial $F \in \mathcal{S}[\Psi][Y_1, \ldots, Y_{e-1}]$ for which

$$E(\Psi_2, \ldots, \Psi_n) = F(S_1, \ldots, S_{e-1}) \in \mathcal{S}[\Psi],$$

as claimed. $\boxed{\odot}$

43.4 Cauchy Modules

Let us use the same notation as in Sections 43.2 and 43.3 and consider a monic separable polynomial

$$f(T) := T^n + a_1 T^{n-1} + \cdots + a_i T^{n-i} + \cdots + a_{n-1} T + a_n \in k[T].$$

Definition 43.4.1 (Ampère). *The n interpolating functions*

$$f_i(T) = f_i(X_1, \ldots, X_{i-1}, T) \in k[X_1, \ldots, X_{i-1}][T], \quad 1 \le i \le n,$$

are recursively defined as

$$f_1(T) := f(T) \quad and \quad f_i(T) := \frac{f_{i-1}(T) - f_{i-1}(X_{i-1})}{T - X_{i-1}}, \quad 1 < i \le n.$$

$\boxed{\odot}$

Definition 43.4.2. *The polynomials $f_i(X_i) = f_i(X_1, \ldots, X_i), 1 \le i \le n$, are called the* Cauchy modules *associated with f.* $\boxed{\odot}$

Example 43.4.3. For $n = 5$ we have

$$\begin{aligned}
f_1(X_1) &= X_1^5 + X_1^4 a_1 + X_1^3 a_2 + X_1^2 a_3 + X_1 a_4 + a_5, \\
f_2(X_2) &= X_2^4 + X_2^3 X_1 + X_2^2 X_1^2 + X_2 X_1^3 + X_1^4 \\
&\quad + a_1(X_2^3 + X_2^2 X_1 + X_2 X_1^2 + X_1^3) \\
&\quad + a_2(X_2^2 + X_2 X_1 + X_1^2) \\
&\quad + a_3(X_2 + X_1) + a_4, \\
f_3(X_3) &= X_3^3 + X_3^2 X_2 + X_3 X_2^2 + X_2^3 + X_3^2 X_1 \\
&\quad + X_3 X_2 X_1 + X_2^2 X_1 + X_3 X_1^2 + X_2 X_1^2 + X_1^3 \\
&\quad + a_1(X_3^2 + X_3 X_2 + X_2^2 + X_3 X_1 + X_2 X_1 + X_1^2) \\
&\quad + a_2(X_3 + X_2 + X_1) + a_3,
\end{aligned}$$

$$f_4(X_4) = X_4^2 + X_4 X_3 + X_3^2 + X_4 X_2 + X_3 X_2$$
$$+ X_2^2 + X_4 X_1 + X_3 X_1 + X_2 X_1 + X_1^2$$
$$+ a_1(X_4 + X_3 + X_2 + X_1) + a_2,$$
$$f_5(X_5) = X_5 + X_4 + X_3 + X_2 + X_1 + a_1.$$

Lemma 43.4.4 (Cauchy). *The Cauchy modules satisfy* $\deg_i(f_i) + i = n + 1$ *and* $\mathrm{lc}(f_i) = 1$ *for each* i.

Moreover, under the further assumption that the roots are all distinct, for each i *the roots of* $f_i(\alpha_1, \ldots, \alpha_{i-1}, X_i)$ *are* $\{\alpha_j, i \le j \le n\}$.

Proof. The claims being true by definition for $i = 1$, we have by induction

$$f_i(\alpha_1, \ldots, \alpha_{i-1}, \alpha_j) = \frac{f_{i-1}(\alpha_1, \ldots, \alpha_{i-2}, \alpha_j) - f_{i-1}(\alpha_1, \ldots, \alpha_{i-2}, \alpha_{i-1})}{\alpha_j - \alpha_{i-1}} = 0$$

for each $j, i \le j \le n$.

Lemma 43.4.5 (Cauchy). *Assume that* $g \in k[X]$ *is such that there is a* $u \in k$ *for which* $g(\alpha) = u$ *for each root* $\alpha \in \mathfrak{R}_f$ *of* f. *Then* **Rem**$(g, f) = u$.

Proof. The polynomial **Rem**$(g, f) - u$ has degree less than $\deg(f)$, vanishes at each root of f and therefore is zero.

Now let $g_n(X_1, \ldots, X_n) \in \mathcal{A} = k[X_1, \ldots, X_{n-1}][X_n]$ be symmetric in the variables $X_1, \ldots, X_n$, and recursively write:

- $g_{n-1} := \mathbf{Rem}(g_n, f_n(X_n)) \in k[X_1, \ldots, X_{n-1}]$, the remainder of the division of g_n by f_n in $k[X_1, \ldots, X_{n-1}][X_n]$;
- $g_{n-2} := \mathbf{Rem}(g_{n-1}, f_{n-1}(X_{n-1})) \in k[X_1, \ldots, X_{n-2}]$, the remainder of the division of g_{n-1} by f_{n-1} in $k[X_1, \ldots, X_{n-2}][X_{n-1}]$;

  $\ldots$

- $g_i := \mathbf{Rem}(g_{i+1}, f_{i+1}(X_{i+1})) \in k[X_1, \ldots, X_i]$, the remainder of the division of g_{i+1} by f_{i+1} in $k[X_1, \ldots, X_i][X_{i+1}]$;

  $\ldots$

- $g_1 := \mathbf{Rem}(g_2, f_2(X_2)) \in k[X_1]$, the remainder of the division of g_2 by f_2 in $k[X_1][X_2]$;
- $g_0 := \mathbf{Rem}(g_1, f_1(X_1)) \in k$, the remainder of the division of g_1 by f_1 in $k[X_1]$.

We remark that we can assume that $g_i \in k[X_1, \ldots, X_i]$ rather than the weaker version $g_i \in k(X_1, \ldots, X_i)$, because $\mathrm{lc}(f_i) = 1$.

Remark 43.4.6 (Cauchy). Each g_i is a symmetric polynomial in the variables $X_1, \ldots, X_i$.

Moreover, repeatedly applying Lemma 43.4.5, if $\#\mathfrak{R}_f = n$, that is, all the roots of f are distinct, we have that, for each i, $g_i(\alpha_1, \ldots, \alpha_i) \in k(\alpha_1, \ldots, \alpha_i)$ satisfies

$$g_i(\alpha_1, \ldots, \alpha_i) = g_{i+1}(\alpha_1, \ldots, \alpha_i, \alpha_j), \quad i < j \le n,$$

so that $g_0 = g_n(\alpha_1, \ldots, \alpha_n)$.

Donc alors la valeur $\lfloor g_0 \rfloor$ *de* $\lfloor g_n(\alpha_1, \ldots, \alpha_n) \rfloor$, *determinèe comme nous l'avons dit ci-dessus, sera une fonction rationelle et même entière, par conséquent une fonction continue des coefficients renfermés dans* $f(x)$. *D'ailleurs chacun de ces coefficients représentera, au signe près, ou la somme des racines de l'èquation* $\lfloor f(x) = 0 \rfloor$, *ou la somme formée avec les produits qu'on obtient en multipliant ces racines deux à deux, trois à troix, etc. Donc la valeur trouvée de* $\lfloor g_0 \rfloor$ *pourra être encore considérée comme une fonction continue des racines de l'èquation* $\lfloor f(x) = 0 \rfloor$; *et dans la formule*

$$[g_n(\alpha_1, \ldots, \alpha_n) = g_0]$$

qui se vérifiera toutes les fois que les racines $\lfloor \alpha_1, \ldots, \alpha_n \rfloor$ *seront inégales, les deux membres varieront par degrée insensible en même temps que ces racines.*

[...]

Il est maintenant facile de s'assurer que [the result] *s'etende, avec la formule* $[g_n(\alpha_1, \ldots, \alpha_n) = g_0]$, *au cas même oú l'èquation* $\lfloor f(x) = 0 \rfloor$ *offre des racines égales. Car des racines égales de l'èquation* $\lfloor f(x) = 0 \rfloor$ *peuvent être considérées des valeurs variables de racines supposées d'abord inégales, mais trés peu différent les unes des autres; et puisque la formule* $[g_n(\alpha_1, \ldots, \alpha_n) = g_0]$, *dont les deux membres varient par dégres insensibles avec les racines, par conséquent avec leurs différences, continuera de subsister pour des valeurs de ces différences aussi rapprochées de zéro que l'on voudra, elle subsistera certainement das le cas même oú ces différences viendront à s'évanouir.*[11]

Thus we have:

THÉORÈME II. *Soient*

$$f(x)$$

une function entière de x, du degré n, et

$$f(a, x) = \frac{f(x) - f(a)}{x - a}, \quad f(a, b, x) = \frac{f(a, x) - f(a, b)}{x - b}, \quad \ldots$$

les functions interpolaires de divers ordres qui renfermant avec la variable x diverses valeurs particulières a, b, c, ... de cette variable. Concevons d'ailleurs que les lettres

$$a, b, c, \ldots, h, k$$

représentent les n racines de l'équation

$$f(x) = 0$$

et désignon par

$$F(a, b, c, \cdots, h, k)$$

[11] Cauchy A., Usage des fonctions interpolaires dans ls determination des fonctions symmetriques des racines d'une équation algébrique donnée, *C.R. Acad. Sci. Paris* **11** (1840), p. 933. In: Cauchy A., *Oeuvres* t. V, Gauthier–Villars, Paris (1882), pp. 476–477.

une function entière mais symmétrique de ces racines. Pour éliminer de cette même fonction les racines

$$k, h, \ldots, c, b, a$$

il suffira de la diviser successivement par les divers terms de la suite

$$f(a, b, c, \ldots, h, k), \; f(a, b, c, \ldots, h,), \ldots, f(a, b, c), \; f(a, b), \; f(a),$$

considérés le premier comme fonction de k, le second comme fonction de h, …, l'avant-dernier comme fonction de b, le dernier comme fonction de a. Le dernier des restes ainsi obtenus sera indépendant de a, b, c, …, h, k, et représentera nécessariment la valeur U de la fonction symmétrique

$$F(a, b, c, \ldots, h, k)$$

exprimée à l'aide des coefficients que renferme le premier membre de l'èquation | f (x) = 0|.[12]

In summary the argument above gives:

Theorem 43.4.7 (Cauchy). *Let $g_n(X_1, \ldots, X_n) \in \mathcal{A}$ be a symmetric polynomial in the variables $X_1, \ldots, X_n$ and let $g_{n-1}, \ldots, g_0$ be obtained as above by successively dividing $g_n(X_1, \ldots, X_n)$ by $f_n(X_n), f_{n-1}(X_{n-1}), \ldots, f_1(X_1)$.*

The last remainder g_0 will be independent of $X_1, \ldots, X_n$ and gives the value $g_n(\alpha_1, \ldots, \alpha_n)$ as a function of the coeffiecients $a_1, \ldots, a_n$ of f. ⊡

Example 43.4.8. For $n = 3$ we have

$$f_1(X_1) = X_1^3 + X_1^2 a_1 + X_1 a_2 + a_3,$$
$$f_2(X_2) = X_2^2 + X_2 X_1 + X_1^2 + a_1(X_2 + X_1) + a_2,$$
$$f_3(X_3) = X_3 + X_2 + X_1 + a_1.$$

For the symmetric polynomial

$$g = X_3^3 X_2 + X_3 X_2^3 + X_3^3 X_1 + X_2^3 X_1 + X_3 X_1^3 + X_2 X_1^3$$

we have

$$\begin{aligned}
g_2(X_1, X_2) = {} & -2X_2^4 - 4X_2^3 X_1 - 6X_2^2 X_1^2 - 4X_2 X_1^3 - 2X_1^4 \\
& + a_1(-4X_2^3 - 9X_2^2 X_1 - 9X_2 X_1^2 - 4X_1^3) \\
& + a_1^2(-3X_2^2 - 6X_2 X_1 - 3X_1^2) - a_1^3(X_2 - X_1), \\
g_1(X_1) = {} & X_1^3 a_1 + X_1^2 a_1^2 + X_1 a_1 a_2 + a_1^2 a_2 - 2a_2^2, \\
g_0 = {} & a_1^2 a_2 - 2a_2^2 - a_1 a_3.
\end{aligned}$$

⊡

Let us now extend the notation of Sections 43.2 and 43.3 by denoting

- $\mathbb{F}$ the prime field of k,
- $\mathsf{J} = \mathbb{I}(\sigma_1 + a_1, \sigma_2 - a_2, \ldots \sigma_n - (-1)^n a_n) \subset \mathcal{A}$,
- $\mathsf{A} := \mathsf{k}[X_1, \ldots, X_n]$,

---

[12] Cauchy A., *op. cit.*, pp. 474–475.

- $J^e := Jk[X_1, \ldots, X_n]$,
- $\mathcal{B} := \{X_1^{a_1} X_2^{a_2} \cdots X_n^{a_n}, 0 \le a_i < i\}$,
- $\mathcal{B}' := \{X_1^{a_1} X_2^{a_2} \cdots X_n^{a_n}, 0 \le a_i < n - i\}$,
- $\Gamma := G(K_f/k) \subset \mathsf{S}_n$,
- $R : \Gamma \hookrightarrow \mathsf{S}_{\mathfrak{R}_f} = \mathsf{S}_n$, the canonical representation of Γ as a permutation group of the roots of f, defined by

$$R(u) = s_u : u(\alpha_i) = \alpha_{s_u(i)},$$

- $\widetilde{} : \mathcal{A} \to K_f \subset \mathsf{k}$, the k-algebra morphism

$$\widetilde{g} = g(\alpha_1, \ldots, \alpha_n), \quad \text{for each } g \in \mathcal{A}.$$

Remark 43.4.9. For each $g \in \mathcal{A}$ and each $s \in \mathsf{S}_n$ we have

$$\widetilde{s(g)} = s(g)(\alpha_1, \ldots, \alpha_n) = s\left(g(\alpha_1, \ldots, \alpha_n)\right) = s(\widetilde{g}),$$

where S_n is interpreted as follows:

$\mathsf{S}_n = G(\mathcal{F}/\mathcal{K})$ on the left-hand side, and
$\mathsf{S}_n = G(K_f/k)$ on the right-hand side.

⊙

Remark 43.4.10. For each $g(X_1, \ldots, X_n) \in \mathcal{A} = k[X_1, \ldots, X_n]$ which is symmetric in the variables $X_1, \ldots, X_n$, we have both

$$\widetilde{g} \in k \quad \text{and} \quad g - \widetilde{g} \in \mathsf{J}.$$

Moreover, with the present notation, Cauchy's Theorem 43.4.7 can be read as

Let $g(X_1, \ldots, X_n) \in \mathcal{A}$ be a symmetric polynomial in the variables $X_1, \ldots, X_n$; set $g_n := g$, and let $g_{n-1}, \ldots, g_0$ be obtained by successively dividing $g_n(X_1, \ldots, X_n)$ by $f_n(X_n), f_{n-1}(X_{n-1}), \ldots, f_1(X_1)$.
The last remainder g_0 satisfies

$$g_0 = \widetilde{g} \in \mathbb{F}(a_1, \ldots, a_n).$$

⊙

Proposition 43.4.11 (Machì–Valibouze). *The reduced Gröbner basis of* J *(and also of* J^e*) w.r.t. any term ordering* $<$ *induced by* $X_1 < X_2 < \cdots < X_n$ *is* $\{f_1, \ldots, f_n\}$. ⊙

Proof. Note that, for each i, we have $\mathbf{T}(f_i) = X_i^{n-i+1}$ for any term ordering $<$ induced by $X_1 < X_2 < \cdots < X_n$. Therefore Buchberger's First Criterion (Lemma 22.5.1) implies that $\{f_1, \ldots, f_n\}$ is the Gröbner basis of the ideal it generates w.r.t. any such term ordering.

Therefore we just have to prove that $\mathsf{J} = \mathbb{I}(f_1, \ldots, f_n)$.

Clearly, Cauchy's Theorem implies that $\mathsf{J} \subseteq \mathbb{I}(f_1, \ldots, f_n)$, and equality is a direct consequence of the remark that $\mathbf{N}(\mathbb{I}(f_1, \ldots, f_n)) = \mathcal{B}'$, so that

$$\dim(\mathsf{A}/\mathsf{J}) = \#\mathcal{B} = n! = \#\mathcal{B}' = \dim\left(\mathsf{A}/\mathbb{I}(f_1, \ldots, f_n)\right).$$

⊙

To complete our argument, we need to consider the "generic" monic polynomial

$$f(T) = T^n + a_1 T^{n-1} + \cdots + a_{n-1} T + a_n \in \mathbb{F}(a_1, \ldots, a_n)[T]$$

and express the associated Cauchy modules f_i, which are symmetric polynomials in $X_1, \ldots, X_i$, as elements in

$$f_i \in \mathbb{F}(a_1, \ldots, a_n)(X_1, \ldots, X_i).$$

Lemma 43.4.12. *With the present notation it holds that:*

(1) $f_n(T) - f_n(X_n) = T - X_n;$

(2) $f(T) = f_{v+1}(T) \prod_{j=1}^{v}(T - X_j) + \sum_{i=1}^{v} f_i(X_i) \prod_{j=1}^{i-1}(T - X_j)$ *for each* v, $1 \le v < n;$

(3) $f(T) = F(T) + \sum_{i=1}^{n} f_i(X_i) \prod_{j=1}^{i-1}(T - X_j);$

(4) $f(T) - F(T) = \sum_{i=1}^{n-1} \left(a_i - (-1)^i \sigma_i\right) T^{n-i} = \sum_{i=1}^{n} f_i(X_i) \prod_{j=1}^{i-1}(T - X_j).$

Proof.

(1) This requires only a trivial verification.

(2) We have $f_i(T)(T - X_{i-1}) = f_{i-1}(T) - f_{i-1}(X_{i-1})$, so that (for $i = 1$) $f(T) = f_1(T) = f_2(T)(T - X_1) + f_1(X_1)$.

Thus, inductively,

$$f(T) = f_v(T) \prod_{j=1}^{v-1}(T - X_j) + \sum_{i=1}^{v-1} f_i(X_i) \prod_{j=1}^{i-1}(T - X_j)$$

$$= \left(f_v(X_v) + f_{v+1}(T)(T - X_v)\right) \prod_{j=1}^{v-1}(T - X_j) + \sum_{i=1}^{v-1} f_i(X_i) \prod_{j=1}^{i-1}(T - X_j)$$

$$= f_{v+1}(T) \prod_{j=1}^{v}(T - X_j) + \sum_{i=1}^{v} f_i(X_i) \prod_{j=1}^{i-1}(T - X_j).$$

(3) We have that (1) implies

$$f_n(T) \prod_{j=1}^{n-1}(T - X_j) = f_n(X_n) \prod_{j=1}^{n-1}(T - X_j) + \prod_{j=1}^{n}(T - X_j)$$

$$= f_n(X_n) \prod_{j=1}^{n-1}(T - X_j) + F(T).$$

The claim then follows by substitution of this result into the formula (2) for $v := n - 1$.

(4) Trivial.

Denoting by $h_d(X_1, \ldots, X_i)$ the dth *complete sum* in $k[X_1, \ldots, X_i]$, that is, (compare Definition 6.3.2) the sum of all terms of degree d in $k[X_1, \ldots, X_i]$, we have

Proposition 43.4.13. *It holds that, setting $a_0 = 1$:*

(1) $$\frac{h_d(X_1, \ldots, X_{i-2}, X_i) - h_d(X_1, \ldots, X_{i-2}, X_{i-1})}{X_i - X_{i-1}} = h_{d-1}(X_1, \ldots, X_i)$$

for each d and each $i \leq n$;

(2) $f_i(X_i) = \sum_{d=0}^{n-i+1} a_{n-i+1-d} h_d(X_1, \ldots, X_i).$

Proof.

(1) Trivial.

(2) From (1) we have

$$
\begin{aligned}
f_i(X_i) &= \frac{f_{i-1}(X_i) - f_{i-1}(X_{i-1})}{X_i - X_{i-1}} \\
&= \sum_{d=0}^{n-i+2} a_{n-i+2-d} \frac{h_d(X_1, \ldots, X_{i-2}, X_i) - h_d(X_1, \ldots, X_{i-2}, X_{i-1})}{X_i - X_{i-1}} \\
&= \sum_{d=1}^{n-i+2} a_{n-i+2-d} h_{d-1}(X_1, \ldots, X_i) \\
&= \sum_{d=0}^{n-i+1} a_{n-i+1-d} h_d(X_1, \ldots, X_i).
\end{aligned}
$$

$\boxdot$

Corollary 43.4.14. *(Compare Proposition 6.3.15 and Fact 6.3.14.)*
The Gröbner basis of $\mathcal{I} = \mathbb{I}(\sigma_1, \ldots, \sigma_n)$ w.r.t. the lex term ordering $<$ induced by $X_1 > X_2 > \cdots > X_n$ is

$$\{h_{n-i+1}(X_1, \ldots, X_i), 1 \leq i \leq n\}.$$

Proof. We have just to apply Propositions 43.4.11 and 43.4.13 with $a_1 = \cdots = a_n = 0$. $\boxdot$

Historical Remark 43.4.15. When stating Proposition 6.3.15 and Fact 6.3.14, I was completely unaware of Proposition 43.4.13 and of Cauchy's Theorem 43.4.7, which imply Proposition 43.4.11.

Only later did I realize that they were reported in Valibouze's *Habilitation*, where the history is also told:

L'idéal |J| des relations symétriques est engendré par les n polynômes |σ₁ − a₁, σ₂ − a₂, …, σₙ − aₙ|. Augustin Cauchy utilise les fonctions interpolaires introduites par Ampère pour calculer un système de générateurs qui se révéle être une base standard réduite, pour l'ordre lexicographique, de l'idéal |J|. Cette base standard sera à tester l'appartenance à l'idéal |J| et a évaluer sur k un polynôme symétrique en les racines de f. Comme elle est réduite, la base standard de |J| permet de retrouver la base naturelle de |A/J|. Cauchy calcule cette base standard pour n = 4 et en 1990 et, avec Antonio Machì, nout la calculons pour tout n en utilizant des séries géneratrices tronquées |⋯|. Alain Lascoux suggère une démonstration plus courte qui fait appel aux Λ-anneaux et aux différences divisées sur les S-fonctions.[13]

43.5 Resolvents and Polynomial Roots

Let us consider again the monic separable polynomial

$$f(T) := T^n + a_1 T^{n-1} + \cdots + a_i T^{n-i} + \cdots + a_{n-1} T + a_n \in k[T]$$

and use freely the same notation as in Sections 43.2 and 43.3.

Definition 43.5.1. *Let $\Psi \in \mathcal{A}_H$ be a resolvent of $H \subset \mathsf{S}_n$ and let*

$$\mathcal{L}_\Psi[\sigma_1, \ldots, \sigma_n, T] \in k[\sigma_1, \ldots, \sigma_n][T] = \mathcal{S}[T]$$

be the Lagrange resolvent of H associated with Ψ.

The (H, Ψ)-Lagrange resolvent of f is the polynomial

$$\mathcal{L}_{\Psi,f}[T] := \mathcal{L}_\Psi[-a_1, \ldots, (-1)^i a_i, \ldots, (-1)^n a_n, T] \in k[T].$$

If $\mathcal{L}_{\Psi,f}$ is separable, one says that Ψ is f-separable.

Proposition 43.5.2. *If k is an infinite field and $A \subset k$ is a ring whose fraction field is k, there is a resolvent $\Psi \in A[X_1, \ldots, X_n]$ of H for which $\mathcal{L}_{\Psi,f}$ is separable; moreover Ψ can be chosen to be homogeneous.*

Proof. Since the roots α_i of f are distinct, the $n!$ polynomials

$$\left(\sum_{i=1}^n U_i \alpha_{s(i)} \right) - 1 \in k[U_1, \ldots, U_n],$$

where s runs over the elements of S_n, are all different.

Thus, writing, for each $H' \in (\mathsf{S}_n/H)_l$,

$$\phi_{H'} := \prod_{s \in H'} \left(\left(\sum_{i=1}^n U_i \alpha_{s(i)} \right) - 1 \right) \in k[U_1, \ldots, U_n],$$

we have $\gcd(\phi_{H'}, \phi_{H''}) = 1$ for each $H', H'' \in (\mathsf{S}_n/H)_l, H' \neq H''$.

Since A is infinite, we can choose $u_1, \ldots, u_n \in A$ such that the elements $\phi_{H'}(u_1, \ldots, u_n), H' \in (\mathsf{S}_n/H)_l$, are all distinct.

For such values, we set

$$\Psi := \prod_{s \in H} \left(\left(\sum_{i=1}^n u_i X_{s(i)} \right) - 1 \right) \in \mathsf{I}(H) \cap A[X_1, \ldots, X_n].$$

[13] Valibouze A., *Théorie de Galois constructive*, Mémoir d'Habilitation, Paris 6 (1998), p. 23.

Its conjugates in $\mathcal{K}$ are the polynomials

$$\Psi_{H'} := \prod_{s \in H'} \left(\left(\sum_{i=1}^{n} u_i X_{s(i)} \right) - 1 \right), \qquad H' \in (S_n/H)_l,$$

which are all distinct, so that $\Psi \in A[X_1, \ldots, X_n]$ is the required resolvent of H.

We have $\mathcal{L}_{\Psi.f} = \prod_{H' \in (S_n/H)_l} (T - \Psi_{H'}(\alpha_1, \ldots, \alpha_n))$, which is therefore separable.

The same argument as in Remark 43.3.4 proves that Ψ can be made homogeneous: denote

$$(\Psi_{H'})_t = (t\sigma_1)^{\deg(\Psi_{H'})} \Psi_{H'}(X_1/t\sigma_1, \ldots, X_n/t\sigma_1),$$
$$P_t(T) := \prod_{H' \in (S_n/H)_l} (T - (\Psi_{H'})_t),$$
$$\mathcal{D}(t) := \mathrm{Disc}(P_t) \in \mathcal{S}[t] = k[\sigma_1, \ldots, \sigma_n, t],$$
$$D(t) := \mathcal{D}(a_1, \cdots, a_n, t) \in A[t].$$

We clearly have $D(t) \neq 0$ so that, A being infinite, there is a $\lambda \in A$ such that $\mathcal{D}(\lambda) \neq 0$; thus

$$\Theta := \Psi_\lambda \in \mathsf{I}(H) \cap A[X_1, \ldots, X_n]$$

is homogeneous of degree $\deg(\Psi) = \#(H)$, and its conjugates in $\mathcal{F}$ are

$$\Theta_{H'} := (\Psi_{H'})_\lambda \in A[X_1, \ldots, X_n], \qquad H' \in (S_n/H)_l.$$

Writing, for each $H' \in (S_n/H)_l$, $\theta_{H'} := \Theta_{H'}(\alpha_1, \ldots, \alpha_n)$ we have $\mathcal{L}_{\Theta.f} = \prod_{H' \in (S_n/H)_l} (T - \theta_{H'})$, whose discriminant satisfies $D(\lambda) \neq 0$.

Therefore

- the $\Theta_{H'}$ are all distinct,
- Θ is a homogeneous resolvent of H,
- $\mathcal{L}_{\Theta.f}$ is separable.

$\boxed{\odot}$

Let us now denote

$\mathbf{Z} := \mathcal{Z}(\mathsf{J}^e) = \{(\beta_1, \ldots, \beta_n) \in \mathsf{k}^n : p(\beta_1, \ldots, \beta_n) = 0 \text{ for each } p \in \mathsf{J}^e\} \subset \mathsf{k}^n;$

$\Gamma := G(K_f/k) \subset S_n,$

$H := (S_n/\Gamma)_r := \{\Gamma s : s \in S_n\}$, the set of the right classes of $\Gamma := G(K_f/k) \subset S_n,$

$N := \#\Gamma.$

Lemma 43.5.3. *It holds that:*

(1) $\mathbf{Z} = \{(\alpha_{s(1)}, \ldots, \alpha_{s(n)}) : s \in S_n\};$
(2) $\mathcal{A}/\mathsf{J} \cong \mathrm{Span}_k(\mathcal{B}'), \mathsf{A}/\mathsf{J}^e \cong \mathrm{Span}_\mathsf{k}(\mathcal{B}');$
(3) *both* J *and* J^e *are radical.*

Proof. (1) is obvious; (2) is a trivial consequence of Proposition 43.4.11; (3) follows from

$$\dim_k(\mathcal{A}/\sqrt{J}) = \#Z = n! = \dim_k(\mathcal{A}/J).$$

Now, for each $\Gamma' \in H$ let us denote

$$\mathsf{W}_{\Gamma'} := \{(\alpha_{s(1)}, \ldots, \alpha_{s(n)}) : s \in \Gamma'\},$$
$$P_{\Gamma'} := \prod_{s \in \Gamma'} \left(T - \sum_{i=1}^{n} U_i \alpha_{s(i)}\right),$$
$$\mathfrak{m}_{\Gamma'} := \mathcal{I}(\mathsf{W}_{\Gamma'}) = \{g \in \mathcal{A} : g(\beta_1, \ldots, \beta_n) = 0 \text{ for each } (\beta_1, \ldots, \beta_n) \in \mathsf{W}_{\Gamma'}\}.$$

We also denote $\mathsf{W} := \mathsf{W}_\Gamma$, $\Psi := P_\Gamma$, $\mathfrak{m} := \mathfrak{m}_\Gamma$.

Lemma 43.5.4. *With the present notation, it holds that:*

(1) *the u-resultant of* Z,

$$\Psi_\mathsf{Z} := \prod_{s \in \mathsf{S}_n} \left(T - \sum_{i=1}^{n} U_i \alpha_{s(i)}\right) \in k[U_1, \ldots, U_n][T],$$

factorizes into irreducible components as $\Psi_\mathsf{Z} = \prod_{\Gamma' \in H} P_{\Gamma'}$;
(2) *the irreducible components of* Z *are the* $\mathsf{W}_{\Gamma'}$: $\mathsf{Z} = \bigcup_{\Gamma' \in H} \mathsf{W}_{\Gamma'}$;
(3) *each* $\mathfrak{m}_{\Gamma'}$ *is maximal in* $\mathcal{A}$;
(4) $\mathsf{J} = \bigcap_{\Gamma' \in H} \mathfrak{m}_{\Gamma'}$;
(5) $\mathfrak{m}_{\Gamma'} = \mathcal{Z}(\mathsf{W}_{\Gamma'})$ *for each* $\Gamma' \in H$;
(6) $K_f \cong \mathcal{A}/\mathfrak{m}_{\Gamma'}$ *for each* $\Gamma' \in H$.

Lemma 43.5.5 (Arnaudiès–Valibouze). *Let* $g \in \mathfrak{m}$, $g \notin \bigcap_{\substack{\Gamma' \in H \\ \Gamma' \neq \Gamma}} \mathfrak{m}_{\Gamma'}$. *Then* $\mathfrak{m}$ *is generated by*

$$\{\sigma_1 + a_1, \sigma_2 - a_2, \ldots, \sigma_n - (-1)^n a_n, g\}.$$

Proof. Writing $\mathfrak{a} := \mathsf{J} + (g)$, by assumption we have $\mathcal{Z}(\mathfrak{a}) = \mathsf{W} = \mathcal{Z}(\mathfrak{m})$, whence

(1) there is a $\rho \in \mathbb{N}$ for which $\mathfrak{m}^\rho \subset \mathfrak{a}$,[14]
(2) $\mathsf{J} \subset \mathfrak{a} \subset \sqrt{\mathfrak{a}} = \mathfrak{m}$.

Since $\left(\bigcap_{\substack{\Gamma' \in H \\ \Gamma' \neq \Gamma}} \mathfrak{m}_{\Gamma'}\right) + \mathfrak{m}^\rho = \mathcal{A}$, there are $u \in \left(\bigcap_{\substack{\Gamma' \in H \\ \Gamma' \neq \Gamma}} \mathfrak{m}_{\Gamma'}\right)$ and $v \in \mathfrak{m}^\rho$ for which $1 = u + v$. Therefore, for each $x \in \mathfrak{m}$, we have $x = xu + xv$ with

$$xu \in \mathfrak{m}^\rho \subset \mathfrak{a}, \quad xv \in \mathfrak{m}\left(\bigcap_{\substack{\Gamma' \in H \\ \Gamma' \neq \Gamma}} \mathfrak{m}_{\Gamma'}\right) = \bigcap_{\Gamma' \in H} \mathfrak{m}_{\Gamma'} = \mathsf{J} \subset \mathfrak{a},$$

so that $x \in \mathfrak{a}$.

---

[14] $\mathcal{A}$ is Noetherian.

Now let Θ be a resolvent of a subgroup H of S_n and denote

$\theta := \widetilde{\Theta} = \Theta(\alpha_1, \ldots, \alpha_n),$
$\Theta_1 = \Theta, \Theta_2, \ldots, \Theta_v,$ the distinct conjugates $s(\Theta), s \in \mathsf{S}_n,$ of Θ in $\mathcal{F} \supset \mathcal{K},$
$H_i := \{s \in \mathsf{S}_n : s(\Theta) = \Theta_i\}, 1 \le i \le v.$

Remarking that θ is a root of the (H, Θ)-Lagrange resolvent $\mathcal{L}_{\Theta, f}[T]$, let us assume that

(1) θ is a simple root of $\mathcal{L}_{\Theta, f}[T]$ and wlog
(2) its conjugates in K_f are $\theta, \widetilde{\Theta}_2, \ldots, \widetilde{\Theta}_r, 1 \le r \le v,$

and denote

$h := \prod_{i=1}^{r}(T - \widetilde{\Theta}_i)$, the monic irreducible factor of $\mathcal{L}_{\Theta, f}[T]$ in $k[T]$,
$O := \{\Theta_1, \ldots, \Theta_r\},$
$S := \{s \in \mathsf{S}_n : s(\Theta_i) \in O, \text{ for each } i, 1 \le i \le r\},$
$g := h(\Theta) \in \mathcal{A} = k[X_1, \ldots, X_n].$

Theorem 43.5.6 (Arnaudiès–Valibouze). *With the present notation we have that:*

(1) *the Γ-orbit of Θ in $\mathcal{F}$ is O;*
(2) $\Gamma \subset S \subset \bigcup_{i=1}^{r} H_i$ *and* $[S : \Gamma] = [S \cap H : \Gamma \cap H]$;
(3) $\Gamma = S = \bigcup_{i=1}^{r} H_i \iff \mathfrak{m} = \mathbb{I}(\sigma_1 + a_1, \sigma_2 - a_2, \ldots \sigma_n - (-1)^n a_n, g).$

Proof.

(1) For $s \in \Gamma$, since $h \in k[T]$ we have

$$h(s(\widetilde{\Theta})) = h(\widetilde{s(\Theta)}) = s(h(\widetilde{\Theta})) = s(0) = 0;$$

therefore $s(\Theta) \in O$.

For each $i, 1 \le i \le r$, as h is irreducible there is an $s \in \Gamma$ for which $s(\widetilde{\Theta}) = \widetilde{s(\Theta)} = \widetilde{\Theta}_i$; since, with the same argument above, we have $h(\widetilde{s(\Theta)}) = 0$ then necessarily $s(\Theta) = \Theta_i$.

(2) The inclusion $S \subset \bigcup_{i=1}^{r} H_i$ being trivial, let us prove $\Gamma \subset S$. For $s \in \Gamma$ we have

$$s\left(\{\widetilde{\Theta}_1, \ldots, \widetilde{\Theta}_r\}\right) = \{\widetilde{\Theta}_1, \ldots, \widetilde{\Theta}_r\};$$

therefore the same argument as above allows us to deduce that

$$
\begin{aligned}
s \in \Gamma &\implies s(\widetilde{\Theta}_i) = \widetilde{s(\Theta_i)} \quad \text{for each } i, 1 \le i \le r, \\
&\implies s(\widetilde{\Theta}_i) \in \{\widetilde{\Theta}_1, \ldots, \widetilde{\Theta}_r\} \quad \text{for each } i, 1 \le i \le r, \\
&\implies s(\Theta_i) \in \{\Theta_1, \ldots, \Theta_r\} \quad \text{for each } i, 1 \le i \le r, \\
&\implies s \in S.
\end{aligned}
$$

Moreover,

- O is the S-orbit of Θ,
- $\{s \in S : s(\Theta) = \Theta\} = S \cap H,$
- $\{s \in \Gamma : s(\Theta) = \Theta\} = \Gamma \cap H,$

whence $r = [S : S \cap H] = [\Gamma : \Gamma \cap H]$. Thus

$$[S : S \cap H][S \cap H : \Gamma \cap H] = [S : \Gamma][\Gamma : \Gamma \cap H]$$

gives, as required, $[S : \Gamma] = [S \cap H : \Gamma \cap H]$.

(3) Clearly $g(\beta_1, \ldots, \beta_n) = 0$ for each $(\beta_1, \ldots, \beta_n) \in \mathsf{W}_\Gamma$.

Therefore, Lemma 43.5.5 reduces the statement to

$$\Gamma = S = \bigcup_{i=1}^{r} H_i \iff g(\beta_1, \ldots, \beta_n) \neq 0 \quad \text{for each } (\beta_1, \ldots, \beta_n) \in \bigcup_{\substack{\Gamma' \in H \\ \Gamma' \neq \Gamma}} \mathsf{W}_{\Gamma'}.$$

We have

$$g(\beta_1, \ldots, \beta_n) \neq 0 \text{ for each } (\beta_1, \ldots, \beta_n) \in \bigcup_{\substack{\Gamma' \in H \\ \Gamma' \neq \Gamma}} \mathsf{W}_{\Gamma'}$$

$$\iff h(\Theta(\alpha_{s(1)}, \ldots, \alpha_{s(n)})) \neq 0 \text{ for each } s \in \mathsf{S}_n \setminus \Gamma$$

$$\iff h(s(\widetilde{\Theta})) \neq 0 \text{ for each } s \in \mathsf{S}_n \setminus \Gamma.$$

Since h is a simple factor of $\mathcal{L}_{\Theta, f}[T]$, the only $s \in \mathsf{S}_n$ for which $h(s(\widetilde{\Theta})) = 0$ are those satisfying $s(\Theta) \in O$, that is, the elements of $\bigcup_{i=1}^{r} H_i$.

Thus

$$\mathfrak{m} = \mathbb{I}\left(\sigma_1 + a_1, \sigma_2 - a_2, \ldots \sigma_n - (-1)^n a_n, g\right)$$

$$\iff g(\beta_1, \ldots, \beta_n) \neq 0 \text{ for each } (\beta_1, \ldots, \beta_n) \in \bigcup_{\substack{\Gamma' \in H \\ \Gamma' \neq \Gamma}} \mathsf{W}_{\Gamma'}$$

$$\iff h(s(\widetilde{\Theta})) \neq 0 \text{ for each } s \in \mathsf{S}_n \setminus \Gamma$$

$$\iff s \notin \bigcup_{i=1}^{r} H_i \text{ implies } s \in \mathsf{S}_n \setminus \Gamma$$

$$\iff \bigcup_{i=1}^{r} H_i \subset \Gamma,$$

which, by (2), is equivalent to $\Gamma = S = \bigcup_{i=1}^{r} H_i$. $\boxed{\odot}$

43.6 Lagrange Resolvent and Galois Group

Let

H be a subgroup of S_n,

$e := [\mathsf{S}_n : H]$,

Θ be a resolvent of H,

$\Theta = \Theta_1, \ldots, \Theta_e$ be its distinct conjugates in $\mathcal{F}$ over $\mathcal{K}$,

$$H_i := \{s \in \mathsf{S}_n : s(\Theta_1) = \Theta_i\}, 1 \leq i \leq e,$$
$$\theta := \widetilde{\Theta},$$

ν be the multiplicity of θ in $\mathcal{L}_{(\cdot), f}[T]$,

so that

(A) $\mathcal{L}_{(\cdot)}[T] = \prod_{i=1}^{e} (T - \Theta_i)$,
(B) $\mathcal{L}_{(\cdot), f}[T] = \prod_{i=1}^{e} (T - \widetilde{\Theta}_i)$,
(C) $(\mathsf{S}_n/H)_l = \{H_i, 1 \leq i \leq e\}, H_1 = H$;

we can wlog assume that

(D) $\widetilde{\Theta}_i = \theta \iff i \leq \nu$,

and write

$$O := \{\Theta_1, \ldots, \Theta_\nu\},$$
$$S := \{s \in \mathsf{S}_n : s(\Theta_i) \in O, 1 \leq i \leq \nu\}.$$

Proposition 43.6.1. *With the present notation:*

(1) *if θ is a simple root of $\mathcal{L}_{(\cdot), f}[T]$ then*

$$G(K_f/k(\theta)) = \{s \in G(K_f/k) : s(\theta) = \theta\} = G(K_f/k) \cap H;$$

(2) *if $\nu > 1$ then*

$$G(K_f/k(\theta) = \{s \in G(K_f/k) : s(\theta) = \theta\} = G(K_f/k) \cap S;$$

moreover,
(a) $[k(\theta) : k] = [G(K_f/k) : G(K_f/k) \cap S]$,
(b) $G(K_f/k) \cap S \supset G(K_f/k) \cap H$ *and*
(c) $[G(K_f/k) : G(K_f/k) \cap H] \leq \nu[k(\theta) : k]$.

Proof.

(1) For each $s \in G(K_f/k)$ and each $p \in \mathcal{A}$ we have (Remark 43.4.9) $\widetilde{s(p)} = s(\widetilde{p})$; so, for each $s \in G(K_f/k) \cap H$, we have

$$\theta = \widetilde{\Theta} = \widetilde{s(\Theta)} = s(\widetilde{\Theta}) = s(\theta),$$

so that $G(K_f/k) \cap H \subset G(K_f/k(\theta))$.

If, instead, $s \in G(K_f/k) \setminus H$ then $\Theta' := s(\Theta)$ is a conjugate, distinct from θ, of Θ in $\mathcal{F}$ over $\mathcal{K}$; since $\mathsf{S}_n = G(\mathcal{F}/\mathcal{K})$ and $\mathcal{L}_{(\cdot), f}[T] = \prod_{s \in \mathsf{S}_n} (T - \widetilde{s(\Theta)})$, we have necessarily

$$s(\theta) = s(\widetilde{\Theta}) = \widetilde{s(\Theta)} = \widetilde{\Theta}' \neq \widetilde{\Theta} = \theta$$

and $s \notin G(K_f/k(\theta))$.

(2) For each $s \in G(K_f/k)$ we have $s \in \{\mathsf{s} \in G(K_f/k) : \mathsf{s}(\theta) = \theta\}$ iff, for each $i, 1 \leq i \leq \nu, \widetilde{s(\Theta_i)} = \theta$, which is equivalent to $s(\Theta_i) \in O$; thus $s \in S$ and the claim follows.

Moreover:

(a) since $[K_f : k(\theta)] = \#G(K_f/k(\theta)) = \#(G(K_f/k) \cap S)$, we have

$$[k(\theta) : k] = \frac{[K_f : k]}{[K_f : k(\theta)]} = \frac{\#G(K_f/k)}{\#(G(K_f/k) \cap S)} = [G(K_f/k) : G(K_f/k) \cap S];$$

(b) $s \in H \implies s(\Theta_1) = \Theta_1 \implies s \in S$, so that

$$G(K_f/k) \cap H \subset G(K_f/k) \cap S;$$

(c) also, we have

$$G(K_f/k) \cap S = \{s \in G(K_f/k) : s(\Theta_i) \in O, 1 \leq i\}$$

and $G(K_f/k) \cap H = \{s \in S : s(\Theta) = \Theta\}$; thus

$$\begin{aligned}[G(K_f/k) \cap S : G(K_f/k) \cap H] &= \#\{s(\Theta) : s \in G(K_f/k) \cap S\} \\ &\leq \#\{s(\Theta) : s \in G(K_f/k) \cap H\} \\ &= \nu \end{aligned}$$

and

$$\begin{aligned}\big[G(K_f/k) &: G(K_f/k) \cap H\big] = \\ \big[G(K_f/k) &: G(K_f/k) \cap S\big]\big[G(K_f/k) \cap S : G(K_f/k) \cap H\big] \\ &\leq \big[k(\theta) : k\big]\nu.\end{aligned}$$

Corollary 43.6.2. *With the present notation:*

(1) *if $\nu = 1$ then the degree of θ over k is $[G(K_f/k) : G(K_f/k) \cap H]$;*
(2) $\theta \in k \iff G(K_f/k) \subseteq S$;
(3) $H_i \cap G(K_f/k) \subset S$ *for each* $i \leq \nu$;
(4) $H_i \cap G(K_f/k) = \emptyset$ *for each* $i > \nu$;
(5) $G(K_f/k) \cap S = \bigcup_{i=1}^{\nu} H_i \cap G(K_f/k).$

$\boxed{\odot}$

Let us denote

$h(T) \in k[T]$, the monic irreducible component of $\mathcal{L}_{\ominus,f}$ for which $h(\theta) = 0$,
$\mathcal{R} := \{\Theta_1, \ldots, \Theta_e\}$,
$S(\Phi) := \{s \in \mathsf{S}_n : s(\Phi) = \Phi\}$ for each $\Phi \in \mathcal{R}$,
$\theta_1 = \theta, \theta_2, \ldots, \theta_d$, the $G(K_f/k)$-conjugates of θ,
$\mathcal{R}_i := \{\Phi \in \mathcal{R} : \widetilde{\Phi} = \theta_i\}, 1 \leq i \leq d,$
$S_i := \{s \in \mathsf{S}_n : s(\Phi) \in \mathcal{R}_i$ for each $\Phi \in \mathcal{R}_i\},$
$\mathcal{O}$ as the set of the $G(K_f/k)$-orbits of $\mathcal{R} = \bigcup_{i=1}^{d} \mathcal{R}_i,$

so that

(E) $h^\nu \mid \mathcal{L}_{\Theta,f}, h^{\nu+1} \nmid \mathcal{L}_{\Theta,f}$,
(F) $h(T) = \prod_{i=1}^{d} (T - \theta_i)$,
(G) $S_1 = S, \mathcal{R}_1 = O$,
(H) $\#\mathcal{R}_i = \nu$ and (Proposition 43.6.1) $d = [G(K_f/k) : G(K_f/k) \cap S_i]$ for each i.

For each $\Omega \in \mathcal{O}$, $\{\Omega \cap \mathcal{R}_i, 1 \le i \le d\}$ is the set, for each j, of the $G(K_f/k) \cap S_j$ orbits of Ω; thus such a set is independent of the choice of j and depends only on Ω. Thus we have

$$m_\Omega := \#(\Omega \cap \mathcal{R}_i) = [G(K_f/k) \cap S_j : G(K_f/k) \cap S(\Phi)]$$
$$\text{for each } i : 1 \le i \le d, j : 1 \le j \le d \text{ and } \Phi \in \Omega \cap \mathcal{R}_j;$$

as a consequence,

(I) $\#\Omega = dm_\Omega$ for each $\Omega \in \mathcal{O}$,
(J) $\sum_{\Omega \in \mathcal{O}} m_\Omega = \#\mathcal{R}_1 = \nu$,
(K) $h^\nu(T) = \prod_{\Omega \in \mathcal{O}} \prod_{\Phi \in \Omega} \left(T - \widetilde{\Phi}\right)$.

Let us now introduce a *second* resolvent Ψ of the *same* subgroup H and set $\Psi_i := s(\Psi), s \in H_i$.

Definition 43.6.3. *Two monic (not necessarily irreducible or squarefree) factors F and G of, respectively, $\mathcal{L}_{\Theta,f}$ and $\mathcal{L}_{\Psi,f}$ are said to be* parallel *iff there is a $J \subset \{1, \dots, e\}$ for which*

$$F(T) = \sum_{j \in J} \left(T - \widetilde{\Theta}_j\right) \quad and \quad G(T) = \sum_{j \in J} \left(T - \widetilde{\Psi}_j\right).$$

$\boxdot$

If F and G are parallel and either $\mathcal{L}_{\Theta,f}$ or $\mathcal{L}_{\Psi,f}$ is f-separable then J is unique. Denoting, for each $\Omega \in \mathcal{O}$,

$$J_\Omega := \{j : 1 \le j \le e, \Theta_j \in \Omega, \},$$
$$P_\Omega(T) := \prod_{j \in J_\Omega} \left(T - \widetilde{\Psi}_j\right)$$

then by definition, the factor in $\mathcal{L}_{\Psi,f}$ parallel to $h^\nu(T)$ is

$$\prod_{\Omega \in \mathcal{O}} \prod_{\Psi \in \Omega} \left(T - \widetilde{\Psi}\right) = \prod_{\Omega \in \mathcal{O}} \prod_{j \in J_\Omega} \left(T - \widetilde{\Psi}_j\right) = \prod_{\Omega \in \mathcal{O}} P_\Omega(T).$$

Moreover, for each $\Omega \in \mathcal{O}$, $\{\Psi_j : j \in J_O\}$ is a $G(K_f/k)$-orbit, so that

(L) $P_\Omega(T) \in k[T]$ for each $\Omega \in \mathcal{O}$,
(M) $\deg(P_\Omega) = \#\Omega = dm_\Omega$.

As the conclusion of this *tour de force*, we have

Theorem 43.6.4 (Arnaudiès–Valibouze). *With the present notation, the factor of $\mathcal{L}_{\Psi,f}$ which is parallel to h^ν is $\prod_{\Omega \in \mathcal{O}} P_\Omega$; moreover, for each $\Omega \in \mathcal{O}$ the polynomial $P_\Omega(T)$ is an irreducible factor in $k[T]$ of $\mathcal{L}_{\Psi,f}$ and we have $\deg(h) \mid \deg(P_\Omega)$.* $\boxdot$

Let us impose on each $\mathbb{N}^e$, $e \in \mathbb{N} \setminus \{0\}$, the ordering $\preceq$ defined by

$$(b_1, \ldots, b_e) \preceq (a_1, \ldots, a_e) \iff b_i \leq a_i \text{ for each } i.$$

Let us fix a subgroup H of S_n and let us denote (using the same notation as Section 43.1)

$\mathcal{E}$ the set of all conjugacy classes of the subgroups of S_n,
$\mathcal{C}_i \in \mathcal{E}$ the conjugacy class to which $G(K_f/k)$ belongs,
$\mathcal{C}_j \in \mathcal{E}$ the conjugacy class to which H belongs,
$e := [\mathsf{S}_n : H]$,
$(a_1, \ldots, a_e)$ the sequence such that $A_i^j := (a_1, \ldots, a_e, 0, \ldots, 0, \ldots)$.

Theorem 43.6.5. *With such notation, let Θ be a resolvent of a subgroup H of S_n; let us assume that each irreducible factor of $\mathcal{L}_{\Theta,f}$ is separable and, for each j, $1 \leq j \leq e$, let us denote by b_j the number of irreducible factors of $\mathcal{L}_{\Theta,f}$. Then*

(1) $(b_1, \ldots, b_e) \preceq (a_1, \ldots, a_e)$,
(2) *if $\mathcal{L}_{\Theta,f}$ is separable, that is, Θ is f-separable,*

$$(b_1, \ldots, b_e) = (a_1, \ldots, a_e).$$

Proof. Let $\{\tau_1, \ldots, \tau_e\} \subset \mathsf{S}_n$ be such that $\{\tau_1 H, \ldots, \tau_e H\}$ is the set of all the left cosets of H, with $\tau_1 = \mathrm{Id}_{\mathsf{S}_n}$. Write, for each i, $H_i := \tau_i H \tau_i^{-1}$, $\Theta_i := \tau_i(\Theta)$ and $\theta_i := \widetilde{\Theta_i}$, so that $H_1 = H$ and $\Theta_1 = \Theta$. We have

$$\mathcal{L}_{\Theta} = \prod_{i=1}^{e} (T - \Theta_i),$$

$$\mathcal{L}_{\Theta,f} = \prod_{i=1}^{e} (T - \theta_i),$$

$$H_i = \{\tau \in \mathsf{S}_n : \tau(\Theta_i) = \Theta_i\}.$$

Let us fix a value j, $1 \leq j \leq e$, and a k-irreducible simple factor p of $\mathcal{L}_{\Theta,f}$, with $\deg(p) = j$; then $p = \prod_{i \in J} (T - \theta_i)$ for some $J \subset \{1, \ldots, e\}$, $\#J = j$.

If $i \in J$ then, by Corollary 43.6.2(1), $j = \deg(p) = [G(K_f/k) : G(K_f/k) \cap H_i]$; let us therefore denote, for each j,

$$n_j := \#\{i : 1 \leq i \leq e, j = [G(K_f/k) : G(K_f/k) \cap H_i]\}.$$

Clearly we have both $jb_j \leq n_j$ and $A_i^j := (n_1, n_2/2, \ldots, n_e/e, 0, \ldots, 0, \ldots)$, which proves (1).

If, moreover, $\mathcal{L}_{\Theta,f}$ is separable then we have

$$\sum_{j=1}^{e} jb_j = \deg(\mathcal{L}_{\Theta,f}) = e = \sum_{j=1}^{e} j\frac{n_j}{j}$$

and, since $jb_j \leq n_j = jn_j/j$ for each j, we obtain (2). $\boxed{\odot}$

43.7 Computing Galois Groups of a Polynomial

Given a value n, let us denote

- $\mathcal{E} := \{\mathcal{C}_1, \ldots, \mathcal{C}_s\}$, the set of all the conjugacy classes of S_n,
- for each j, $1 \leq j \leq s$,
  - a subgroup $H_j \in \mathcal{C}_j$,
  - $e_j := [\mathsf{S}_n : H_j]$,
  - $\tau_{j1} = \mathrm{Id}_{\mathsf{S}_n}, \ldots, \tau_{je_j} \in \mathsf{S}_n$, elements such that $\{\tau_{j1} H_j, \ldots, \tau_{je_j} H_j\}$ is the set of all the left cosets of H_j,
  - Θ_j as a resolvent of H_j,
  - $\Theta_{jm} := \tau_{jm}(\Theta_j)$, $1 \leq m \leq e_j$,
  - $\mathcal{L}_{\Theta_j} := \prod_{m=1}^{e_j} \left(T - \Theta_{jm} \right)$.

If $\mathcal{L}_{\Theta_j, f}$ is f-separable, we denote, for each $j \leq s$,

- a_{jm}, the number of irreducible factors of $\mathcal{L}_{\Theta_j, f}$ whose degree is m, $1 \leq m \leq e_j$;
- $\pi(\Theta_j, f) := (a_{j1}, \ldots, a_{je_j})$.

Then, combining Proposition 43.1.13 and Theorem 43.6.4 we obtain

Theorem 43.7.1. *With the present notation, and under the assumption that all resolvents Θ_j, $1 \leq j \leq s$, are f-separable, then the conjugacy class of $G(K_f/k)$ is $\mathcal{C}_r$, where r denotes the index of the row of the partition array B_i^j coinciding with the array*

$$(\pi(\Theta_1, f), \pi(\Theta_2, f), \ldots, \pi(\Theta_s, f)).$$

$\boxed{\odot}$

This gives an effective method for computing the conjugacy class of $G(K_f/k)$; it requires one to

(1) determine the partition array B_i^j,
(2) choose suitable f-separable resultants Θ_j,
(3) compute $\mathcal{L}_{\Theta_j, f}$,
(4) factorize them, deducing the values $\pi(\Theta_j, f)$, and
(5) deduce the conjugacy class of $G(K_f/k)$ by comparing the values of the partition array B_i^j.

Let us remark that:

- steps (4) and (5) do not require any special comment;
- we will briefly discuss step (3) in Algorithm 43.7.2;
- the hard task is steps (1) and (2), which have been solved systematically by Arnaudiès and Valibouze for $n \leq 11$.[15]

---

[15] Here I report their result for $n = 4$; for $5 \leq n \leq 11$, I refer to the survey Valibouze A., Computation of the galois groups of the resolvent factors for the direct and inverse Galois problems, *L.N. Comp. Sci.* **948** (1995), 456–468, and to the LITP reports quoted therein.

Algorithm 43.7.2. The computation of $\mathcal{L}_{\Theta_j, f}$ is a direct application of the results of Proposition 43.4.11; it is sufficient to compute the Gröbner basis G of the ideal generated by

$$\{f_1, \ldots, f_n, T - \Theta_j\} \subset k[T, X_1, \ldots, X_n]$$

w.r.t. the lex ordering induced by $T < X_1 < X_2 < \cdots < X_n$; then we have

$$\{\mathcal{L}_{\Theta_j, f}\} = G \cap k[T].$$

$\boxed{\odot}$

We report here the results of Arnaudiès and Valibouze for $n = 4$; the following table lists the 11 conjugacy classes of S_4, reporting for a chosen element $H_j \in \mathcal{C}_j$ their structure, generators and order:

H_1	I_4	[]	1
H_2	S_2	$[(3, 4)]$	2
H_3	S_2	$[(1, 2)(3, 4)]$	2
H_4	A_3	$[(1, 2, 3)]$	3
H_5	$\mathsf{S}_2 \times \mathsf{S}_2$	$[(1, 2), (3, 4)]$	4
H_6	V_4	$[(1, 2)(3, 4), (1, 3)(2, 4)]$	4
H_7	$\mathbb{Z}_4$	$[(1, 2)(3, 4), (1, 3, 2, 4)]$	4
H_8	S_3	$[(2, 3, 4), (3, 4)]$	6
H_9	D_4	$[(3, 4), (1, 2)(3, 4), (1, 3)(2, 4)]$	8
H_{10}	A_4	$[(1, 2)(3, 4), (1, 3)(2, 4), (2, 3, 4)]$	12
H_{11}	S_4	$[(1, 4), (2, 4), (3, 4)]$	24

Here A_n denotes the alternating group, D_n the dihedral group, V_4 the *Viergruppe* and $\mathsf{I}_n := \{\mathrm{Id}_{\mathsf{S}_n}\}$. The corresponding resultants are[16]

$$\Theta_1 = u_1 X_1 + u_2 X_2 + u_3 X_3,$$
$$\Theta_2 = X_1 + X_3 X_4,$$
$$\Theta_3 = X_1 X_2 + X_1 X_3 + X_2 X_4,$$
$$\Theta_4 = (X_2 - X_3)(X_3 - X_4)(X_4 - X_2),$$
$$\Theta_5 = X_1 X_2,$$
$$\Theta_6 = X_1 X_3 + X_2 X_4 - X_1 X_2 - X_3 X_4,$$
$$\Theta_7 = X_1 X_2^2 + X_2 X_3^2 + X_3 X_4^2 + X_4 X_1^2,$$
$$\Theta_8 = X_1,$$
$$\Theta_9 = X_1 X_2 + X_3 X_4,$$
$$\Theta_{10} = \prod_{1 \le i < j \le 4} (X_j - X_i),$$
$$\Theta_{11} = 1.$$

---

[16] Here u_1, u_2, u_3 are distinct and non-zero values.

	1	2	3	4	5	6
1	[(24, 1)]	[(12, 1)]	[(12, 1)]	[(8, 1)]	[(6, 1)]	[(6, 1)]
2	[(12, 2)]	[(2, 1), (5, 2)]	[(6, 2)]	[(4, 2)]	[(2, 1), (2, 2)]	[(3, 2)]
3	[(12, 2)]	[(6, 2)]	[(4, 1), (4, 2)]	[(4, 2)]	[(2, 1), (2, 2)]	[(6, 1)]
4	[(8, 3)]	[(4, 3)]	[(4, 3)]	[(2, 1), (2, 3)]	[(2, 3)]	[(2, 3)]
5	[(6, 4)]	[(2, 2), (2, 4)]	[(2, 2), (2, 4)]	[(2, 4)]	[(2, 1), (1, 4)]	[(3, 2)]
6	[(6, 4)]	[(3, 4)]	[(6, 2)]	[(2, 4)]	[(3, 2)]	[(6, 1)]
7	[(6, 4)]	[(3, 4)]	[(2, 2), (2, 4)]	[(2, 4)]	[(1, 2), (1, 4)]	[(3, 2)]
8	[(4, 6)]	[(2, 3), (1, 6)]	[(2, 6)]	[(1, 2), (1, 6)]	[(2, 3)]	[(1, 6)]
9	[(3, 8)]	[(1, 4), (1, 8)]	[(3, 4)]	[(8, 1)]	[(1, 2), (1, 4)]	[(3, 2)]
10	[(2, 12)]	[(1, 12)]	[(2, 6)]	[(2, 4)]	[(1, 6)]	[(2, 3)]
11	[(1, 24)]	[(1, 12)]	[(1, 12)]	[(1, 8)]	[(1, 6)]	[(1, 6)]

	7	8	9	10	11
1	[(6, 1)]	[(4, 1)]	[(3, 1)]	[(2, 1)]	[(1, 1)]
2	[(3, 2)]	[(2, 1), (1, 2)]	[(1, 1), (1, 2)]	[(1, 2)]	[(1, 1)]
3	[(2, 1), (2, 2)]	[(2, 2)]	[(3, 1)]	[(2, 1)]	[(1, 1)]
4	[(2, 3)]	[(1, 1), (1, 3)]	[(1, 3)]	[(2, 1)]	[(1, 1)]
5	[(1, 2), (1, 4)]	[(2, 2)]	[(1, 1), (1, 2)]	[(1, 2)]	[(1, 1)]
6	[(3, 2)]	[(1, 4)]	[(3, 1)]	[(2, 1)]	[(1, 1)]
7	[(2, 1), (1, 4)]	[(1, 4)]	[(1, 1), (1, 2)]	[(1, 2)]	[(1, 1)]
8	[(1, 6)]	[(1, 1), (1, 3)]	[(1, 3)]	[(1, 2)]	[(1, 1)]
9	[(1, 2), (1, 4)]	[(1, 4)]	[(1, 1), (1, 2)]	[(1, 2)]	[(1, 1)]
10	[(1, 6)]	[(1, 4)]	[(1, 3)]	[(2, 1)]	[(1, 1)]
11	[(1, 6)]	[(1, 4)]	[(1, 3)]	[(1, 2)]	[(1, 1)]

Figure 43.1. Partition Array for S_4

The partition array is given in Figure 43.1.

This table is applied as follows.

- Compute the discriminant $\mathcal{L}_{\Theta_{10}.f}$ of f and check whether $\mathcal{L}_{\Theta_{10}.f}$ is a square; if so, we have

$$G(K_f/k) = H_j, \quad j \in \{1, 3, 4, 6, 10\}.$$

- Then one computes a factorization of $\mathcal{L}_{\Theta_8.f}$; if $\pi(\Theta_8, f)$ is

  $[(4, 1)] \implies G(K_f/k) = H_1,$

  $[(2, 1), (1, 2)] \implies G(K_f/k) = H_2,$

  $[(2, 2)]$ and $\mathcal{L}_{\Theta_{10}.f}$ is a square $\implies G(K_f/k) = H_3,$

  $[(2, 2)]$ and $\mathcal{L}_{\Theta_{10}.f}$ is not a square $\implies G(K_f/k) = H_5,$

  $[(1, 1), (1, 3)]$ and $\mathcal{L}_{\Theta_{10}.f}$ is a square $\implies G(K_f/k) = H_4,$

  $[(1, 1), (1, 3)]$ and $\mathcal{L}_{\Theta_{10}.f}$ is not a square $\implies G(K_f/k) = H_8,$

  $[(1, 4)]$ then compute a factorization of $\mathcal{L}_{\Theta_9.f}$; if $\pi(\Theta_9, f)$ is

  ◦ $[(1, 3)]$ and $\mathcal{L}_{\Theta_{10}.f}$ is a square $\implies G(K_f/k) = H_{10},$

  ◦ $[(1, 3)]$ and $\mathcal{L}_{\Theta_{10}.f}$ is not a square $\implies G(K_f/k) = H_{11},$

  ◦ $[(3, 1)] \implies G(K_f/k) = H_6,$

  ◦ $[(1, 1), (1, 2)]$ then compute a factorization of $\mathcal{L}_{\Theta_4.f}$;[17] if $\pi(\Theta_4, f)$ is

  ◇ $[(1, 8)] \implies H_9,$

  ◇ $[(2, 4)] \implies H_7.$

---

[17] One could similarly use $\mathcal{L}_{\Theta_i.f}, i \in \{1, 2, 3, 7\}$.

44

Kronecker IV

The main effort in the research on solving techniques has always been devoted to "practical" complexity, namely smooth and fast software tools,[1] while "theoretical" complexity has never been deeply considered. A noteworthy exception is the work of the TERA group, based in École Polytechnique, Buenos Aires and Santander, around Marc Giusti, Joos Heintz and Luis M. Pardo, which in a series of papers produced in the 1990s[2] devised a solver with good complexity. The input is assumed to be a finite set of polynomials generating a zero-dimensional ideal $\mathsf{J} \subset \mathcal{Q}$ and given by a straight-line program, the output being

- a system of coordinates in noetherian position for the ideal,
- a primitive element of $\mathcal{Q}/\mathsf{J}$,
- its minimal polynomial $q(T) = g_0(T) \in K[T]$ and
- either

    an *allgemeine* basis $(g_0(T), Z_1 - g_1(T), \ldots, Z_r - g_r(T))$ or
    a Kronecker/RUR presentation

$$q(T), \quad \frac{\partial q}{\partial T}(T)Z_1 - w_1(Y_1, \ldots, Y_d, T), \quad \ldots, \quad \frac{\partial q}{\partial T}(T)Z_r - w_r(T).$$

[1] The main effort within the **Polynomial System Solving (PoSSo)** group was devoted to efficient memory management!

[2] Of which here and in the Bibliography I quote only the most relevant papers:

- Giusti M., Heintz J., Morais J.E., Pardo L.M., When polynomial equation systems can be "solved" fast?, in *L. N. Comp. Sci.* **948** (1995), pp. 205–231, Springer.
- Giusti M., Heintz J., Morais J.E., Morgensten J., Pardo L.M., Straight-line programs in geometric elimination theory, *J. Pure Appl. Algebra* **124** (1998), 101–146.
- Giusti M., Heintz J., Hägele K., Morais J.E., Pardo L.M., Montaña J.L., Lower bounds for diophantine approximation, *J. Pure Appl. Algebra* **117–118** (1997), 277–311.
- Giusti M., Heintz J., Morais J.E., Pardo L.M., Le rôle des structures de données dans les problèmes d'élimination, *C.R. Acad. Sci. Paris* **325** (1997), 1223–1228.
- Morais J.E., *Resolución eficaz de systemas de ecuaciones polinomiales*, Ph.D. thesis, University of Cantabria, Santander (1997).
- Giusti M., Lecerf G., Salvy B., A Gröbner free alternative for polynomial system solving, *J. Complexity* **17** (2001), 154–211.
- Lecerf G., *Une alternative aux méthodes de réécriture pour résolution des systémes algébriques*, Ph.D. thesis, École Polytechnique (2001).

The relevant result is that such an algorithm has (low) polynomial complexity w.r.t. the natural measure of the data (the number of variables, the degrees of the input polynomials, the number of roots, the size of the straight-line program).

What is amazing is that this good-complexity theoretical result has produced a software solver whose practical performances compare with the best available Gröbner-based solvers.

After posing the problem tackled by the TERA group (Section 44.1), discussing the technical tools, mainly an appropriate Newton–Hensel lifting (Sections 44.2 and 44.3) and presenting the general structure of the Kronecker package (Section 44.4), I discuss in depth its three steps (Sections 44.5–44.7), its genericity conditions, showing that the "good" choices live in an open Zariski set (Section 44.8), and finally sketch its complexity analysis (Section 44.9).

44.1 Kronecker Parametrization

Let:

- $\mathsf{I} \subset k[X_1, \ldots, X_n]$ be an unmixed radical ideal;
- $d := \dim(\mathsf{I})$, $r := n - d = r(\mathsf{I})$ be the dimension and rank of I;
- $\mathsf{M} := (c_{ij}) \in GL(n, k)$ be an invertible $n \times n$ square matrix with entries in k such that, writing $Y_i := \sum_j c_{ij} X_j$ for each i,

$$\{Y_1, \ldots, Y_n\} = \{V_1, \ldots, V_d, Z_1, \ldots, Z_r\}$$

is a Noether position (Definition 27.9.4) for I;
- $K := k(V_1, \ldots, V_d)$, and let $K \subset \Omega(k)$ be its algebraic closure;
- $\mathsf{J} := \mathsf{I}K[Z_1, \ldots, Z_r]$, the zero-dimensional extension of I;
- $\mathsf{s} := \deg(\mathsf{J})$ be the multiplicity (Definitions 27.12.9 and 27.13.7) of I.

Recall (Section 34.2) that a K-linear form

$$U := \lambda_1 Z_1 + \cdots + \lambda_r Z_r, \quad \lambda_i \in K, \quad \lambda_1 \neq 0,$$

is a primitive element of $K[Z_1, \ldots, Z_r]/\mathsf{J}$ iff

$$\operatorname{Span}_K\{1, U, U^2, \ldots, U^{\mathsf{s}-1}\} \cong K[Z_1, \ldots, Z_r]/\mathsf{J}.$$

Definition 44.1.1 (Giusti–Heintz–Morales–Pardo). *With the notation above the assignment of*

- *a matrix* $\mathsf{M} := (c_{ij}) \in GL(n, k)$, *such that, setting* $Y_i := \sum_j c_{ij} X_j$,

$$\{Y_1, \ldots, Y_n\} = \{V_1, \ldots, V_d, Z_1, \ldots, Z_r\}$$

is a Noether position for I,

- *a primitive element*

$$U := \lambda_1 Z_1 + \cdots + \lambda_r Z_r, \quad \lambda_i \in K, \quad \lambda_1 \neq 0$$

  of $K[Z_1, \ldots, Z_r]/J$,
- *the minimal polynomial* $q(T) := g_0(T) \in k[V_1, \ldots, V_d][T]$ *of* U,
- *the parametrization* $(g_1(T), \ldots, g_r(T))$, $g_i \in k(V_1, \ldots, V_d)[T]$ *of the variety* $\mathcal{Z}(\mathsf{I})$ *such that* $Z_i - g_i(U) \in \mathsf{J}$ *for each* i

is called a geometric resolution *of the variety* $\mathcal{Z}(\mathsf{I})$.   ⊙

Remark 44.1.2 (Giusti–Lecerf–Salvy).   Recalling Kronecker's result (41.3) and Proposition 42.9.3, one can remark that for each polynomial $p(T) \in K[T]$ which is relatively prime with $q(T)$, and thus invertible in $K[T]/q(T)$, one obtains, on setting

$$w_i(T) := \mathbf{Rem}(p(T)g_i(T), g_0(T)) \in k(V_1, \ldots, V_d)[T], \qquad 1 \leq i \leq r,$$

another parametrization $\left(\dfrac{w_1(T)}{p(T)}, \ldots, \dfrac{w_r(T)}{p(T)} \right)$ of the variety $\mathcal{Z}(\mathsf{I})$, with

$$p(U)Z_i - w_i(U) \in \mathsf{J} \qquad \text{for each } i.$$

In particular the result (41.3) of Kronecker (where g_0 is assumed irreducible) and that of Proposition 42.9.3 (where g_0 is assumed squarefree) are obtained on setting $p := \partial q/\partial T$.   ⊙

Definition 44.1.3 (Giusti–Lecerf–Salvy).   *A parametrization*

$$\begin{cases} q(V_1, \ldots, V_d, T) &= 0, \\ \dfrac{\partial q}{\partial T}(V_1, \ldots, V_d, T)Z_1 &= w_1(V_1, \ldots, V_d, T), \\ &\vdots \\ \dfrac{\partial q}{\partial T}(V_1, \ldots, V_d, T)Z_r &= w_r(V_1, \ldots, V_d, T) \end{cases}$$

of a radical and equidimensional ideal $\mathsf{I} \subset \mathcal{P}$, $\dim(\mathsf{I}) = d$, *in "generic" position is called a* Kronecker parametrization *of* I.   ⊙

The discussion above allows us to state:

Proposition 44.1.4 (Giusti–Lecerf–Salvy).   *With the notation above one can wlog assume that:*

(1) $q(T) := g_0(T) \in k[V_1, \ldots, V_d][T]$;

(2) *for each* $i \leq r$, $w_i \in k[V_1, \ldots, V_d][T]$;

(3) *for each* $i \leq r$, $\dfrac{\partial q}{\partial T} g_i(T) \equiv w_i(T) \bmod q(T)$;

(4) $q(U) \in \mathsf{I}$;

(5) *for each* $i \leq r$; $\dfrac{\partial q}{\partial T}(U)Z_i - w_i(U) \in \mathsf{I}$;

(6) $\deg(q) = \deg_T(q) = \deg(\mathsf{J})$;

(7) $\deg(w_i) < \deg_T(q) = \deg(\mathsf{J})$ *for each* $i \le r$.

Moreover, for any radical ideal I, *given any system of coordinates*

$$\{Y_1, \ldots, Y_n\} = \{V_1, \ldots, V_d, Z_1, \ldots, Z_r\}$$

which is in Noether position for I *and any primitive element, there is a unique geometric resolution of the variety* $\mathcal{Z}(\mathsf{I})$.

Proof. To prove that, we consider (compare Sections 41.8 and 41.9)

- new variables $\Lambda_{d+1}, \ldots, \Lambda_n$,
- the field $K_\Lambda := K(\Lambda_{d+1}, \ldots, \Lambda_n)$,
- the extension $\mathsf{I}_\Lambda := \mathsf{I}K_\Lambda[Z_1, \ldots, Z_r]$ of I in $K_\Lambda[Z_1, \ldots, Z_r]$,
- the $K(\Lambda_{d+1}, \ldots, \Lambda_n)$-linear form

$$U_\Lambda := \Lambda_{d+1}Z_1 + \cdots + \Lambda_n Z_r$$

which is a primitive element of I_Λ, and

- its characteristic polynomial $q_\Lambda(T) \in (K_\Lambda[Z_1, \ldots, Z_r]/\mathsf{I}_\Lambda)[T]$, which is square-free, monic and of degree $\deg(\mathsf{I}^e)$.

Differentiating $q_\Lambda(T)$ with respect to each Λ_{d+i} we deduce the geometric resolution

$$\begin{cases} q_\Lambda(V_1, \ldots, V_d, T) &= 0, \\ \dfrac{\partial q_\Lambda}{\partial T}(V_1, \ldots, V_d, T)Z_1 &= -\dfrac{\partial q_\Lambda}{\partial \Lambda_{d+1}}(V_1, \ldots, V_d, T), \\ &\vdots \\ \dfrac{\partial q_\Lambda}{\partial T}(V_1, \ldots, V_d, T)Z_r &= -\dfrac{\partial q_\Lambda}{\partial \Lambda_n}(V_1, \ldots, V_d, T). \end{cases}$$

$\boxed{\odot}$

On the basis of these considerations, the TERA group aimed to solve the following:

Problem 44.1.5 (Giusti–Heintz–Morales–Pardo). *Let*

$$f_1, \ldots, f_r, g \in K[Z_1, \ldots, Z_r]$$

and, for each ρ, *write*

$\mathsf{Z}_\rho := \{\alpha \in \mathsf{K}^r : f_1(\alpha) = \cdots = f_\rho(\alpha) = 0 \ne g(\alpha)\}$,
$\mathsf{J}_\rho := \mathbb{I}(f_1, \ldots, f_\rho) : g^\infty$,
$\mathsf{L}_\rho := \sqrt{\mathsf{J}_\rho}$,
$\mathsf{V}_\rho := \mathcal{Z}(\mathsf{J}_\rho)$,

so that

$$\mathcal{I}(\mathsf{Z}_\rho) = \mathcal{I}(\mathsf{V}_\rho) = \mathsf{L}_\rho, \quad \mathsf{V}_\rho = \mathcal{Z}(\mathsf{L}_\rho) = \mathcal{Z}\mathcal{I}(\mathsf{Z}_\rho).$$

Assuming that:

(1) Z_r *is finite;*

(2) *for each ρ, $\dim(\mathsf{L}_\rho) = r - \rho$;*

(3) *the Jacobian matrix $\left(\dfrac{\partial f_i}{\partial Z_j}\right)$ of $f_1, \ldots, f_\rho$ w.r.t. $Z_1, \ldots, Z_r$ has rank ρ at each point of V_ρ,*

compute the parametrization

$$\begin{cases} q(U) &=& 0, \\ Z_1 &=& w_1(U), \\ & \vdots & \\ Z_r &=& w_r(U), \end{cases}$$

of Z_r, where $q(U) \in K[U]$ and $w_i(U) \in K(U)$ for each i. ◉

Remark 44.1.6. The interest of this problem is easily justified by the considerations made in Remark 35.3.9, Theorem 35.6.8 and Remark 35.6.9 on the ARGH scheme; at each step of the computation one obtains an ideal $\mathfrak{f} \subset \mathbb{Q}[X_1, \ldots, X_n]$ and a polynomial $g \in \mathbb{Q}[X_1, \ldots, X_n]$, and one needs to compute the roots of the ideal

$$\left(\sqrt{\mathfrak{f} : g^\infty}\right)^e = \sqrt{\mathfrak{f}^e : g^\infty} \subset \mathbb{Q}(V_1, \ldots, V_d)[Z_1, \ldots, Z_r],$$

where we wlog assume that $\{V_1, \ldots, V_d, Z_1, \ldots, Z_r\}$ is in Noether position for $\mathfrak{f}$ and $d = \dim(\mathfrak{f})$. Denoting by $(g_1, \ldots, g_s)$ any basis of $\mathfrak{f}^e$ it is sufficient (compare Corollary 36.1.6) to perform a generic linear combination $f_i := \sum_{j=1}^{s} \lambda_{ij} g_j$ in order to obtain a regular sequence $f_1, \ldots, f_r$.

Thus the setting relating to the ARGH scheme coincides with that of Problem 44.1.5; thus such a problem can be interpreted as *solving an ARGH-component of a given ideal by producing its Kronecker parametrization.*

Moreover, in this setting:

(1) $\mathfrak{f}^e$ is zero-dimensional;

(2) each L_ρ has rank ρ since $f_1, \ldots, f_r$ is a regular sequence;

(3) it has been proved[3] that the Jacobian condition is satisfied by any generic combination of the basis elements. ◉

44.2 Lifting Points

The variety $\mathcal{Z}(f_1, \ldots, f_r)$ is a subvariety of the δ-dimensional variety

$$\mathsf{V}_\rho \supset \mathcal{Z}(f_1, \ldots, f_r), \quad \delta = r - \rho,$$

[3] Krick T., Pardo L.M., Une approache informatique pour l'approximation diophantienne, *C.R. Acad. Sci. Paris* **318** (1994), 407–412.

defined by the polynomials $f_1, \ldots, f_\rho$ which satisfy conditions (2), (3) of Problem 44.1.5. We call this sequence of polynomials a *lifting system* of V_ρ. Let us now consider a new system of coordinates $\{Y_1, \ldots, Y_r\}$ which is a Noether position for L_ρ and the projection $\phi : K^r \mapsto K^\delta$ defined by $\phi(a_1, \ldots, a_r) = (a_1, \ldots, a_\delta)$.

Definition 44.2.1. *A point* $\mathsf{p} := (p_1, \ldots, p_\delta) \in K^\delta$ *is called a* lifting point *of* V_ρ *w.r.t. the lifting system* $f_1, \ldots, f_\rho$ *(and the frame* $\{Y_1, \ldots, Y_r\}$*) if the Jacobian matrix of* $f_1, \ldots, f_\rho$ *w.r.t.* $Y_{\delta+1}, \ldots, Y_r$ *is invertible at each point of the variety* $V_\mathsf{p} := V_\rho \cap \phi^{-1}(p_1, \ldots, p_\delta)$ *whose zero-dimensional ideal* $\mathcal{I}(V_\mathsf{p})$ *we denote as* L_p. $\quad\boxed{\odot}$

Definition 44.2.2. *With the notation above, the assignment of*

- *a lifting system* $f_1, \ldots, f_\rho$ *of* V_ρ,
- *a matrix* $M := (c_{ij}) \in GL(r, K)$ *such that* $\{Y_1, \ldots, Y_r\}$, $Y_i := \sum_j c_{ij} Z_j$, *is a Noether position for* V_ρ,
- *a lifting point* $\mathsf{p} := (p_1, \ldots, p_\delta)$ *of* V_ρ *w.r.t. the lifting system* $f_1, \ldots, f_\rho$ *and the frame* $\{Y_1, \ldots, Y_r\}$,
- *a primitive element* $U := \lambda_{\delta+1} Y_{\delta+1} + \cdots + \lambda_r Y_r, \lambda_i \in K, \lambda_{\delta+1} \neq 0,$ *of* $K[Y_{\delta+1}, \ldots, Y_r]/L_\mathsf{p}$,
- *the minimal polynomial* $q(T)$ *of* U,
- *the parametrization* $(g_{\delta+1}(T), \ldots, g_r(T))$ *of* V_p

such that

- $Y_j - g_j(U) \in L_\mathsf{p}$ *for each* $j, \delta < j \leq r$,
- $V_\mathsf{p} = \{(p_1, \ldots, p_\delta, g_{\delta+1}(\alpha), \ldots, g_r(\alpha)) : \alpha \in \mathcal{R}\}, \mathcal{R} := \{\alpha \in K : q(\alpha) = 0\}$,

is called a lifting fiber *of* V_ρ. $\quad\boxed{\odot}$

Remark 44.2.3.   Writing

$$M^{-1} := (d_{ij}) \in GL(r, K), \text{ the inverse of } M,$$
$$h_i(Y_1, \ldots, Y_r) := f_i\left(\sum_j d_{1j} Y_j, \ldots, \sum_j d_{rj} Y_j\right),$$

the following relations are satisfied by the data above:

(1) $U(g_{\delta+1}(T), \ldots, g_r(T)) = \lambda_{\delta+1} g_{\delta+1}(T) + \cdots + \lambda_r g_r(T) = T,$
(2) $h_i(p_1, \ldots, p_\delta, g_{\delta+1}(T), \ldots, g_r(T)) \in \mathbb{I}(q(U)),$
(3) $\mathsf{s} := \deg(L_\mathsf{p}) = \deg(L_r).$

Moreover, by Proposition 44.1.4, for any lifting fiber of V_ρ there exists a unique geometric resolution of V_ρ for the same Noether position and primitive element.

$\boxed{\odot}$

Proposition 44.2.4. *The specialization of the minimal polynomial and the parametrization of this geometric resolution on the lifting point* p *gives exactly the minimal polynomial and the parametrization of the lifting fiber.*

Proof.   Assume that U is not a primitive element of V_ρ; we can then choose a primitive element U' of V_ρ which is also a primitive element for V_p. The specialization of the corresponding Kronecker parametrization of V_ρ gives a parametrization of V_p. By linear algebra on

$$\mathrm{Span}_K\{1, U', U'^2, \ldots, U'^{\mathsf{s}-1}\} \equiv K[Y_{\delta+1}, \ldots, Y_r]/\mathsf{L_p}$$

one can compute the minimal polynomial of U, whose degree is necessarily less than s, giving the required contradiction. $\boxdot$

Lemma 44.2.5.   *With the notation and assumptions above, the set of points* $(p_1, \ldots, p_\delta, \lambda_{\delta+1}, \ldots, \lambda_r) \in K^r$ *such that either*

    $\mathsf{p} := (p_1, \ldots, p_\delta)$ *is not a lifting point or*
    $U := \lambda_{\delta+1} Y_{\delta+1} + \cdots + \lambda_r Y_r$ *is not a primitive element of* V_p

is contained in an algebraic proper subset of K^r.

Proof.   Let J be the Jacobian matrix of $f_1, \ldots, f_\rho$ w.r.t. $Y_{\delta+1}, \ldots, Y_r$. The integral dependency relation of $\det(J)$ modulo L_ρ is given by a monic polynomial $F(U) \in K[Y_1, \ldots, Y_\delta][U]$. By assumption, $\det(J)$ is not a zero-divisor of $K[Y_1, \ldots, Y_r]/\mathsf{L}_\rho$, so that $A(Y_1, \ldots, Y_\delta) := F(0) \neq 0$ satisfies $A \in \mathsf{L}_\rho + \det(J)$. Each point $\mathsf{p} := (p_1, \ldots, p_\delta)$ such that $A(p_1, \ldots, p_\delta) \neq 0$ is a lifting point.

Let now fix a lifting point $\mathsf{p} := (p_1, \ldots, p_\delta)$ and consider (compare Proposition 44.1.4)

- the ideal $\mathsf{L_\Lambda} := \mathsf{L_p} K(\Lambda_{\delta+1}, \ldots, \Lambda_r)[Y_{\delta+1}, \ldots, Y_r]$,
- $Y_\Lambda := \Lambda_{\delta+1} Y_{\delta+1} + \cdots + \Lambda_r Y_r \in K(\Lambda_{\delta+1}, \ldots, \Lambda_n)[Y_{\delta+1}, \ldots, Y_n]/\mathsf{L}$,
- $q_\Lambda(T) \in (K(\Lambda_{\delta+1}, \ldots, \Lambda_r)[Y_{\delta+1}, \ldots, Y_r]/\mathsf{L_\Lambda})[T]$, its minimal polynomial,
- $\mathrm{Disc}(q) \in K(\Lambda_{\delta+1}, \ldots, \Lambda_r)$, its discriminant (compare Theorem 10.6.5).

Then any point $(\lambda_{\delta+1}, \ldots, \lambda_r)$ such that $\mathrm{Disc}(q)(\lambda_{\delta+1}, \ldots, \lambda_r) \neq 0$ gives a primitive element $Y := \lambda_{\delta+1} Y_{\delta+1} + \cdots + \lambda_r Y_r$ for V_p. $\boxdot$

44.3 Newton–Hensel Lifting

Proposition 44.3.1.   *Let*

    R be an integral domain,
    I be an ideal of R,
    $I^\star \subset R[T]$ be its extension,
    $\mathbf{f} := (f_1, \ldots, f_r)$, $f_i \in R[Z_1, \ldots, Z_r]$,
    $U := \lambda_1 Z_1 + \cdots + \lambda_r Z_r$, $\lambda_i \in R$, be a linear form,
    $q(T) \in R[T]$ be a monic polynomial, $\mathsf{s} := \deg(q) > 1$,
    $\mathbf{v} := (v_1(T), \ldots, v_r(T))$, $v_i(T) \in R[T]$, $deg(v_i) < \mathsf{s}$,
    $J = \big(\partial f_i/\partial Z_j\big)$ be the Jacobian matrix of $f_1, \ldots, f_r$ w.r.t. $Z_1, \ldots, Z_r$

and assume that the following relations,

(A) $f_j(v_1(T), \ldots, v_r(T)) \equiv 0 \bmod I^\star + \mathbb{I}(q)$, *for each* j,
(B) $T \equiv \lambda_1 v_1(T) + \cdots + \lambda_r v_r(T) \bmod I^\star + \mathbb{I}(q)$,
(C) $J(v_1(T), \ldots, v_r(T))$ *is invertible modulo* $I^\star + \mathbb{I}(q)$,

hold in $R[T]$; *then the following objects exist and can be computed:*

$-$ *a monic polynomial* $Q(T) \in R[T]$,
$-$ $\mathbf{V} := (V_1(T), \ldots, V_r(T))$, $V_i(T) \in R[T]$.

Thus in $R[T]$ *the following hold:*

(1) $\deg(Q) = \mathsf{s}$;
(2) $Q(T) \equiv q(T) \bmod I^\star$;
(3) $\deg(V_i) < \mathsf{s}$, *for each* i;
(4) $V_i(T) \equiv v_i(T) \bmod I^\star$, *for each* i;
(5) $f_j(V_1(T), \ldots, V_r(T)) \equiv 0 \bmod (I^\star)^2 + \mathbb{I}(Q)$, *for each* j;
(6) $T \equiv \lambda_1 V_1(T) + \cdots + \lambda_r V_r(T) \bmod (I^\star)^2 + \mathbb{I}(Q)$.

Proof.   Consider a generic vector

$$\mathbf{w} := (w_1(T), \ldots, w_r(T)), \quad w_i(T) \in R[T], \quad \deg(w_i) < \mathsf{s},$$

and write the Taylor expansion of $\mathbf{f}$ between $\mathbf{w}$ and $\mathbf{v}$

$$\mathbf{f}(\mathbf{w}) = \mathbf{f}(\mathbf{v}) + J(\mathbf{v}) \cdot (\mathbf{w} - \mathbf{v}) + \cdots.$$

Since the aim is to find such a vector $\mathbf{w}$ which moreover satisfies

$$\mathbf{w} \equiv \mathbf{v} \bmod I^\star \quad \text{and} \quad \mathbf{f}(\mathbf{w}) \equiv 0 \bmod ((I^\star)^2 + \mathbb{I}(q)),$$

we use these conditions in the expansion above, obtaining

$$0 \equiv \mathbf{f}(\mathbf{w}) \equiv \mathbf{f}(\mathbf{v}) + J(\mathbf{v}) \cdot (\mathbf{w} - \mathbf{v}) \bmod ((I^\star)^2 + \mathbb{I}(q))$$

and thus deducing, thanks to assumption (C) above, the existence and the uniqueness of a solution

$$\mathbf{w} := \mathbf{v} - J^{-1}(\mathbf{v}) \cdot \mathbf{f}(\mathbf{v}) \bmod ((I^\star)^2 + \mathbb{I}(q)).$$

The polynomial

$$\Delta(T) := U(\mathbf{w}) - T = \lambda_1 w_1(T) + \cdots + \lambda_r w_r(T) - T \in R[T]$$

is such that $\deg(\Delta) < \mathsf{s}$ and assumptions (B), (C) allow us to deduce that all its coefficients are member of I.
    Setting $Y := T + \Delta(T)$ we have

$$\Delta(Y) = \Delta(T) + \Delta'(T)(Y - T) + \cdots = \Delta(T) + \Delta'(T)\Delta(T) + \cdots;$$

hence $\Delta(Y) \equiv \Delta(T) = Y - T \bmod I^2$.

Therefore, defining $p(T), u_i(T)$ as the unique polynomials such that

$$p - q'\Delta, u_i - w_i'\Delta \in \mathbb{I}(q), \quad \deg(p) < \mathsf{s}, \quad \deg(u_i) < \mathsf{s},$$

we define, $\mathrm{mod}((I^\star)^2 + \mathbb{I}(q))$,

$$Q(Y) := q(Y) - p(Y) \equiv q(Y) - \Delta(Y)q'(Y) \equiv q(T) - (Y - T)q'(T) \equiv q(T)$$

and

$$V_i(Y) := w_i(Y) - u_i(Y) \equiv w_i(Y) - \Delta(Y)w'(Y) \equiv w_i(T) - (Y - T)w'(T) \equiv w_i(T).$$

We have therefore the ideal equality

$$\begin{aligned}
\mathsf{H} &:= \mathbb{I}(q(T), Y - T - \Delta(T), Z_1 - w_1(T), \dots, Z_r - w_r(T)) \\
&= \mathbb{I}(Q(Y), T - Y - \Delta(Y), Z_1 - V_1(Y), \dots, Z_r - V_r(Y))
\end{aligned}$$

in the ring $(R/I^2)[U, Y, Z_1, \dots, Z_r]$.

Therefore Q and $\mathbf{V}$ satisfy the required conditions:

(1) $\deg(p) < \mathsf{s} = \deg(q) \implies \deg(Q) = \mathsf{s}$;
(2) it is sufficient to remark that $p \in I$;
(3) $\deg(u_i) < \mathsf{s}, \deg(w_i) < \mathsf{s} \implies \deg(V_i) < \mathsf{s}$, for each i;
(4) it is sufficient to remark that $u_i \in I$;
(5) we have

$$f_j(V_1(Y), \dots, V_r(Y)) \equiv f_j(Z_1, \dots, Z_r) \equiv f_j(w_1(T), \dots, w_r(T)) \bmod \mathsf{H}$$

and $f_j(w_1(T), \dots, w_r(T)) \in (I^\star)^2 + \mathbb{I}(q)$, hence

$$f_j(V_1(Y), \dots, V_r(Y)) \in \left((I^\star)^2 + \mathsf{L}\right) \cap R[Y] = (I^\star)^2 + \mathbb{I}(Q(Y));$$

(6) since, $\mathrm{mod}((I^\star)^2 + \mathsf{H})$,

$$\begin{aligned}
Y - U(V_1(Y), \dots, V_r(Y)) &\equiv Y - U(Z_1, \dots, Z_r) \\
&\equiv Y - U(w_1(T), \dots, w_r(T)) \\
&= Y - T - \Delta(T) \\
&= 0,
\end{aligned}$$

we have $Y - U(V_1(Y), \dots, V_r(Y)) \in \left((I^\star)^2 + \mathsf{L}\right) \cap R[Y] = (I^\star)^2 + \mathbb{I}(Q(Y))$.
$\boxed{\odot}$

Corollary 44.3.2. *With the same notation and assumptions as in Proposition 44.3.1, for any $\epsilon \in \mathbb{N}$ the following objects exist and can be computed,*

- *a monic polynomial $Q(T) \in R[T]$,*
- $\mathbf{V} := (V_1(T), \dots, V_r(T)), V_i(T) \in R[T],$

such that

(1) $\deg(Q) = \mathsf{s}$,
(2) $Q(T) \equiv q(T) \bmod I^\star$,

(3) $\deg(V_i) < \mathsf{s}$, *for each i,*
(4) $V_i(T) \equiv v_i(T) \bmod I^\star$, *for each i,*
(5) $f_j(V_1(T), \ldots, V_r(T)) \equiv 0 \bmod ((I^\star)^{\epsilon+1} + \mathbb{I}(Q))$, *for each j,*
(6) $T \equiv \lambda_1 V_1(T) + \ldots \lambda_r V_r(T) \bmod ((I^\star)^{\epsilon+1} + \mathbb{I}(Q))$

hold in $R[T]$. ◎

44.4 Kronecker Package: Description

The solution of Problem 44.1.5 is incremental on the number of equations to be solved: each system V_ρ is solved iteratively, each resolution being encoded by means of a lifting fiber.

Thus, at each step the algorithm depends on the choice of a Noether position for V_ρ, a lifting point and a primitive element, Zariski-openness granting that such choices can be made randomly.

Let us therefore assume that we have

- $\delta = r - \rho$,
- the δ-dimensional variety $V_\rho \subset \mathcal{Z}(f_1, \ldots, f_r)$,
- the projection $\phi : \mathsf{K}^r \mapsto \mathsf{K}^\delta$ defined by $\phi(a_1, \ldots, a_r) = (a_1, \ldots, a_\delta)$,
- the lifting system $f_1, \ldots, f_\rho$ of V_ρ,
- a frame of coordinates which is in Noether position for V_ρ and which, for simplicity, we assume to be $\{Z_1, \ldots, Z_r\}$,
- a lifting point $\mathsf{p} := (p_1, \ldots, p_\delta)$ of V_ρ w.r.t. the lifting system $f_1, \ldots, f_\rho$ and the frame $\{Z_1, \ldots, Z_r\}$,
- the primitive element $U := \lambda_{\delta+1} Z_{\delta+1} + \cdots + \lambda_r Z_r, \lambda_i \in k, \lambda_{\delta+1} \neq 0$, of $K[Z_{\delta+1}, \ldots, Z_r]/\mathsf{L_p}$,
- the minimal polynomial $q(T)$ of U,
- the parametrization $(v_{\delta+1}(T), v_{\delta+2}(T), \ldots, v_r(T))$ of both V_p and V_ρ.

Up to now we have simply assumed that

the Noether position $\{Z_1, \ldots, Z_r\}$,
the lifting point $\mathsf{p} := (p_1, \ldots, p_\delta)$ and
the primitive element $\lambda_{\delta+1} Z_{\delta+1} + \cdots + \lambda_r Z_r$

are sufficiently generic to satisfy all the conditions of genericity required by the algorithm; we will discuss such conditions more deeply in Section 44.8.

lifting step: Thus we are assuming to have a geometric resolution

$$\begin{cases} q(T) &= 0, \\ Z_{\delta+1} &= v_{\delta+1}(T), \\ &\vdots \\ Z_r &= v_r(T) \end{cases}$$

for the primitive element

$$U := \lambda_{\delta+1} Z_{\delta+1} + \cdots + \lambda_r Z_r \in K[Z_{\delta+1}, \ldots, Z_r]/\mathsf{L_p}$$

of the variety $\mathsf{V_p}$ defined by

$$\mathbf{a} := (p_1, \ldots, p_\delta, \alpha_{\delta+1}, \ldots, \alpha_r) \in \mathsf{V_p} \iff f_1(\mathbf{a}) = \cdots = f_\rho(\mathbf{a}) = 0 \neq g(\mathbf{a})$$

and the zero-dimensional radical ideal

$$\mathsf{L_p} := \mathcal{I}(\mathsf{V_p}) = \mathsf{L}_\rho + \mathbb{I}(Z_1 - p_1, \ldots, Z_\delta - p_\delta).$$

We compute a geometric resolution

$$\left\{ \begin{array}{rcl} Q(Z_\delta, T) & = & 0, \\ Z_{\delta+1} & = & V_{\delta+1}(Z_\delta, T), \\ & \vdots & \\ Z_r & = & V_r(Z_\delta, T) \end{array} \right.$$

for the primitive element

$$U := \sum_{i=1}^{\rho} \lambda_{\delta+i} Z_{\delta+i} \in k(Z_\delta)[Z_{\delta+1}, \ldots, Z_r]/\mathsf{L}_D^e$$

of the variety V_D defined by

$$\mathbf{a} = (p_1, \ldots, p_{\delta-1}, \alpha_\delta, \ldots, \alpha_r) \in \mathsf{V}_D \iff f_1(\mathbf{a}) = \cdots = f_\rho(\mathbf{a}) = 0 \neq g(\mathbf{a})$$

and the one-dimensional radical ideal

$$\mathsf{L}_D := \mathcal{I}(\mathsf{V}_D) = \mathsf{L}_\rho + \mathbb{I}(Z_1 - p_1, \ldots, Z_{\delta-1} - p_{\delta-1}).$$

intersection step: From these data we compute a geometric resolution

$$\left\{ \begin{array}{rcl} q(Z) & = & 0, \\ Z_\delta & = & v_\delta(Z), \\ & \vdots & \\ Z_r & = & v_r(Z) \end{array} \right.$$

for the primitive element

$$U := \sum_{j=0}^{\rho} \lambda_{\delta+j} Z_{\delta+j} \in K[Z_\delta, \ldots, Z_r]/\mathsf{L}'$$

of the zero-dimensional radical ideal

$$\mathsf{L}' := \sqrt{\mathsf{L}_r + \mathbb{I}(Z_1 - p_1, \ldots, Z_{\delta-1} - p_{\delta-1}, f_{\rho+1})}.$$

cleaning step: We now remove the points $\mathbf{a} \in \mathcal{Z}(\mathsf{L}')$ such that $g(\mathbf{a}) = 0$, thus obtaining the required geometric resolution

$$\begin{cases} q'(T) & = & 0, \\ Z_\delta & = & v'_\delta(T), \\ & \vdots & \\ Z_r & = & v'_r(T) \end{cases}$$

for the primitive element

$$U := \lambda_\delta Z_\delta + \cdots + \lambda_r Z_r \in K[Z_\delta, \ldots, Z_r]/\mathsf{L_{p'}}$$

and the lifting point $\mathbf{p}' := (p_1, \ldots, p_{\delta-1})$ of the variety $\mathsf{V_{p'}}$ defined by

$$\mathbf{a} := (p_1, \ldots, p_{\delta-1}, \alpha_\delta, \ldots, \alpha_r) \in \mathsf{V_{p'}} \iff f_1(\mathbf{a}) = \cdots = f_{\rho+1}(\mathbf{a}) = 0 \neq g(\mathbf{a})$$

and the zero-dimensional radical ideal

$$\mathsf{L_{p'}} := \mathcal{I}(\mathsf{V_p}) = \mathsf{L}_{\rho+1} + (Z_1 - p_1, \ldots, Z_{\delta-1} - p_{\delta-1}).$$

44.5 Kronecker Package: Lifting Step

Defining

$$\epsilon := \deg(\mathsf{L_p}) + 1 = \deg(\mathsf{L}_\rho) + 1,$$
$$h_i(Z_\delta, Z_{\delta+1}, \ldots, Z_r) := f_i(p_1, \ldots, p_{\delta-1}, Z_\delta, Z_{\delta+1}, \ldots, Z_r),$$

we can apply Corollary 44.3.2 to the data

$$R := K[Z_\delta],$$
$$I := \mathbb{I}(Z_\delta) \subset R,$$
$$I^\star := \mathbb{I}(Z_\delta) \subset K[Z_\delta, T],$$
$$\mathbf{f} := (h_1, \ldots, h_r), h_i \in R[Z_{\delta+1}, \ldots, Z_r],$$
$$U := \lambda_{\delta+1} Z_{\delta+1} + \cdots + \lambda_r Z_r,$$
$$q(T) \in R[T] \text{ the minimal polynomial of } U,$$
$$\mathbf{v} := (v_{\delta+1}(T), v_{\delta+2}(t), \ldots, v_r(T)),$$

thus obtaining

a monic polynomial $Q(T) \in K[Z_\delta][T]$,
$$\mathbf{V} := (V_{\delta+1}(T), \ldots, V_r(T)), V_i(T) \in K[Z_\delta][T],$$

such that

(1) $h_j(V_{\delta+1}(T), \ldots, V_r(T)) \equiv 0 \bmod (I^{\epsilon+1} + \mathbb{I}(Q))$, for each j,

(2) $T \equiv \lambda_{\delta+1} V_1(T) + \cdots + \lambda_r V_r(T) \bmod (I^{\epsilon+1} + \mathbb{I}(Q))$.

That is, we obtain

a polynomial $Q(Z_\delta, T) \in K[Z_\delta, T]$,
$\mathbf{V} := (V_{\delta+1}(Z_\delta, T), \ldots, V_r(Z_\delta, T)), V_i(Z_\delta, T) \in K[Z_\delta, T]$,

such that

(1) $f_j(p_1, \ldots, p_{\delta-1}, Z_\delta, V_{\delta+1}(Z_\delta, T), \ldots, V_r(Z_\delta, T)) \equiv 0 \bmod Q$, for each j,
(2) $T \equiv \lambda_{\delta+1} V_{\delta+1}(Z_\delta, T) + \cdots + \lambda_r V_r(Z_\delta, T) \bmod Q$.

We therefore have the parametrization

$$\left\{ \begin{array}{rcl} Q(Z_\delta, T) & = & 0, \\ Z_{\delta+1} & = & V_{\delta+1}(Z_\delta, T), \\ & \vdots & \\ Z_r & = & V_r(Z_\delta, T) \end{array} \right.$$

of

$$\mathsf{V}_D := \{\mathbf{a} = (p_1, \ldots, p_{\delta-1}, \alpha_\delta, \ldots, \alpha_r) \in \mathsf{K}^r : f_1(\mathbf{a}) = \cdots = f_\rho(\mathbf{a}) = 0 \neq g(\mathbf{a})\}$$

and the one-dimensional radical ideal

$$\mathsf{L}_D := \mathcal{I}(\mathsf{V}_D) = \mathsf{L}_\rho + \mathbb{I}(Z_1 - p_1, \ldots, Z_{\delta-1} - p_{\delta-1}).$$

44.6 Kronecker Package: Intersection Step

Let us therefore assume that we have a one-dimensional ideal

$$\mathsf{I} \subset K[Z_\delta, Z_{\delta+1}, \ldots, Z_r],$$

and its radical $\mathsf{L} := \sqrt{\mathsf{I}}$, given by means of the geometric resolution

$$\left\{ \begin{array}{rcl} Q(Z_\delta, T) & = & 0, \\ Z_{\delta+1} & = & V_{\delta+1}(Z_\delta, T), \\ & \vdots & \\ Z_r & = & V_r(Z_\delta, T) \end{array} \right.$$

for

- the Noether position $\{Z_\delta, Z_{\delta+1}, \ldots, Z_r\}$ for L,
- the primitive element $U := \lambda_{\delta+1} Z_{\delta+1} + \cdots + \lambda_r Z_r, \lambda_i \in K \subset K(Z_\delta)$,

and a polynomial $f \in K[Z_\delta, Z_{\delta+1}, \ldots, Z_r]$ such that $\mathsf{L} + \mathbb{I}(f)$ is zero-dimensional.[4]
Let us consider $Q(Z_\delta, T)$ and $f(Z_\delta, V_{\delta+1}(Z_\delta, T), \ldots, V_r(Z_\delta, T))$ as polynomials in $K[Z_\delta][T]$ and their resultant:

---

[4] We can apply the results of this section to the case $\mathsf{L} := \mathsf{L}_D$ and then obtain
$$f(Z_\delta, Z_{\delta+1}, \ldots, Z_r) := f_{\rho+1}(p_1, \ldots, p_{\delta-1}, Z_\delta, Z_{\delta+1}, \ldots, Z_r).$$

Lemma 44.6.1. *Denoting*

$A(Z_\delta) := \mathrm{Res}(Q, f(Z_\delta, V_{\delta+1}, \ldots, V_r) \in K[Z_\delta]$,

$B := K[Z_\delta, Z_{\delta+1}, \ldots, Z_r]/\mathsf{L}$,

$B' := K(Z_\delta)[Z_{\delta+1}, \ldots, Z_r]/\mathsf{L}^e$,

$F(T) \in K[Z_\delta][T]$ *as the integral dependency relation of* f *modulo* L,

Φ_f *as the endomorphism of multiplication by* f *in* B',

$\chi_f(T) \in K(Z_\delta)[T]$ *as the characteristic polynomial of* Φ_f,

$m_f(T) \in K(Z_\delta)[T]$ *as the monic polynomial of* Φ_f,

$\mathsf{W} := \mathcal{Z}(\mathsf{L}^e) = \{\mathsf{b}_1, \ldots \mathsf{b}_t\}$, $\sigma_i := \mathrm{mult}(\mathsf{b}_i, \mathsf{L}^e)$,

$\phi : \mathsf{K}^{\rho+1} \mapsto \mathsf{K}$ *as the projection* $\phi(\alpha, \beta_1, \ldots, \beta_\rho) = \alpha$,

for each $\alpha \in \mathsf{K}$, $\mathsf{W}_\alpha := \{\mathsf{a}_1, \ldots, \mathsf{a}_s\} = \pi^{-1}(\alpha) \cap \mathcal{Z}(\mathsf{L})$, *each* a_i *being counted with the proper multiplicity* $s_i := \mathrm{mult}(\mathsf{a}_i, \mathsf{L})$, $\sum_i s_i = \deg(\mathsf{L}^e)$,

we have:

(1) $m_f(T), \chi_f(T) \in K[Z_\delta][T]$;

(2) $C_i(Z_\delta) \in K[Z_\delta]$ *and* $\deg(C_i) \leq (\mathsf{s} - i)\deg(f)$, *setting* $\chi_f := \sum_i C_i(Z_\delta)T^i$ *and* $\mathsf{s} := \deg(\mathsf{L}^e)$;

(3) $C_0(Z_\delta) \in \mathsf{L} + \mathbb{I}(f)$;

(4) $C_0(\alpha) = \prod_{\mathsf{a}_i \in \mathsf{W}_\alpha} f(\mathsf{a}_i)^{s_i}$;

(5) $C_0(Z_\delta)$ *and* $A(Z_\delta)$ *coincide up to a sign*;

(6) $\deg(A) \leq \mathsf{s}\deg(f)$;

(7) $\{\alpha \in \mathsf{K} : A(\alpha) = 0\} = \{\phi(\mathsf{b}) : \mathsf{b} \in \mathsf{W}, f(\mathsf{b}) = 0\}$.

Proof. Since $F(\Phi_f) = 0$ we can deduce that $m_f \mid F$ and, since both are monic, Gauss's Lemma (Corollary 6.1.5) implies (1). Moreover we know (Corollary 40.5.2) that $\chi(Z_\delta, T) = \prod_{i=1}^{t} (T - f(Z_\delta, \mathsf{b}_i))^{\sigma_i}$, so that we can deduce (2).

We remark that B is a finite $K[Z_\delta]$-module of rank $\mathsf{s} := \deg(\mathsf{L}^e)$. Since any $K[Z_\delta]$-basis of B induces a $K(Z_\delta)$-basis of B' (compare Section 36.3), the characteristic polynomials of Φ_f in B and B' coincide, so that the Cayley–Hamilton Theorem in B implies that $\chi(Z_\delta, f) \in \mathsf{L}$ and hence (3).

Moreover, for each $\alpha \in \mathsf{K}$, writing $B_0 := \mathsf{K}[Z_\delta, Z_{\delta+1}, \ldots, Z_r]/\mathsf{L} + \mathbb{I}(Z_\delta - \alpha)$ and remarking that the specialization at α of the $K[Z_\delta]$-basis of B gives a K-basis of B_0, we deduce that $C_0(\alpha)$ is the constant coefficient of the characteristic polynomial $\chi(T) = \prod_{i=1}^{s} (T - f(\mathsf{a}_i))^{s_i}$ of multiplication by f in B_0, whence (4).

Since $\mathsf{a} \in \mathcal{Z}(\mathsf{L} + \mathbb{I}(f)) \implies C_0(\pi(\mathsf{a})) = 0$ for each $\mathsf{a} \in \mathsf{K}^{\rho+1}$, as a consequence of (3), and, for each $\alpha \in \mathsf{K}$ which annihilates C_0, (4) implies the existence of $\mathsf{a} \in \pi^{-1}(\alpha) \cap \mathcal{Z}(\mathsf{L})$ which annihilates f, we obtain (5), of which (6), (7) are direct consequences. $\quad\boxed{\odot}$

Since Z_δ is probably not a primitive element for

$$K[Z_\delta, Z_{\delta+1}, \ldots, Z_r]/\sqrt{\mathsf{L} + \mathbb{I}(f)},$$

although this is true for a generic element $\lambda_\delta Z_\delta + U$, let us therefore introduce a new variable Z, denote by

$$\widehat{\cdot} : K[Z_\delta, Z_{\delta+1}, \ldots, Z_r][T] \mapsto K[Z, Z_{\delta+1}, \ldots, Z_r][T]$$

the substitution

$$\widehat{g} := g(\lambda_\delta^{-1}(Z - T), T, Z_{\delta+1}, \ldots, Z_r) \text{ for each } g(Z_\delta, T, Z_{\delta+1}, \ldots, Z_r)$$

and assume that $Z = \lambda_\delta \widehat{Z_\delta} + T$ has the required properties; this is true for almost all choices of λ_δ.

Definition 44.6.2. *A point* $\lambda_\delta \in k$ *is called a* Liouville *point w.r.t. the above geometric resolution of* L *if*

(1) $\lambda_\delta \neq 0$,
(2) $\widehat{Q}$ *is monic in* T *and* $\deg_T(\widehat{Q}) = \deg_T(Q) = \mathsf{s} = \deg(\mathsf{L})$,
(3) $\widehat{Q}$ *is squarefree and relatively prime with* $\widehat{P}$, $P := \partial Q/\partial T$.

Lemma 44.6.3. *With the above notation, if* λ_δ *is a Liouville point then the variables* $\{Z, Z_{\delta+1}, \ldots, Z_r\}$ *are in Noether position w.r.t.* $\widehat{\mathsf{L}} := \{\widehat{f} : f \in \mathsf{L}\}$ *and*

$$\begin{cases} \widehat{Q}(Z, T) &= 0, \\ Z_{\delta+1} &= \widehat{V}_{\delta+1}(Z, T), \\ &\vdots \\ Z_r &= \widehat{V}_r(Z, T) \end{cases}$$

is a geometric resolution of $\widehat{\mathsf{L}}$ *for the primitive element* U.

Proof. First of all, we note that $\widehat{\mathsf{L}} \cap K[Z] = \{0\}$ since for each $h(Z) \in \widehat{\mathsf{L}} \cap K[Z]$ we have $Q(Z_\delta, T) \mid h(\lambda_\delta Z_\delta + T)$ and $\widehat{Q}(Z, T) \mid h(Z)$; since $\widehat{Q}(Z, T)$ is monic in T this implies that $h(Z) = 0$ as required.

Thus, in order to prove that $\{Z, Z_{\delta+1}, \ldots, Z_r\}$ is in Noether position it is sufficient to prove that each Z_i is dependent on Z. Set $\mathsf{L}_1 := \widehat{\mathsf{L}} + \mathbb{I}(T - U) \subset K[Z, Z_{\delta+1}, \ldots, Z_r, T]$ and consider a bivariate polynomial $h(Z_\delta, Z_i) \in \mathsf{L}$, monic and whose total degree is bounded by $\deg_{Z_i}(h)$, whose existence is implied by the assumption that $\{Z_\delta, Z_{\delta+1}, \ldots, Z_r\}$ is in Noether position w.r.t. I. Then $h(\lambda_\delta^{-1}(Z - T), Z_i) \in \mathsf{L}$ and, since $\widehat{Q}(Z, T) \in \mathsf{L}$ and its total degree is bounded by s, we can deduce the existence of a polynomial $H(Z, Z_i) \in \mathsf{L}$, monic and whose total degree is bounded by $\deg_{Z_i}(h)$, that is, the dependency of Z_i on Z.

Since $\widehat{Q}$ remains squarefree, U remains primitive. $\qquad\qquad\boxed{\odot}$

Lemma 44.6.4. *Almost each* $\lambda_\delta \in K$ *is a Liouville point.*

Proof. Setting $W := \lambda_\delta^{-1} Z$ we have

$$\widehat{Q}(Z, T) = Q(\lambda_\delta^{-1}(Z - T), T) = Q(W - \lambda_\delta^{-1}T, T).$$

Both the discriminant of $Q(W - \Lambda T, T)$ and the resultant, in $K[W, \Lambda][T]$, of $Q(W - \Lambda T, T)$ and $\frac{\partial Q}{\partial T}(W - \Lambda T, T)$ are polynomials in $K[W, \Lambda]$ and do not vanish for

$\Lambda = 0$; hence almost all choices for $\lambda_\delta \neq 0$ give (3). Also, writing as $h(Z, T) := H(Q)$ (Definition 23.2.1) the homogeneous part of maximal degree s of Q, the coefficient of T^{s} in $\widehat{Q}(Z, T)$ is $h(-\lambda_\delta^{-1}, 1)$; since again $h(0, 1) \neq 0$, almost all choices for $\lambda_\delta \neq 0$ give (2). $\quad\boxed{\odot}$

Remark 44.6.5.   If $\lambda_\delta \in K$ is a Liouville point w.r.t. the above geometric resolution of L and $\widehat{f}$ denotes the polynomial $\widehat{f} := f(\lambda_\delta^{-1}(Z - T), Z_{\delta+1}, \ldots, Z_r)$ then the resultant $A(Z) \in K[Z]$ of the polynomials (in $K[Z][T]$)

$$\widehat{Q}(Z, T) \quad \text{and} \quad f(\lambda_\delta^{-1}(Z - T), \widehat{V}_{\delta+1}(Z, T), \ldots, \widehat{V}_r(Z, T))$$

satisfies $A(\lambda_\delta Z_\delta + U) \in \mathsf{L}$. In fact we have already proved (in Lemma 44.6.1) that $A(Z) \in \widehat{\mathsf{L}} + \mathbb{I}(\widehat{f})$; thus replacing Z with $\lambda_\delta Z_\delta + U$ we obtain $A(\lambda_\delta Z_\delta + U) \in \mathsf{L}$.

Moreover, each root $(\alpha, \beta_{\delta+1}, \ldots, \beta_r) \in \pi^{-1}(\alpha)$ of $\widehat{\mathsf{L}}$, where $A(\alpha) = 0$, corresponds to the root $(\beta_\delta, \beta_{\delta+1}, \ldots, \beta_r)$ of $\mathsf{L} + \mathbb{I}(f)$, where

$$\beta_\delta = \lambda_\delta^{-1}\left(\alpha - \sum_{j=1}^{n} \lambda_{\delta+j}\beta_{\delta+j}\right) \quad \text{or, equivalently,} \quad \alpha = \sum_{j=0}^{\rho} \lambda_{\delta+j}\beta_{\delta+j}.$$

This is not yet sufficient to describe $\mathcal{Z}(\mathsf{L} + \mathbb{I}(f))$, because we are still missing the parametrization of the coordinates. Letting

$T_\delta, T_{\delta+1}, \ldots, T_r$ be new variables,
$K_t := K(T_\delta, T_{\delta+1}, \ldots, T_r)$,
$\mathsf{L}_t := \mathsf{L}K_t[Z_\delta, Z_{\delta+1}, \ldots, Z_r]$,
$U_t := U + T_{\delta+1}Z_{\delta+1} + \cdots + T_r Z_r = \sum_{j=1}^{\rho}(\lambda_{\delta+j} + T_{\delta+j})Z_{\delta+j}$,

let us assume that we have the geometric resolution

$$\begin{cases} q_t(Z_\delta, T) &= 0, \\ Z_{\delta+1} &= V_{t,\delta+1}(Z_\delta, T), \\ &\vdots \\ Z_r &= V_{t,r}(Z_\delta, T) \end{cases}$$

of L_t for the primitive element U_t. Since, for a Liouville point λ_δ for L, $\lambda_\delta + T_\delta$ is a Liouville point for L_t, in this setting the resultant computation returns a polynomial $A_t(Z) \subset K_t[Z]$ such that $A_t((\lambda_\delta + T_\delta)Z_\delta + T) \in \mathsf{L}_t$. More precisely, if we express $A_t(Z)$ as

$$A(Z) + T_\delta A_\delta(Z) + T_{\delta+1}A_{\delta+1}(Z) + \cdots + T_r A_r(Z) + B,$$

where $A_i(Z) \in K[Z]$ and $B(T_\delta, T_{\delta+1}, \ldots, T_r, Z) \in \mathbb{I}(T_\delta, T_{\delta+1}, \ldots, T_r)^2$ and we evaluate it in

$$(\lambda_\delta + T_\delta)Z_\delta + U_t = (\lambda_\delta Z_\delta + U) + T_\delta Z_\delta + T_{\delta+1}Z_{\delta+1} + \cdots + T_r Z_r$$

then a Taylor expansion allows us to deduce that both $A(\lambda_\delta Z_\delta + U)$ and

$$A'(\lambda_\delta Z_\delta + U) + Z_{\delta+j}A_{\delta+j}(\lambda_\delta Z_\delta + U), \quad 0 \leq j \leq \rho,$$

are members of L.

The roots of the polynomial $A_t(Z)$ are the values of the linear form

$$U_t := \sum_{j=0}^{\rho} \lambda_{\delta+j}(Z_{\delta+j} + T_{\delta+j})$$

at the roots of $\mathsf{L} + \mathbb{I}(f)$. Thus, setting $\mathcal{Z}(\mathsf{L} + \mathbb{I}(f)) =: \{\mathsf{a}_1, \ldots \mathsf{a}_s\}$ and a linear form $U := \sum_{j=0}^{\rho} \lambda_{\delta+j} Z_{\delta+j}$, we have $A_t(Z) = \prod_{j=1}^{s} \left(Z - U_t(\mathsf{a}_j)\right)^{s_j} \in \mathsf{L}_t$ and

$$\prod_{j=1}^{s}(Z - U_t(\mathsf{a}_j))^{s_j} \equiv A(Z) + \sum_{i=0}^{r} A_{\delta+i}(Z)T_{\delta+i} \mod (\mathsf{L}_t + \mathbb{I}(T_\delta, T_{\delta+1}, \ldots, T_r)^2).$$

Thus, by expansion, we obtain

$$A(Z) := \prod_{j=1}^{s}(Z - U(\mathsf{a}_j))^{s_j},$$

$$A_{\delta+i}(Z) := -\sum_{h=1}^{s} Z_{\delta+i}(\mathsf{a}_h)s_h(Z - U(\mathsf{a}_h))^{s_h - 1} \prod_{\substack{j=1 \\ j \neq h}}^{s}(Z - U(\mathsf{a}_j))^{s_j}.$$

Thus $D(Z) := \gcd(A, A') = \prod_{j=1}^{s}(Z - U(\mathsf{a}_j))^{s_j - 1}$ divides each $A_{\delta+i}$, so that

$$\begin{cases} A(Z)/D(Z) &=& 0, \\ A'(Z)/D(Z)Z_\delta &=& A_\delta(Z)/D(Z), \\ A'(Z)/D(Z)Z_{\delta+1} &=& A_{\delta+1}(Z)/D(Z), \\ &\vdots& \\ A'(Z)/D(Z)Z_r &=& A_r(Z)/D(Z) \end{cases}$$

is the required geometric resolution of $\mathsf{L} + \mathbb{I}(f)$ for the primitive element $U := \sum_{j=0}^{\rho} \lambda_{\delta+j} Z_{\delta+j}$.

Finally, euclidean arithmetic allows us to compute the inverse of A'/D modulo A/D and gives the representation

$$\begin{cases} q(Z) &=& 0, \\ Z_\delta &=& v_\delta(Z), \\ &\vdots& \\ Z_r &=& v_r(Z). \end{cases}$$

▣

Now, in order to apply Remark 44.6.5 successfully, we need to compute the geometric resolution

$$\begin{cases} q_t(Z_\delta, T) &=& 0, \\ Z_{\delta+1} &=& V_{t,\delta+1}(Z_\delta, T), \\ &\vdots& \\ Z_r &=& V_{t,r}(Z_\delta, T) \end{cases}$$

of L_t for the primitive element U_t starting with our data, that is, the parametrization

$$\begin{cases} Q(Z_\delta, T) &= 0, \\ Z_{\delta+1} &= V_{\delta+1}(Z_\delta, T), \\ &\vdots \\ Z_r &= V_r(Z_\delta, T) \end{cases}$$

of $L := L_D$ and the primitive element $U := \lambda_{\delta+1} Z_{\delta+1} + \cdots + \lambda_r Z_r$.

Since in $K_t[Z_\delta, Z_{\delta+1}, \ldots, Z_r]/L_t$ we have $Z_i \equiv V_i(Z_\delta, U) \bmod L_t$, we obtain

$$U_t \equiv U + T_{\delta+1} V_{\delta+1}(Z_\delta, U) + \cdots + T_r V_r(Z_\delta, U)$$
$$\equiv U + T_{\delta+1} V_{\delta+1}(Z_\delta, U_t) + \cdots + T_r V_r(Z_\delta, U_t)$$

modulo $(L_t + \mathbb{I}(T_\delta, T_{\delta+1}, \ldots, T_r)^2)$, whence

$$U \equiv U_t - T_{\delta+1} V_{\delta+1}(Z_\delta, U_t) - \cdots - T_r V_r(Z_\delta, U_t).$$

Therefore replacing U in the parametrization and applying a Taylor expansion we obtain

$$q_t(Z_\delta, T) := Q(Z_\delta, T) - \frac{\partial Q}{\partial T} \sum_{i=1}^{\rho} T_{\delta+i} V_{\delta+i}(Z_\delta, T)$$
$$\equiv Q(Z_\delta, T) - T_{\delta+1} W_{\delta+1}(Z_\delta, T) - \cdots - T_r W_r(Z_\delta, T)$$

and

$$V_{t.\delta+i}(Z_\delta, T) := V_{\delta+i}(Z_\delta, T) - \frac{\partial Q}{\partial T} \sum_{i=1}^{\rho} T_{\delta+i} V_{\delta+i}(Z_\delta, T)$$

modulo $(L_t + \mathbb{I}(T_\delta, T_{\delta+1}, \ldots, T_r)^2)$.

Algorithm 44.6.6. In conclusion we have to compute

- the data q_t and $V_{t.\delta+i}$ with the formulas above;
- the resultant $A_t(Z)$ of the polynomials

$$q_t(\lambda_\delta^{-1}(Z - T), T)$$

and

$$f_{\rho+1}\left(p_1, \ldots, p_{\delta-1}, \lambda_\delta^{-1}(Z - T), \widehat{V}_{t.\delta+1}(Z_\delta, T), \ldots, \widehat{V}_{t.r}(Z_\delta(Z_\delta, T))\right)$$

- and, by expansion, the data $A, D, A_\delta, \ldots, A_r$.

This computation, which mainly amounts to that of $A_t(Z)$, for complexity reasons (see the discussion in Section 44.9) is not performed in $K(Z_\delta)[T]$ but in

$$R[T], \quad R := K[T_\delta, \ldots, T_r][Z_\delta]/\mathbb{I}(Z_\delta - p)^{\mathsf{ds}+1},$$

where $p \in k$ is a "generic" value, $\mathsf{d} := \deg(f)$ and $\mathsf{s} := \deg(Q)$, so that ds counts the roots, with multiplicity, of $L + \mathbb{I}(f)$.

44.7 Kronecker Package: Cleaning Step

So now we have the geometric resolution

$$\begin{cases} q(Z) &= 0, \\ Z_\delta &= v_\delta(Z), \\ &\vdots \\ Z_r &= v_r(Z) \end{cases}$$

of $\mathsf{L}+\mathbb{I}(f)$ and we need to remove from $\mathsf{W} := \mathcal{Z}(\mathsf{L}+\mathbb{I}(f)) =: \{\mathsf{a}_1,\ldots\mathsf{a}_s\}$ the roots such that $g(\mathsf{a}_i) = 0$.

This is easily performed by computing

$$G(Z) := g(p_1,\ldots,p_{\delta-1}, v_\delta(Z),\ldots,v_r(Z)),$$
$$e(Z) := \gcd(q, G),$$
$$q' := q/e,$$
$$w'_i := v_i \bmod q'.$$

We need however to be sure that the lifting point $\mathsf{p} := (p_1,\ldots,p_{\delta-1})$ is not "bad": it is sufficient to be sure that

$$\mathsf{p} \notin \pi(\mathsf{W}'' \cap \mathcal{Z}(g)),$$

where

$$\mathsf{W}' := \{\mathsf{a} \in \mathsf{W} : g(\mathsf{a}) \neq 0\},$$
$$\mathsf{W}'' := \mathcal{ZI}(\mathsf{W}'),$$
$$\pi : \mathsf{K}^r \mapsto \mathsf{K}^{\delta-1} \text{ is the projection } \pi(\alpha_1,\ldots,\alpha_r) = (\alpha_1,\ldots,\alpha_{\delta-1}).$$

Such a point is called a *cleaning point*. Clearly almost all lifting points are cleaning points: the bad points are the projections of the intersection of a variety with dimension $\delta - 1$ and the hypersuface g; such a projection over $\mathsf{K}^{\delta-1}$ has dimension $\delta - 2$.

44.8 Genericity Conditions

With the same notation and assumptions as in Section 44.4 we are now ready to discuss the genericity conditions; we therefore begin with

a Noether position $\{Z_1,\ldots,Z_r\}$,
a lifting point $\mathsf{p} := (p_1,\ldots,p_\delta)$,
a primitive element $U := \lambda_{\delta+1} Z_{\delta+1} + \cdots + \lambda_r Z_r$ of $K[Z_{\delta+1},\ldots,Z_r]/\mathsf{L}_\mathsf{p}$

for V_ρ.

The computation we sketched in Section 44.4 and discussed in the following sections returns a lifting fiber of $\mathsf{V}_{\rho+1}$ for the lifting point $(p_1,\ldots,p_{\delta-1})$ and the primitive element $\lambda_\delta Z_\delta + \lambda_{\delta+1} Z_{\delta+1} + \cdots + \lambda_r Z_r$, provided that the following conditions hold:

- $\{Z_1, \ldots, Z_r\}$ is in Noether position for $\mathsf{V}_{\rho+1}$;
- $\mathsf{p} := (p_1, \ldots, p_{\delta-1})$ is a lifting point for $\mathsf{V}_{\rho+1}$;
- λ_δ is a Liouville point for L_D;
- $\lambda_\delta Z_\delta + \lambda_{\delta+1} Z_{\delta+1} + \cdots + \lambda_r Z_r$ is a primitive element for $\mathsf{V}_{\rho+1}$;
- $\mathsf{p} := (p_1, \ldots, p_{\delta-1})$ is a cleaning point for W''.

We need moreover three further assumptions:

(1) $\{Z_0, \ldots, Z_r\}$ is in Noether position for each homogeneous ideal ${}^h\mathsf{L}_\rho \subset K[Z_0, \ldots, Z_r]$;

(2) p_δ is "lucky" for the truncated computation;

(3) U and λ_δ are "lucky" for the resultant computation of Remark 44.6.5.

The reasons for these assumptions are as follows.

(1) Consider, as an example, the one-dimensional prime ideal $\mathfrak{p}$ generated by $f(Z_1, Z_2) := Z_1^2 - Z_2 \in K[Z_1, Z_2]$. While the variables are in Noether position, any specialization of Z_1 returns a single point, notwithstanding $\deg(f) = 2$.

However, this does not happen for a truly generic frame of coordinates: if we perform such a generic change of coordinates, obtaining

$$f(Y_1, Y_2) := d_{11}^2 Y_1^2 + d_{11} d_{12} Y_1 Y_2 + d_{12}^2 Y_2^2 - d_{21} Y_1 + d_{22} Y_2,$$

then each evaluation of Y_1 returns two points.

More generally, we need to avoid a degeneration in which

$$\deg(f(p_1, \ldots, p_\delta, Z_{\delta+1}, \ldots, Z_r)) < \deg(f);$$

this cannot occur if $\{Z_0, Z_1, \ldots, Z_r\}$ is in Noether position for the homogeneous ideal ${}^h\mathsf{L}_r$.

(2) Computation of the gcd and resultant for polynomials has good complexity if performed over $L[T]$, where L is a field, but this implies the ability to execute zero-testing and inverting, which is beyond the present model.

The computation of, e.g., the resultant

$$A(Z_\delta) := \mathrm{Res}(Q, f(Z_\delta, V_{\delta+1}, \ldots, V_r)),$$

which satisfies $\deg(A) \le \eta := \deg(\mathsf{L}^e)) \deg(f)$, costs

$$\mathsf{M}(\eta) := \mathcal{O}(\eta \log^2(\eta) \log \log(\eta))$$

arithmetical operations in $K(Z_\delta)$ if it is performed in this *field* where we do not have a good complexity model for zero-testing and inverting; the result of this computation gives a polynomial $A(Z_\delta) \in K[Z_\delta]$ of degree η.

Let us now fix a value $p \in K$ and, remarking that $K(Z_\delta) \subset K[[Z_\delta]]$, consider the *ring*

$$R := K[[Z_\delta]]/(Z_\delta - p)^{\eta+1} \cong \operatorname{Span}_K \{1, \ldots, Z_\delta^\eta\}$$

where we can perform both

(0) zero-testing: an element $g = \sum_{i=0}^\eta c_i Z_\delta^i \in R$ is zero iff $c_i = 0$ for each i iff $g(p) = 0$,

(1) inverting: the inverse in R of the invertible polynomial g, $g(p) \neq 0$, is the polynomial $s(Z_\delta)$, $\deg(s) \le \eta$, which satisfies

$$s(Z_\delta)g(Z_\delta) + t(Z_\delta)(Z_\delta - p)^{\eta+1} = \gcd(g(Z_\delta), (Z_\delta - p)^{\eta+1} = 1$$

for a suitable $t(Z_\delta)$, $\deg(t) < \deg(g)$.

Thus any element

$$g(Z_\delta) := d(Z_\delta)/r(Z_\delta) \in K(Z_\delta), d(Z_\delta), r(Z_\delta) \in K[Z_\delta]$$

can be canonically represented by an element $\dot{g} \in \operatorname{Span}_K \{1, \ldots, Z_\delta^\eta\}$ such that

$$\dot{g}(Z_\delta)r(Z_\delta) \equiv d(Z_\delta) \bmod (Z_\delta - p)^{\eta+1}.$$

Thus the same algorithm which, if performed on $K(Z_\delta)$, returns $A(Z_\delta)$ with complexity $\mathsf{M}(\eta)$ can be performed also on R with the same complexity $\mathsf{M}(\eta)$, returning some polynomial $B(Z_\delta) := \sum_{i=0}^\eta c_i Z_\delta^i \in K[Z_\delta]$ of degree bounded by η.

Can we assume that such a polynomial $B(Z_\delta)$ is the *true* resultant $A(Z_\delta)$, that is, that $B(Z_\delta) = \dot{A}(Z_\delta) = A(Z_\delta)$? The answer is obvious: the solution is correct iff in each step of the computation the algorithm in R gives the same answer as the algorithm in $K(Z_\delta)$, if

(0) when zero-testing $g(Z_\delta) \in K(Z_\delta)$, $g = 0 \iff \dot{g}(p) = 0$,

(1) when computing the inverse $h(Z_\delta) := g^{-1}(Z_\delta)$ of $g(Z_\delta) \in K(Z_\delta)$, the polynomial $s(Z_\delta)$ such that $s(Z_\delta)\dot{g}(Z_\delta) \equiv 1 \bmod (Z_\delta - p)^{\eta+1}$ satisfies $s = \dot{h}$.

This behavior depends on the choice of $p \in K$. There are some values p for which the wrong answer is returned, and this approach fails; however, almost all choices are "lucky", thus allowing one to produce the correct answer $B(Z_\delta) = \dot{A}(Z_\delta) = A(Z_\delta)$ with good complexity $\mathsf{M}(\eta)$ while computing in the ring R rather than in the field $K(Z_\delta)$.

(3) In a similar way a better complexity is obtained if the computation (see Remark 44.6.5) of the resultant $A(Z) \in K[Z]$ of the polynomials $\widehat{q}_i(Z, T)$ and $f(\lambda_\delta^{-1}(Z - T), \widehat{V}_{\delta+1}(Z, T), \ldots, \widehat{V}_r(Z, T))$ in $K[Z][T]$ can be performed with the ring arithmetic of $K[Z]$ rather than the field arithmetic of $K(Z)$. Such a possibility depends on a "lucky" choice of the Liouville point λ_δ and of the primitive U.

44.9 Complexity Considerations

We record here the following.[5]

Fact 44.9.1. *Let R be an integral domain and K be a field. Writing*

$$\mathsf{M}(n) := \mathcal{O}(n \log^2(n) \log \log(n)),$$

the following hold.

(1) *The bit-complexity of the arithmetic operations (addition, multiplication, finding the quotient, remainder and gcd) of integers of bit-size[6] n costs $\mathsf{M}(n)$.*

(2) *The multiplication and division of polynomials in $R[T]$ whose degree is bounded by n costs $\mathcal{O}(n \log(n) \log \log(n))$ arithmetical operations in R.*

(3) *The computation of the gcd and resultant of polynomials in $K[T]$ whose degree is bounded by n costs $\mathsf{M}(n)$ arithmetical operations in K.*

(4) *The multiplication of two n-square matrices in R costs $\mathcal{O}(n^{\omega})$ arithmetical operations in R, with $\omega < 2.39$.*

(5) *The inversion of an n-square matrix in K costs $\mathcal{O}(n^{\omega})$ arithmetical operations in K.*

(6) *If $D = k[T]/q(T)$, where k is a field and q is a squarefree monic polynomial, the inversion of an n-square matrix in D costs $\mathcal{O}(n^{\Omega})$ arithmetical operations (addition, multiplication, finding the determinant and the adjoint matrix) in D, where $\Omega < 4$.*

(7) *Performing a linear substitution into a polynomial $p \in K[X]$ of degree d costs $\mathsf{M}(d)$ arithmetical operations in K.* ◉

Definition 44.9.2. *A polynomial $f \in k[X_1, \ldots, X_n]$ is said to be* given by a *straight-line program of size L if there is a sequence $\{Q_1, \ldots, Q_L\} \subset k[X_1, \ldots, X_n]$ where $f \in \{Q_1, \ldots, Q_L\}$ and, for each i, $1 \le i \le L$, either*

$$Q_i \in \{X_1, \ldots, X_n\}, \quad Q_i \in k,$$

or there are $j_1, j_2 < i$ such that

$$Q_i = Q_{j_1} + Q_{j_2} \ or$$
$$Q_i = Q_{j_1} - Q_{j_2} \ or$$
$$Q_i = Q_{j_1} Q_{j_2}.$$

◉

---

[5] Compare

- Aho A.V., Hopcroft J.E., Ullman J.D., *The Design and Analysis of Computer Algorithms*, Addison–Wesley (1974).
- Bini D., Pan V., *Polynomial and Matrix Computations*, Birkhäuser (1994).
- Bürgisser P., Clausen M., Shorolahi M.A., *Algebraic Complexity Theory*, Springer (1997).

[6] That is, integers m such that $\log(m) \le n$.

Theorem 44.9.3. *Let*

$$f_1, \ldots, f_r, g \in K[Z_1, \ldots, Z_r]$$

be polynomials of degree bounded by D and given by a straight-line program of size at most L; with the same notation and assumptions as in Problem 44.1.5, a geometric resolution of Z_r can be computed with

$$\mathcal{O}(r(rL + r^\Omega)\mathsf{M}^2(DS))$$

arithmetic operations in K, where

$$S := \max(\deg(Z_\rho), 1 \le \rho < r) \le D^{r-1},$$

by means of a probabilistic algorithm. Its probability of returning correct results relies on the choices of elements of K; choices which give an incorrect result are contained in a closed Zariski set.

Proof (sketch):   Let us remark that

- The computation of the objects in Proposition 44.3.1 costs $\mathcal{O}\left((rL + r^\Omega)\mathsf{M}(S)\mathsf{a}(2)\right)$, where $\mathsf{a}(j)$ is the cost of the arithmetical operations in R/I^j :

  the evaluation of $\mathbf{f}$ and J has complexity $\mathcal{O}(rL)$,

  the inversion of J costs $\mathcal{O}(r^\Omega)$,

  the updating of Q and $\mathbf{V}$ costs $\mathcal{O}(r^2)$;

  all these costs are evaluated in terms of arithmetical operations in $R/I^2[T]$, each such operation costing $\mathsf{M}(S)\mathsf{a}(2)$.
- The computation of the objects in Corollary 44.3.2 costs $\mathcal{O}\left((rL + r^\Omega)\mathsf{M}(S) \sum_{j=0}^{\log_2(\epsilon)} \mathsf{a}(2^j)\right)$ operations.
- The lifting step costs $\mathcal{O}\left((rL + r^\Omega)\mathsf{M}^2(S)\right)$ operations since $\mathsf{a}(j) = \mathsf{M}(j)$, so that

$$\sum_{j=0}^{\log_2(S+1)} \mathsf{M}(2^j) \le \mathsf{M}(S) \sum_{j=0}^{\log_2(S+1)} \frac{1}{2^j} \in \mathcal{O}(\mathsf{M}(S)).$$

- The intersection step costs $\mathcal{O}\left(r(L + r^2)\mathsf{M}(S)\mathsf{M}(dS)\right)$ operations as follows:
  - If $p \in K[Z_\delta, U]$ is stored in a two-dimensional array of size $\mathcal{O}(S^2)$, $S := \deg(p)$, the computation of $\widehat{p}$ costs $\mathcal{O}(S\mathsf{M}(S))$ arithmetical operations in K. It is sufficient to compute $\widehat{p_i}$ for each homogeneous component p_i of p of degree i; in this setting, since also $\widehat{p_i}$ is homogeneous of degree i, it is sufficient to compute $\widehat{p_i}(Z_\delta, 1) = p_i(\lambda_\delta^{-1}(Z_\delta - 1), 1)$ that is, to perform a linear tranformation over each univariate polynomial $p_i(Z_\delta, 1) \in K[Z_\delta]$. Thus the cost is $\mathcal{O}\left(\sum_{i=0}^{S} \mathsf{M}(i)\right) \subset \mathcal{O}(S\mathsf{M}(S))$.
  - The computation of $\mathrm{Res}(Q, f(Z_\delta, V_{\delta+1}, \ldots, V_r))$ costs

$$\mathcal{O}\left((L + r^2)\mathsf{M}(\mathsf{s})\mathsf{M}(ds)\right) \le \mathcal{O}\left((L + r^2)\mathsf{M}(S)\mathsf{M}(DS)\right)$$

arithmetical operations in K, where L is the size of f, $\mathsf{d} := \deg(f) \leq D$ and $\mathsf{s} := \deg(Q) \leq S$. The computation is in fact performed in $k[[Z_\delta]]/((Z_\delta - p)^{\mathsf{ds}+1})$.

○ The computation of $A_t(Z)$ costs $\mathcal{O}\left((r - D)(L + r^2)\mathsf{M}(S)\mathsf{M}(DS)\right)$ arithmetical operations in K: apply the result above to

$$K[T_\delta, T_{\delta+1}, \ldots, T_r]/\mathbb{I}(T_\delta, T_{\delta+1}, \ldots, T_r)^2.$$

○ The computation of the geometric resolution of L_t costs $\mathcal{O}\left(r^2\mathsf{M}(S)\mathsf{M}(DS)\right)$ arithmetical operations in K:

the arithmetics in $L := K[Z_\delta]/(Z_\delta - p)^{\mathsf{ds}+1}$ costs

$$\mathcal{O}(\mathsf{M}(\mathsf{ds})) \leq \mathcal{O}(\mathsf{M}(DS));$$

the computation of the polynomials $W_{\delta+i}$ requires $\mathcal{O}(r\mathsf{M}(S))$ arithmetical operations in L;

the computation of the polynomials $V_{t,d+i}$ requires $\mathcal{O}(r)$ arithmetical operations in

$$K[T_\delta, T_{\delta+1}, \ldots, T_r]/\mathbb{I}(q_t) + \mathbb{I}(T_\delta, T_{\delta+1}, \ldots, T_n)^2,$$

which means $\mathcal{O}(r^2\mathsf{M}(S))$ arithmetical operations in L and $\mathcal{O}\left(r^2\mathsf{M}(S)\mathsf{M}(DS)\right)$ in K;

○ the removal of multiplicity costs $\mathcal{O}\left(r\mathsf{M}(\mathsf{s})\right)$, $\mathsf{s} := \deg(A_t) \leq S$.

• The cleaning step costs $\mathcal{O}\left((L + r^2)\mathsf{M}(S)\right)$ operations.

$\boxed{\odot}$

Remark 44.9.4.   Therefore "generic" choices give the complete answer.

Moreover, the result can be checked by evaluating the input polynomial: if they satisfy the required equations, at most the algorithm failed to recover *all* the roots.

In the special case in which $g = 1$, that is, the case in which we want to compute the roots $\mathcal{Z}(f_1, \ldots, f_r)$, since Bezout's theorem informs us that the number of solutions is $\prod_{i=1}^r \deg(f_i)$ it is therefore possible to check whether the algorithm recovered all solutions.

$\boxed{\odot}$

It is worthwhile to compare the complexity of this algorithm with a Gröbner-basis approach; let us therefore assume that each polynomial is given by a dense representation:

Corollary 44.9.5.   *Let*

$$f_1, \ldots, f_r, \in K[Z_1, \ldots, Z_r]$$

be polynomials of degree bounded by $D \geq n$ and let $g := 1$; with the same notation and assumptions as in Problem 44.1.5, a geometric resolution of Z_r can be computed with $\mathcal{O}(D^{3(r+\mathcal{O}(1))})$ arithmetic operations in K by means of the probabilistic algorithm of Theorem 44.9.3.

$\boxed{\odot}$

Proof.   By Bezout's theorem $DS \leq D^r$, so that $\mathrm{M}(DS))$ is in $D^{r+\mathcal{O}(1)}$ and $L \leq r\binom{D+r}{r}$, which is in $D^{r+\mathcal{O}(1)}$ too. $\boxdot$

Remark 44.9.6.   We recall that the degree bound of a zero-dimensional Gröbner basis is in the present notation $\mathcal{O}(D^r)$ (compare Sections 38.3 and 38.4), so that the dense representation of each polynomial costs $\gamma := \mathcal{O}(D^{r^2})$ and a rough evaluation of the cost of a Gröbner-basis computation returns $\mathcal{O}(D^{4r^2})$ (see the final comments of Chapter 22). $\boxdot$

45

Duval II

This concluding chapter of Part six on algebraic solving, must be devoted to an application to multivariate systems of the Kronecker Philosophy expounded in the first volume: "solving" does not mean *producing programs which compute the roots of polynomial equation systems*; it means *producing programs which compute* **with** *their roots*.

In the multivariate case, given a finitely generated (zero-dimensional) ideal $J \subset K[Z_1, \ldots, Z_r]$, such a philosophy will be put into effect if we can consider each of its root $(a_1, \ldots, a_r) \in \mathcal{Z}(J)$ as *given*, by means of a proper suitable representation of the ideal itself.

The considerations performed for the univariate case in Part one (see Chapters 8 and 11 and in particular Sections 8.2, 8.3 and 11.4) and Gröbner's reinterpretation of both Kronecker's theory and the Primitive Element Theorem (Section 8.4) in terms of *Allgemeine Basissätze* (Section 34.2), which explicitly link root representations with lex Gröbner bases and triangular sets, give the *leitmotif* of this chapter (see also Section 34.5): I will consider the roots $(a_1, \ldots, a_r) \in \mathcal{Z}(J)$ *given*

- if the ideal $J \subset K[Z_1, \ldots, Z_r]$ is represented by means of the Gröbner-related techniques discussed in the second volume, so that

$$K[Z_1, \ldots, Z_r]/J \rightarrow \bigoplus_{(a_1,\ldots,a_r)\in\mathcal{Z}(J)} K[a_1, \ldots, a_r],$$

- and if procedures on $K[Z_1, \ldots, Z_r]/J$ are given which allow one to perform (via Duval splitting) the four operations and zero-testing on each $K[a_1, \ldots, a_r]$.

This approach will be illustrated by the following instances:

(1) the Kronecker–Duval model, where all that is needed is to quote the approach endorsed by the project **PoSSo**;[1]
(2) the representation of $\sqrt{J}$ by means of an *allgemeine* basis (Definition 34.2.2, Equation (42.1));

---

[1] **Polynomial System Solving. ESPRIT-BRA 6846.**

(3) the representation of J by means of Kronecker's parametrization (Section 41.9, Chapter 44) and Rational Univariate Representation (Proposition 42.9.3, Definition 42.9.16);

(4) the representation of J by means of a Gröbner representation (Definition 29.3.3);

(5) the representation of J by means of the linear representation w.r.t. the term ordering $\prec$ (Definition 29.3.3).

45.1 Kronecker–Duval Model

In relation to the Kronecker–Duval Model, I suggest the rereading of Chapters 5, 6, 12, Section 34.5 and Chapter 42 in the light of the following quotation:[2]

The standard method for computing with algebraic numbers consists in working in a *tower* of fields, each field being defined by the minimal polynomial of an algebraic number, defined over the preceding field. The computation in such a field needs addition, multiplication and extended gcd of univariate polynomials, as well as Euclidean division, and, recursively, similar operations in the smaller fields.

   [...]

[Duval's] dynamic evaluation may be viewed as a lazy factorization: in representation of algebraic numbers, the reducibility of a polynomial may only pose a problem when testing equalities or when inverting elements.

   When such an operation is needed, a gcd computation (already needed for inverting) allows to detect if there is a reducibility problem and, in this case, to get a partial factorization of the reducible polynomial.

   It follows that, for dynamic evaluation, an algebraic number is represented as a root of a (possibly reducible) square free polynomial. When a factorization occurs from a non trivial gcd, the computation splits in two cases, depending on which factor has the algebraic number as a root. Sometimes, one of the cases is irrelevant, but frequently both cases are of interest, and one needs to carry on two independent computations. Thus the domain of computation is a reduced artinian ring which is implemented as a family of towers. The evaluation is dynamic in the sense that the towers change during the computation.

   [...]

Towers are equivalent with special algebraic systems, the *triangular systems* where each polynomial introduces a new variable. More generally, any finite set of algebraic numbers may be represented as a solution of a zero dimensional algebraic system.

PoSSo Report **R2**

45.2 Allgemeine Representation

Let us consider

- K, an infinite, perfect field, where, if $p := \mathrm{char}(K) \neq 0$, it is possible to extract pth roots,

- $\overline{K}$, its algebraic closure,

- $\mathcal{Q} = K[Z_1, \ldots, Z_r]$,

---

[2] Whose author is Daniel Lazard.

- $\mathcal{W} := \{Z_1^{a_1} \cdots Z_r^{a_r} : (a_1, \ldots, a_r) \in \mathbb{N}^r\}$,
- $\mathsf{J} \subset \mathcal{Q}$, a zero-dimensional ideal,
- $\mathsf{J} = \bigcap_{i=1}^{r} \mathsf{q}_i$, its irredundant primary representation in $\mathcal{Q}$,
- for each i, $1 \le i \le \mathsf{r}$,
  - $\mathsf{m}_i = \sqrt{\mathsf{q}_i}$, the associated maximal prime,
  - $K_i := \mathcal{Q}/\mathsf{m}_i$, $K \subset K_i \subset \mathsf{K}$,
  - $\mathcal{Q}_i := K_i[Z_1, \ldots, Z_r]$,
  - the irredundant primary representations $\mathsf{q}_i = \bigcap_{j=1}^{r_i} \mathsf{q}_{ij}$ and $\mathsf{m}_i = \bigcap_{j=1}^{r_i} \mathsf{m}_{ij}$ in $\mathcal{Q}_i$,
  - the roots $\mathsf{b}_{ij} := (b_1^{(ij)}, \ldots, b_r^{(ij)}) \in K_i^r \subset \mathsf{K}^r$, $1 \le j \le r_i$,
  - $d_{ij} := \operatorname{mult}(\mathsf{b}_{ij}, \mathsf{J}) = \deg(\mathsf{q}_{ij})$ for each j, $1 \le j \le r_i$,

  which satisfy

  (1) $\mathsf{m}_{ij} = \mathbb{I}(Z_1 - b_1^{(ij)}, \ldots, Z_r - b_r^{(ij)})$,

  (2) the b_{ij}, $1 \le j \le r_i$, are K-conjugate for each i,

  (3) up to a renumeration, $\sqrt{\mathsf{q}_{ij}} = \mathsf{m}_{ij}$,

  (4) $\mathsf{m}_i = \mathsf{m}_{ij} \cap \mathcal{Q}$;

  (5) $\mathsf{q}_i = \mathsf{q}_{ij} \cap \mathcal{Q}$;

  (6) for each j, l, $1 \le j, l \le r_i$, $d_{ij} = d_{il} =: d_i$,

  (7) $r_i = \deg(\mathsf{m}_i) = [K_i : K]$,

  (8) $\deg(\mathsf{q}_i) = d_i r_i$,

  (9) $\mathsf{J} = \bigcap_{i=1}^{r} \bigcap_{j=1}^{r_i} \mathsf{q}_{ij}$, $\sqrt{\mathsf{J}} = \bigcap_{i=1}^{r} \bigcap_{j=1}^{r_i} \mathsf{m}_{ij}$ are the irredundant primary representations in $\mathsf{K}[Z_1, \ldots, Z_r]$,

  (10) $\mathcal{Z}(\mathsf{J}) = \{\mathsf{b}_{ij} : 1 \le i \le \mathsf{r}, 1 \le j \le r_j\}$.
- $Y := Z_1 + \sum_{i=2}^{r} c_i Z_i$, an *allgemeine* coordinate (Definition 34.4.7) for J,
- $\mathsf{J}^+ = \mathsf{J} + \mathbb{I}\left(Y - Z_1 - \sum_{l=2}^{r} c_l Z_l\right) \subset K[Y, Z_1, \ldots, Z_r]$,
- g_0, the monic primitive generator of $\mathsf{J}^+ \cap K[Y]$,
- for each i,
  - $\mathsf{m}_i^+ = \mathsf{m}_i + \mathbb{I}\left(Y - Z_1 - \sum_{l=2}^{r} c_l Z_l\right)$,
  - $\mathsf{q}_i^+ = \mathsf{q}_i + \mathbb{I}\left(Y - Z_1 - \sum_{l=2}^{r} c_l Z_l\right)$,
  - $h_i \in K[Z_1]$, the monic polynomial such that $\mathbb{I}(h_i) = \mathsf{m}_i^+ \cap K[Z_1]$,
  - for each j, $1 \le j \le r_i$,

    $\mathsf{m}_{ij}^+ = \mathsf{m}_{ij} + \mathbb{I}\left(Y - Z_1 - \sum_{l=2}^{r} c_l Z_l\right)$,

    $\mathsf{q}_{ij}^+ = \mathsf{q}_{ij} + \mathbb{I}\left(Y - Z_1 - \sum_{li=2}^{r} c_l Z_l\right)$,

    $\beta_{ij} = b_1^{(ij)} + \sum_{l=2}^{r} c_l b_l^{(ij)}$,

  which satisfy

  (11) $h_i = \prod_{j=1}^{r_j} \left(Y - \beta_{ij}\right)$ for each i,

  (12) $g_0(Y) = \prod_{i=1}^{r} h_i^{d_i} = \prod_{i=1}^{r} \prod_{j=1}^{r_i} \left(Y - \beta_{ij}\right)^{d_i}$,

  (13) $R := \deg(g_0) = \sum_{i=1}^{r} r_i d_i = \deg(\mathsf{J})$,

  (14) $f_0 := \operatorname{SQFR}(g_0) = \prod_{i=1}^{r} h_i = \prod_{i=1}^{r} \prod_{j=1}^{r_i} \left(Y - \beta_{ij}\right)$.

As a consequence we have the isomorphisms (defined by canonical projection and Chinese Remaindering):

$$
\begin{array}{ccccc}
\mathcal{Q}/\mathsf{J} & \cong & \mathcal{Q}[Y]/\mathsf{J}^+ & \cong & K[Y]/(g_0) \\
\updownarrow & & \updownarrow & & \updownarrow \\
\bigoplus_{i=1}^{r} \mathcal{Q}/\mathsf{q}_i & \cong & \bigoplus_{i=1}^{r} \mathcal{Q}[Y]/\mathsf{q}_i^+ & \cong & \bigoplus_{i=1}^{r} K[Y]/(h_i^{d_i}) \\
\updownarrow & & \updownarrow & & \updownarrow \\
\bigoplus_{i=1}^{r}\bigoplus_{j=1}^{r_j} \mathcal{Q}_i/\mathsf{q}_{ij} & \cong & \bigoplus_{i=1}^{r}\bigoplus_{j=1}^{r_j} \mathcal{Q}_i[Y]/\mathsf{q}_{ij}^+ & \cong & \bigoplus_{i=1}^{r} K_i[Y]/\left(Y-\beta_{ij}\right)^{d_i}
\end{array}
$$

and

$$
\begin{array}{ccccc}
\mathcal{Q}/\sqrt{\mathsf{J}} & \cong & \mathcal{Q}[Y]/\sqrt{\mathsf{J}^+} & \cong & K[Y]/(f_0) \\
\updownarrow & & \updownarrow & & \updownarrow \\
\bigoplus_{i=1}^{r} \mathcal{Q}/\mathsf{m}_i & \cong & \bigoplus_{i=1}^{r} \mathcal{Q}[Y]/\mathsf{m}_i^+ & \cong & \bigoplus_{i=1}^{r} K[Y]/(h_i) \\
\updownarrow & & \updownarrow & & \updownarrow \\
\bigoplus_{i=1}^{r}\bigoplus_{j=1}^{r_j} \mathcal{Q}_i/\mathsf{m}_{ij} & \cong & \bigoplus_{i=1}^{r}\bigoplus_{j=1}^{r_j} \mathcal{Q}_i[Y]/\mathsf{m}_{ij}^+ & \cong & \bigoplus_{i=1}^{r} K_i[Y]/\left(Y-\beta_{ij}\right)
\end{array}
$$

Let us now assume that J is radical, so that $g_0 = f_0$ and the reduced Gröbner basis w.r.t. the lex ordering induced by $Y < Z_1 < \cdots < Z_r$ is the *allgemeine* basis (see Theorem 34.2.1)

$$(g_0(Y), Z_1 - g_1(Y), \ldots, Z_r - g_r(Y))$$

of J^+. Let us show how to apply it in order to perform arithmetical manipulation over each root b_{ij}.

canonical representation: All arithmetical expressions

$$p(\mathsf{b}_{ij}) = p(b_1^{(ij)}, \ldots, b_r^{(ij)}), \quad p \in \mathcal{Q},$$

of each root $\mathsf{b}_{ij} \in \mathcal{Z}(\mathsf{J})$ have a canonical representation

$$p(\mathsf{b}_{ij}) = \widehat{p}(\beta_{ij}),$$

where $\widehat{p}(Y) := \mathbf{Rem}\left(p(g_1(Y), \ldots, g_r(Y)), g_0(Y)\right) \in K[Y]$.

Remark 45.2.1.   For each $q(Y) \in K[Y]$, $\deg(q) < \deg(g_0)$, the polynomial $p = \check{q} \in \mathcal{Q}$ defined by

$$\check{q}(Z_1, \ldots, Z_r) := q\left(Z_1 + \sum_{i=1}^{r} c_i Z_i\right)$$

satisfies the relations $\widehat{\check{q}} = \widehat{p} = q$ and $\check{q}(\mathsf{b}_{ij}) = q(\beta_{ij})$.   ◉

vector space arithmetics: Given two arithmetical expressions $p_1, p_2 \in \mathcal{Q}$ of the roots $\mathsf{b}_{ij} \in \mathcal{Z}(\mathsf{J})$ and values $c_1, c_2 \in K$, the arithmetical expression $p(\mathsf{b}_{ij}) := c_1 p_1(\mathsf{b}_{ij}) + c_2 p_2(\mathsf{b}_{ij})$ has the canonical representation $p(\mathsf{b}_{ij}) = \widehat{p}(\beta_{ij})$, where

$$\begin{aligned}
\widehat{p}(Y) &:= c_1 \widehat{p_1}(Y) + c_2 \widehat{p_2}(Y) \\
&= c_1 \, \mathbf{Rem}\,(p_1(g_1(Y), \ldots, g_r(Y)), g_0(Y)) \\
&\quad + c_2 \, \mathbf{Rem}\,(p_2(g_1(Y), \ldots, g_r(Y)), g_0(Y)) \,.
\end{aligned}$$

multiplication: With the same notation, the arithmetical expression

$$p(\mathsf{b}_{ij}) := p_1(\mathsf{b}_{ij}) p_2(\mathsf{b}_{ij})$$

has the canonical representation $p(\mathsf{b}_{ij}) = \widehat{p}(\beta_{ij})$, where

$$\widehat{p}(Y) := \mathbf{Rem}\,((\widehat{p_1}(Y)\widehat{p_2}(Y)), g_0(Y)) \,.$$

zero-testing: Given an arithmetical expression $p \in \mathcal{Q}$ we have

$$p(\mathsf{b}_{ij}) = 0 \iff \widehat{p}(Y) := \mathbf{Rem}\,(p(g_1(Y), \ldots, g_r(Y)), g_0(Y)) = 0.$$

inverse and division: Given an arithmetical expression $p \in \mathcal{Q}$, our aim is to produce

- a factorization $g_0 = g_0^{(0)} g_0^{(1)}$,
- a polynomial $q(Y) \in K[Y], \deg(q) < \deg(g_0^{(1)})$,

such that:

- if $g_0^{(0)} = g_0, 1 = g_0^{(1)}$ then, for each $\mathsf{b}_{ij} \in \mathcal{Z}(\mathsf{J})$, $p(\mathsf{b}_{ij}) = \widehat{p}(\beta_{ij}) = 0$;
- if $g_0^{(1)} = g_0, 1 = g_0^{(0)}$, then, for each $\mathsf{b}_{ij} \in \mathcal{Z}(\mathsf{J})$,

$$p(\mathsf{b}_{ij}) = \widehat{p}(\beta_{ij}) \neq 0 \qquad \text{and} \qquad p^{-1}(\mathsf{b}_{ij}) = \check{q}(\mathsf{b}_{ij}) = q(\beta_{ij});$$

- otherwise, for each $\mathsf{b}_{ij} \in \mathcal{Z}(\mathsf{J})$, we have
  - $p(\mathsf{b}_{ij}) = \widehat{p}(\beta_{ij}) = 0 \iff g_0^{(0)}(\beta_{ij}) = 0$,
  - $p(\mathsf{b}_{ij}) = \widehat{p}(\beta_{ij}) \neq 0 \iff g_0^{(1)}(\beta_{ij}) = 0$, in which case

$$p^{-1}(\mathsf{b}_{ij}) = \check{q}(\mathsf{b}_{ij}) = q(\beta_{ij}).$$

Thus, writing

$$\begin{aligned}
I_0 &:= \{i : 1 \leq i \leq \mathsf{r}, h_i \mid g_0^{(0)}\} = \{i : h_i \mid \widehat{p}\} \subset \{1, \ldots, \mathsf{r}\}, \\
I_1 &:= \{i : 1 \leq i \leq \mathsf{r}, i \notin I_0\} = \{i : h_i \mid g_0^{(1)}\} = \{i : h_i \nmid \widehat{p}\} \subset \{1, \ldots, \mathsf{r}\}, \\
\mathsf{J}_\iota &:= \cap_{i \in I_\iota} \mathsf{q}_i, \iota \in \{0, 1\}, \\
\mathsf{Z}_\iota &:= \{\mathsf{b}_{ij} : i \in I_\iota\}, \iota \in \{0, 1\}, \\
g_j^{(\iota)} &:= \mathbf{Rem}(g_j, g_0^{(\iota)}) \in K[Y], \iota \in \{0, 1\}, 1 \leq j \leq r,
\end{aligned}$$

one has that:

(a) $g_0^{(\iota)} = \prod_{i \in I_\iota} h_i = \prod_{i \in I_\iota} \prod_{j=1}^{r_i} (Y - \beta_{ij}), \iota \in \{0, 1\}$;

(b) for $\iota \in \{0, 1\}$, $\left(g_0^{(\iota)}(Y), Z_1 - g_1^{(\iota)}(Y), \ldots, Z_r - g_r^{(\iota)}(Y) \right)$ is the *allgemeine* basis of J_ι^+, that is, its reduced Gröbner basis w.r.t. the lex ordering induced by $Y < Z_1 < \cdots < Z_r$;

(c) $\mathbf{Z}_\iota = \mathcal{Z}(\mathsf{J}_\iota), \iota \in \{0, 1\}$;

(d) $\mathbf{Z}_0 = \{\mathbf{b}_{ij} \in \mathcal{Z}(\mathsf{I}) : p(\mathbf{b}_{ij}) = 0\}$;

(e) $\mathbf{Z}_1 = \{\mathbf{b}_{ij} \in \mathcal{Z}(\mathsf{J}) : p(\mathbf{b}_{ij}) \neq 0\}$;

(f) $\mathsf{J} = \mathsf{J}_0 \cap \mathsf{J}_1$;

(g) the isomorphisms

$$
\begin{array}{ccccc}
\mathcal{Q}/\mathsf{J} & \cong & \mathcal{Q}[Y]/\mathsf{J}^+ & \cong & K[Y]/(g_0) \\
\big\updownarrow & & \big\updownarrow & & \big\updownarrow
\end{array}
$$

$$
\mathcal{Q}/\mathsf{J}_0 \oplus \mathcal{Q}/\mathsf{J}_1 \cong \mathcal{Q}[Y]/\mathsf{J}_0^+ \oplus \mathcal{Q}[Y]/\mathsf{J}_1^+ \cong K[Y]/(g_0^{(0)}) \oplus K[Y]/(g_0^{(1)})
$$

hold;

(h) $g_0^{(\iota)}$ is squarefree, $\iota \in \{0, 1\}$;

(i) $\mathsf{J}_\iota = \sqrt{\mathsf{J}_\iota}, \iota \in \{0, 1\}$.

In order to produce both the required

- factorization $g_0 := g_0^{(0)} g_0^{(1)}$ and
- polynomial $q(Y) \in K[Y], \deg(q) < \deg(g_0^{(1)})$,

having the properties listed above, we simply apply Lazard's Theorem 11.3.2 and compute:

$$
g_0^{(0)} := \gcd(g_0, \widehat{p}) \in K[Y];
$$
$$
s, t \in K[Y] \text{ such that } s\widehat{p} + t g_0 = g_0^{(0)};
$$
$$
g_0^{(1)} := g_0/g_0^{(0)};
$$
$$
u, v \in K[Y] \text{ such that } u g_0^{(0)} + v g_0^{(1)} = 1;
$$
$$
q := \mathbf{Rem}(su, g_0^{(1)}).
$$

In fact this computation is simply a reformulation of Lazard's Theorem 11.3.2: writing $p_1(Y) := \widehat{p}(Y)/g_0^{(0)}(Y)$ we have:

- if $g_0^{(0)}(\beta_{ij}) = 0$ then $\widehat{p}(\beta_{ij}) = g_0^{(0)}(\beta_{ij}) p_1(\beta_{ij}) = 0$;
- if $g_0^{(1)}(\beta_{ij}) = 0$ then

$$
\begin{aligned}
q(\beta_{ij})\widehat{p}(\beta_{ij}) &= s(\beta_{ij})u(\beta_{ij})\widehat{p}(\beta_{ij}) \\
&= u(\beta_{ij})s(\beta_{ij})\widehat{p}(\beta_{ij}) + u(\beta_{ij})t(\beta_{ij})g_0(\beta_{ij}) \\
&= u(\beta_{ij})g_0^{(0)}(\beta_{ij}) \\
&= u(\beta_{ij})g_0^{(0)}(\beta_{ij}) + v(\beta_{ij})g_0^{(1)}(\beta_{ij}) \\
&= 1.
\end{aligned}
$$

We remark that $g_0^{(0)} = 1$ implies that

$$g_0^{(1)} = g_0, \, v = 0, \, u = 1 \text{ and } \mathbf{Rem}(su, g_0^{(1)}) = \mathbf{Rem}(s, g_0^{(1)}).$$

45.3 Kronecker Parametrization and Rational Univariate Representation

While using the same notation as in the previous section, let us now assume that we have a rational univariate representation

$$(\chi(Y), \gamma_0(Y), \gamma_1(Y), \ldots, \gamma_r(Y))$$

of J such that

$$\mathcal{Z}(\mathsf{J}) = \{\mathsf{b}_{ij} : 1 \le i \le \mathsf{r}, 1 \le j \le r_j\} = \left\{\left(\frac{\gamma_1(\alpha)}{\gamma_0(\alpha)}, \ldots, \frac{\gamma_r(\alpha)}{\gamma_0(\alpha)}\right) : \alpha \in \mathsf{K}, \chi(\alpha) = 0\right\}.$$

For each i, j, we denote by α_{ij} the root of $\chi(T)$ for which

$$\mathsf{b}_{ij} = \left(\frac{\gamma_1(\alpha_{ij})}{\gamma_0(\alpha_{ij})}, \ldots, \frac{\gamma_r(\alpha_{ij})}{\gamma_0(\alpha_{ij})}\right)$$

and $\psi_i := \prod_j (T - \alpha_{ij})$ for each i, observing that we have

$$\chi(T) = \prod_{ij}(T - \alpha_{ij})^{d_i} = \prod_i \psi_i^{d_i}.$$

We also denote by $\gamma_{-1}(Y) \in K[Y]$ the unique polynomial which satisfies

$$\gamma_0(Y)\gamma_{-1}(Y) \equiv 1 \bmod \chi, \quad \deg(\gamma_{-1}) < \deg(\chi).$$

If J is radical then χ is squarefree, $\gamma_0 = \chi'$, $d_i = 1$ and the representation is a Kronecker parametrization.

With the present notation, if the given RUR is associated to the *allgemeine* coordinate $Z_1 + \sum_{i=2}^r c_i Z_i$ then we have $\chi = g_0$, $\alpha_{ij} = \beta_{ij}$ for each i, j and $\psi_i = h_i$ for each i.

Let us now show how to adapt the considerations of the previous section to this RUR setting:

canonical representation: All arithmetical expressions

$$p(\mathsf{b}_{ij}) = p(b_1^{(ij)}, \ldots, b_r^{(ij)}), \quad p \in \mathcal{Q},$$

of each root $\mathsf{b}_{ij} \in \mathcal{Z}(\mathsf{J})$ have a canonical representation

$$p(\mathsf{b}_{ij}) = \frac{\widehat{p}(\alpha_{ij})}{\gamma_0(\alpha_{ij})},$$

where

$$\widehat{p}(Y) = \mathbf{Rem}\left(p(\gamma_1(Y), \ldots, \gamma_r(Y)), \chi(Y)\right) \in K[Y].$$

vectorspace arithmetics: Given two arithmetical expressions $p_1, p_2 \in \mathcal{Q}$ of the roots $\mathsf{b}_{ij} \in \mathcal{Z}(\mathsf{J})$ and values $c_1, c_2 \in K$, the arithmetitical expression $p(\mathsf{b}_{ij}) := c_1 p_1(\mathsf{b}_{ij}) + c_2 p_2(\mathsf{b}_{ij})$ has the canonical representation $p(\mathsf{b}_{ij}) := \widehat{p}(\alpha_{ij})/\gamma_0(\alpha_{ij})$, where

$$
\begin{aligned}
\widehat{p}(Y) &:= c_1 \widehat{p_1}(Y) + c_2 \widehat{p_2}(Y) \\
&= c_1 \, \mathbf{Rem}\,(p_1(\gamma_1(Y), \ldots, \gamma_r(Y)), \chi(Y)) \\
&\quad + c_2 \, \mathbf{Rem}\,(p_2(\gamma_1(Y), \ldots, \gamma_r(Y)), \chi(Y)).
\end{aligned}
$$

multiplication: The arithmetical expression $p_1(\mathsf{b}_{ij}) p_2(\mathsf{b}_{ij})$ has the canonical representation $q(\alpha_{ij})/\gamma_0(\alpha_{ij})$, where

$$
q(Y) := \mathbf{Rem}\,(\widehat{p_1}(Y)\widehat{p_2}(Y)\gamma_{-1}(Y), \chi(Y)),
$$

so that

$$
\begin{aligned}
\frac{q(\alpha_{ij})}{\gamma_0(\alpha_{ij})} &= \frac{\widehat{p_1}(\alpha_{ij})\widehat{p_2}(\alpha_{ij})\gamma_{-1}(\alpha_{ij})}{\gamma_0(\alpha_{ij})} \\
&= \frac{\left(p_1(\mathsf{b}_{ij})\gamma_0(\alpha_{ij})\right)\left(p_2(\mathsf{b}_{ij})\gamma_0(\alpha_{ij})\right)\gamma_{-1}(\alpha_{ij})}{\gamma_0(\alpha_{ij})} \\
&= p_1(\mathsf{b}_{ij}) p_2(\mathsf{b}_{ij}).
\end{aligned}
$$

zero testing: Given an arithmetical expression $p \in \mathcal{Q}$, we have

$$
p(\mathsf{b}_{ij}) = 0 \iff \widehat{p}(Y) = \mathbf{Rem}\,(p(\gamma_1(Y), \ldots, \gamma_r(Y), \chi(Y)) = 0.
$$

inverse and division: Even if χ is not necessarily squarefree, Lazard's Theorem 11.3.2 can be applied in essentially the same way in order to compute the required factorization $\chi := \chi^{(0)}\chi^{(1)}$ and the polynomial $q(Y) \in K[Y]$, $\deg(q) < \deg(\chi^{(1)})$, having the required properties.

We begin by remarking that of all the data related to the multiplicity of the primary components of J, namely r, r_i, d_i, the only data available to us is[3]

$$
R = \deg(\chi) = \sum_{i=1}^{r} r_i d_i;
$$

however that is all we need, since we have the following.

Lemma 45.3.1. *With the present notation, it holds that*

$$
\gcd(\chi, \widehat{p}^R) = \prod_{i \in I_0} \psi_i^{d_i},
$$

where $I_0 = \{i : 1 \le i \le \mathsf{r}, \psi_i \mid \widehat{p}\}$.

---

[3] See property (13) in Section 45.2.

Proof. In fact, for each i, since $\psi_i^{d_i} \mid \chi$ and $\psi_i^{d_i+1} \nmid \chi$, we have that $\psi_i \mid p \iff \psi_i^{d_i} \mid p^R \iff \psi_i^{d_i} \mid \gcd(\chi, p^R)$. ◉

Corollary 45.3.2. *We write:*

$$\chi^{(0)} := \gcd(\chi, \gamma_{-1}^2 \widehat{p}^R) \in K[Y];$$
$$s, t \in K[Y] \text{ such that } s\gamma_{-1}^2 \widehat{p}^R + t\chi = \chi^{(0)};$$
$$\chi^{(1)} := \chi/\chi^{(0)};$$
$$u, v \in K[Y] \text{ such that } u\chi^{(0)} + v\chi^{(1)} = 1;$$
$$q := \mathbf{Rem}(us\widehat{p}^{R-1}, \chi^{(1)}).$$

Then

- *if $\chi^{(0)} = \chi, 1 = \chi^{(1)}$ then for each $\mathsf{b}_{ij} \in \mathcal{Z}(\mathsf{J})$, $p(\mathsf{b}_{ij}) = 0$;*
- *if $\chi^{(1)} = \chi, 1 = \chi^{(0)}$ then for each $\mathsf{b}_{ij} \in \mathcal{Z}(\mathsf{J})$,*

$$p(\mathsf{b}_{ij}) = \frac{\widehat{p}(\alpha_{ij})}{\gamma_0(\alpha_{ij})} \neq 0 \quad \text{and} \quad p^{-1}(\mathsf{b}_{ij}) = \frac{q(\alpha_{ij})}{\gamma_0(\alpha_{ij})};$$

- *otherwise, for each $\mathsf{b}_{ij} \in \mathcal{Z}(\mathsf{J})$, we have*

$$p(\mathsf{b}_{ij}) = 0 \iff \chi^{(0)}(\alpha_{ij}) = 0,$$
$$p(\mathsf{b}_{ij}) = \frac{\widehat{p}(\beta_{ij})}{\gamma_0(\alpha_{ij})} \neq 0 \iff \chi^{(1)}(\alpha_{ij}) = 0, \text{ in which case}$$

$$p^{-1}(\mathsf{b}_{ij}) = \frac{q(\alpha_{ij})}{\gamma_0(\alpha_{ij})}.$$

Proof. Since for each $\mathsf{b}_{ij} \in \mathcal{Z}(\mathsf{J})$, $\gamma_0(\alpha_{ij}) \neq 0$ and $\gamma_{-1}(\alpha_{ij}) \neq 0$, Lazard's Theorem 11.3.2 implies that:

- denoting $p_1 := \dfrac{\gamma_{-1}^2 \widehat{p}^R}{\chi_0^{(0)}}$ we have

$$\chi_0^{(0)}(\alpha_{ij}) = 0 \implies \widehat{p}^R(\alpha_{ij})\gamma_{-1}^2(\alpha_{ij}) = \chi_0^{(0)}(\alpha_{ij})p_1(\alpha_{ij}) = 0 \implies \widehat{p}^R(\alpha_{ij}) = 0.$$

- $\chi^{(1)}(\alpha_{ij}) = 0$ implies that $\widehat{p}^R(\alpha_{ij}) \neq 0$ and

$$\frac{\widehat{p}(\alpha_{ij})}{\gamma_0(\alpha_{ij})} \frac{q(\alpha_{ij})}{\gamma_0(\alpha_{ij})} = \widehat{p}(\alpha_{ij})q(\alpha_{ij})\gamma_{-1}^2(\alpha_{ij})$$
$$= \widehat{p}(\alpha_{ij})\left(u(\alpha_{ij})s(\alpha_{ij})\widehat{p}^{R-1}(\alpha_{ij})\right)\gamma_{-1}^2(\alpha_{ij})$$
$$= u(\alpha_{ij})s(\alpha_{ij})\gamma_{-1}^2(\alpha_{ij})\widehat{p}^R(\alpha_{ij}) + u(\alpha_{ij})t(\alpha_{ij})\chi(\alpha_{ij})$$
$$= u(\alpha_{ij})\chi^{(0)}(\alpha_{ij})$$
$$= u(\alpha_{ij})\chi^{(0)}(\alpha_{ij}) + v(\alpha_{ij})\chi^{(1)}(\alpha_{ij})$$
$$= 1.$$

◉

Therefore, writing

$$I_0 := \{i : 1 \leq i \leq \mathsf{r}, \psi_i \mid \chi^{(0)}\} = \{i : \psi_i \mid \widehat{p}\} \subset \{1, \dots, \mathsf{r}\},$$

256   *Duval II*

$$I_1 := \{i : 1 \leq i \leq r, i \notin I_0\} = \{i : \psi_i \mid \chi^{(1)}\} = \{i : \psi_i \nmid \widehat{p}\} \subset \{1, \cdots, r\},$$
$$J_\iota := \cap_{i \in I_\iota} \mathfrak{q}_i, \iota \in \{0, 1\},$$
$$Z_\iota := \{\mathbf{b}_{ij} : i \in I_\iota\}, \iota \in \{0, 1\},$$
$$\gamma_j^{(\iota)} := \mathbf{Rem}(\gamma_j, \chi^{(\iota)}) \in K[Y], \iota \in \{0, 1\}, 0 \leq j \leq r,$$

one has

Corollary 45.3.3.   *With the present notation, the following hold:*

(a) $\chi^{(\iota)} = \prod_{i \in I_\iota} \psi_i^{d_i} = \prod_{i \in I_\iota} \prod_{j=1}^{r_i} \left(Y - \alpha_{ij}\right)^{d_i}, \iota \in \{0, 1\}$;

(b) *for $\iota \in \{0, 1\}$, $\left(\chi^{(\iota)}(Y), \gamma_0^{(\iota)}(Y), \gamma_1^{(\iota)}(Y), \ldots, \gamma_r^{(\iota)}(Y)\right)$ is the rational univariate representation of J_ι;*

(c) $Z_\iota = \mathcal{Z}(J_\iota), \iota \in \{0, 1\}$;

(d) $Z_0 = \{\mathbf{b}_{ij} \in \mathcal{Z}(I) : p(\mathbf{b}_{ij}) = 0\}$;

(e) $Z_1 = \{\mathbf{b}_{ij} \in \mathcal{Z}(J) : p(\mathbf{b}_{ij}) \neq 0\}$;

(f) $J = J_0 \cap J_1$;

(g) $Z_\iota = \left\{ \left(\dfrac{\gamma_1^{(\iota)}(\alpha)}{\gamma_0^{(\iota)}(\alpha)}, \ldots, \dfrac{\gamma_r^{(\iota)}(\alpha)}{\gamma_0^{(\iota)}(\alpha)}\right) : \alpha \in K, \chi^{(\iota)}(\alpha) = 0\right\}, \iota \in \{0, 1\}$.

If, moreover, J is radical and the given rational univariate representation is a Kronecker parametrization, setting

$$\xi_j^{(\iota)} := \mathbf{Rem}\left(\gamma_j \frac{\partial \chi^{(\iota)}}{\partial Y} \gamma_{-1}, \chi^{(\iota)}\right) \in K[Y], \iota \in \{0, 1\}, 1 \leq j \leq n,$$

it holds that

(h) $\chi^{(\iota)}$ *is squarefree,* $\iota \in \{0, 1\}$;

(i) $J_\iota = \sqrt{J_\iota}, \iota \in \{0, 1\}$;

(j) $\chi^{(\iota)} = \prod_{i \in I_\iota} \psi_i = \prod_{i \in I_\iota} \prod_{j=1}^{r_i} \left(Y - \beta_{ij}\right), \iota \in \{0, 1\}$;

(k) *for $\iota \in \{0, 1\}$,*

$$\left(\chi^{(\iota)}(Y), \frac{\partial \chi^{(\iota)}}{\partial Y} Z_1 - \xi_1^{(\iota)}(Y), \ldots, \frac{\partial \chi^{(\iota)}}{\partial Y} Z_r - \xi_r^{(\iota)}(Y)\right)$$

is the Kronecker parametrization of J_ι.

$$\boxed{\odot}$$

Proof.   The only non-trivial statements are (b), which is Remark 42.9.18, and (k), for which it is sufficient to remark that, by construction, for each $\mathbf{b}_{ij} \in Z_\iota$ and each $l, 1 \leq l \leq n$, one has

$$b_l^{(ij)} = \frac{\gamma_l(\alpha_{ij})}{\gamma_0(\alpha_{ij})} = \gamma_l(\alpha_{ij})\gamma_{-1}(\alpha_{ij}) = \frac{\gamma_l^{(\iota)}(\alpha_{ij})\dfrac{\partial \chi^{(\iota)}}{\partial Y}(\alpha_{ij})\gamma_{-1}(\alpha_{ij})}{\dfrac{\partial \chi^{(\iota)}}{\partial Y}(\alpha_{ij})} = \frac{\xi_l^{(\iota)}(\alpha_{ij})}{\dfrac{\partial \chi^{(\iota)}}{\partial Y}(\alpha_{ij})}.$$

$$\boxed{\odot}$$

45.4 Gröbner Representation

Using the same notation as in Section 45.2, let us begin by assuming that the zero-dimensional ideal J, $\deg(J) = R$, is given by means of a Gröbner representation (Definition 29.3.3) of its quotient algebra

$$\mathbf{q} = \{q_1, \ldots, q_R\}, \quad q_1 = 1, \qquad \mathcal{M} = \mathcal{M}(\mathbf{q}) := \left\{ \left(a_{lj}^{(h)} \right) \in K^{R^2}, 1 \le h \le r \right\},$$

so that

(1) $Q/J \cong \mathrm{Span}_K(\mathbf{q})$,
(2) $Z_h q_l = \sum_j a_{lj}^{(h)} q_j$ for each l, j, h, $1 \le l, j \le R$, $1 \le h \le n$, in Q/J,

and by recalling that, for each $f \in Q$, its Gröbner description (Definition 29.3.3)[4]

$$\mathbf{Rep}(f, \mathbf{q}) := (\gamma(f, q_1, \mathbf{q}), \ldots, \gamma(f, q_R, \mathbf{q})) \in K^R$$

in terms of this Gröbner representation, which satisfies

$$f - \sum_j \gamma(f, q_j, \mathbf{q}) q_j \in J,$$

can be efficiently computed either by representing f as a linear combination of terms in $\mathcal{W}$ or by using a recursive Horner representation.

In other words, as we observed in Historical Remark 29.3.4, we can efficiently compute structure constants

$$\gamma_{ij}^{(l)} := \gamma(q_i q_j, q_l, \mathbf{q})$$

which satisfy

(3) $q_i q_j = \sum_l \gamma_{ij}^{(l)} q_l$ for each l, j, h, $1 \le i, j, l \le R$.

As a consequence we can adapt Definition 29.3.3 as follows.

Definition 45.4.1. *A Gröbner representation of* J *(or, better, of the algebra* Q/J*) is the assignment of*

(a) *a K-linearly independent set* $\mathbf{q} = \{q_1, \ldots, q_R\}$,
(b) *the set* $\mathcal{M} = \mathcal{M}(\mathbf{q}) := \left\{ \left(a_{lj}^{(h)} \right) \in K^{R^2}, 1 \le h \le n \right\}$ *of n square matrices,*
(c) R^3 *values* $\gamma_{ij}^{(l)} \in K$,

which satisfy

(1) $Q/J \cong \mathrm{Span}_K(\mathbf{q})$,
(2) $Z_h q_l \equiv \sum_j a_{lj}^{(h)} q_j \bmod$ J, *for each l, j, h, $1 \le l, j \le R$, $1 \le h \le n$,*
(3) $q_i q_j \equiv \sum_l \gamma_{ij}^{(l)} q_l \bmod$ J, *for each l, j, h, $1 \le i, j, l \le R$.*

---

[4] Compare the discussion in Section 29.3 and in particular Algorithm 29.3.8.

The values $\gamma_{ij}^{(l)} := \gamma(q_i q_j, q_l, \mathbf{q})$ *are called the* structural constants *of the K-algebra* $\mathcal{Q}/\mathsf{J} = \mathrm{Span}_K(\mathbf{q})$.

The linear representation *of* J *w.r.t. the term ordering* $\prec$ *is the Gröbner representation* $(\mathbf{N}_\prec(\mathsf{J}), \mathcal{M}, \gamma_{ij}^{(l)})$, *where* $\mathbf{q} = (\mathbf{N}_\prec(\mathsf{J}))$. ⊙

The application of Gröbner representations as a tool for performing Kronecker's Philosophy effectively requires the solution of the following.

Problem 45.4.2. *Given*

- *a zero-dimensional ideal* $\mathsf{J}' \supset \mathsf{J}$ *and*
- *a K-basis* $\mathbf{q}' = \{q'_1, \ldots, q'_S\} \subset \mathrm{Span}_K(\mathbf{q})$

such that $\mathcal{Q}/\mathsf{J}' = \mathrm{Span}_K(\mathbf{q}')$, *compute a Gröbner representation of* J'. ⊙

We postpone discussing the solution of this problem to the end of the section (see Algorithm 45.4.8), after we have expounded our application.

Since a Gröbner representation of J gives the natural arithmetics of a K-algebra, in order to apply it for carrying into effect Kronecker's Philosophy, we just need to focus on inversion and division.

On the basis of the discussion in Section 29.3 and in particular Algorithm 29.3.8 we can wlog assume that each arithmetical expression given via a polynomial p is represented either via a recursive Horner representation or as a linear combination of terms in $\mathcal{W}$; thus it can be easily expressed as $p = \sum_\iota \gamma(p, q_\iota, \mathbf{q})q_\iota$.

canonical representation: All arithmetical expressions

$$p(\mathbf{b}_{ij}) = p(b_1^{(ij)}, \ldots, b_r^{(ij)}), \quad p \in \mathcal{Q},$$

of each root $\mathbf{b}_{ij} \in \mathcal{Z}(\mathsf{J})$ have the canonical representation

$$p(\mathbf{b}_{ij}) = \sum_\iota \gamma(p, q_\iota, \mathbf{q})q_\iota(\mathbf{b}_{ij}).$$

vector space arithmetics: Given two such arithmetical expressions $p_1, p_2 \in \mathcal{Q}$ of the roots $\mathbf{b}_{ij} \in \mathcal{Z}(\mathsf{J})$ and values $c_1, c_2 \in K$, the arithmetical expression $p(\mathbf{b}_{ij}) := c_1 p_1(\mathbf{b}_{ij}) + c_2 p_2(\mathbf{b}_{ij})$ has the canonical representation

$$\sum_\iota \Big(c_1\gamma(p_1, q_\iota, \mathbf{q}) + c_2\gamma(p_2, q_\iota, \mathbf{q})\Big)q_\iota(\mathbf{b}_{ij}).$$

multiplication: With the same notation, the arithmetical expression

$$p(\mathbf{b}_{ij}) := p_1(\mathbf{b}_{ij})p_2(\mathbf{b}_{ij})$$

has the canonical representation

$$\sum_\lambda \left(\sum_\iota \sum_\kappa \gamma(p_1, q_\iota, \mathbf{q})\gamma_{\iota\kappa}^{(\lambda)}\gamma(p_2, q_\kappa, \mathbf{q})\right) q_\lambda(\mathbf{b}_{ij}),$$

since

$$\sum_{\lambda}\left(\sum_{\iota}\sum_{\kappa}\gamma(p_1,q_\iota,\mathbf{q})\gamma_{\iota\kappa}^{(\lambda)}\gamma(p_2,q_\kappa,\mathbf{q})\right)q_\lambda(\mathbf{b}_{ij})$$

$$=\sum_{\iota}\sum_{\kappa}\gamma(p_1,q_\iota,\mathbf{q})\gamma(p_2,q_\kappa,\mathbf{q})\left(\sum_{\lambda}\gamma_{\iota\kappa}^{(\lambda)}q_\lambda(\mathbf{b}_{ij})\right)$$

$$=\sum_{\iota}\sum_{\kappa}\gamma(p_1,q_\iota,\mathbf{q})\gamma(p_2,q_\kappa,\mathbf{q})q_\iota(\mathbf{b}_{ij})q_\kappa(\mathbf{b}_{ij})$$

$$=\left(\sum_{\iota}\gamma(p_1,q_\iota,\mathbf{q})q_\iota(\mathbf{b}_{ij})\right)\left(\sum_{\kappa}\gamma(p_2,q_\kappa,\mathbf{q})q_\kappa(\mathbf{b}_{ij})\right)$$

$$=p_1(\mathbf{b}_{ij})p_2(\mathbf{b}_{ij}).$$

zero testing: Given an arithmetical expression $p \in \mathcal{Q}$ we have

$$p(\mathbf{b}_{ij}) = 0 \iff \gamma(p, q_i, \mathbf{q}) = 0 \text{ for each } i.$$

inverse and division: Linear algebra, namely a Gaussian algorithm, is all we need to deal with division.

Let us consider any $p \in \mathcal{Q}$ given by the canonical representation

$$p \equiv \sum_{\iota=1}^{R}\gamma(p,q_\iota,\mathbf{q})q_\iota \bmod \mathsf{J},$$

and let us remark that, for each $i \le r$, setting

$$\delta := \begin{cases} 1 & \mathsf{J} = \sqrt{\mathsf{J}}, \\ R = \deg(\mathsf{J}) & \mathsf{J} \ne \sqrt{\mathsf{J}}, \end{cases}$$

we have:

$$p(\mathbf{b}_{ij}) = 0 \iff p \in \mathfrak{m}_i \iff p^\delta \in \mathfrak{q}_i \iff \mathfrak{q}_i : p^\delta = 1;$$
$$p(\mathbf{b}_{ij}) \ne 0 \iff p \notin \mathfrak{m}_i \iff p^\delta \notin \mathfrak{q}_i \iff \mathfrak{q}_i : p^\delta = \mathfrak{q}_i;$$
$$p(\mathbf{b}_{ij}) = 0 \iff \mathfrak{q}_i + (p^\delta) = \mathfrak{q}_i \iff \mathfrak{m}_i + (p) = \mathfrak{m}_i;$$
$$p(\mathbf{b}_{ij}) \ne 0 \iff \mathfrak{q}_i + (p^\delta) = (1) \iff \mathfrak{m}_i + (p) = (1).$$

Therefore, writing

- $A := \left(\gamma(p^{\delta-1}q_\kappa, q_\lambda, \mathbf{q})\right)_{\lambda\kappa}$,
- $I_0 := \{i : 1 \le i \le \mathsf{r}, p^\delta \in \mathfrak{q}_i\} = \{i : p \in \mathfrak{m}_i\} = \{i : p(\mathbf{b}_{ij}) = 0\} \subset \{1,\ldots,\mathsf{r}\}$,
- $I_1 := \{i : 1 \le i \le \mathsf{r}, i \notin I_0\} = \{i : p^\delta \notin \mathfrak{q}_i\} = \{i : p \notin \mathfrak{m}_i\} = \{i : p(\mathbf{b}_{ij}) \ne 0\}$
  $\subset \{1,\ldots,\mathsf{r}\}$,
- $\mathsf{J}_\iota := \cap_{i\in I_\iota}\mathfrak{q}_i, \iota \in \{0,1\}$,
- $Z_\iota := \{\mathbf{b}_{ij} : i \in I_\iota\}, \iota \in \{0,1\}$,
- $\mathbf{b} = \{b_1,\ldots,b_R\}$, another K-basis of $\mathcal{Q}/\mathsf{J} = \mathrm{Span}_K(\mathbf{b})$,

- $\phi : \mathcal{Q}/\mathsf{J} \cong \mathrm{Span}_K(\mathbf{q}) \to \mathcal{Q}/\mathsf{J} \cong \mathrm{Span}_K(\mathbf{b})$, the morphism defined by $\phi(g) = p^\delta g$ for each $g \in \mathrm{Span}_K(\mathbf{q})$,
- $C := (c_{\iota\kappa})$, the matrix representing the morphism ϕ,
- $\pi : \mathcal{Q} \to \mathcal{Q}/\mathsf{J} \cong \mathrm{Span}_K(\mathbf{q})$, the canonical projection,
- $S := \deg(\mathsf{J}_0)$,

we have:

    (a) $\mathsf{Z}_\iota = \mathcal{Z}(\mathsf{J}_\iota), \iota \in \{0, 1\}$;

    (b) $\mathsf{Z}_0 = \{\mathbf{b}_{ij} \in \mathcal{Z}(\mathsf{I}) : p(\mathbf{b}_{ij}) = 0\}$;

    (c) $\mathsf{Z}_1 = \{\mathbf{b}_{ij} \in \mathcal{Z}(\mathsf{J}) : p(\mathbf{b}_{ij}) \neq 0\}$;

    (d) $\mathsf{J} = \mathsf{J}_0 \cap \mathsf{J}_1$;

    (e) $\mathsf{J}_0 + \mathsf{J}_1 = (1)$;

    (f) $\mathsf{J}_0 = \mathsf{J} + (p^\delta), \mathsf{J}_1 = \mathsf{J} : p^\delta$;

    (g) $\phi(q_\kappa) = p^\delta q_\kappa = \sum_{\iota=1}^R b_\iota c_{\iota\kappa}, 1 \leq \kappa \leq R$;

    (h) $\mathrm{Im}(\phi) \subset \mathcal{Q}/\mathsf{J}$ is the principal ideal generated by $\pi(p^\delta)$;

    (i) $\pi^{-1}(\mathrm{Im}(\phi)) = \mathsf{J}_0 = \mathsf{J} + (p^\delta)$;

    (j) $\ker(\phi) = \{\pi(g) : g \in \mathcal{Q}, \pi(gp^\delta) = 0\} \subset \mathcal{Q}/\mathsf{J}$;

    (k) $\ker(\phi\pi) = \{g \in \mathcal{Q} : gp^\delta \in \mathsf{J}\} = \mathsf{J} : p^\delta = \mathsf{J}_1$;

    (l) S is the rank of C.

Remark 45.4.3.   If J is radical, that is, $\delta = 1$, then both J_0 and J_1 are radical. In this case, the procedure outlined here, which requires the evaluation of $p^{\delta-1}$ and $p^{\delta-1}q'_\iota$ in order to compute both p^δ and $p(p^{\delta-1}q'_\iota)$, is simplified since we need just to evaluate pq'_ι. $\boxdot$

If we now perform Gaussian column reduction on $C := (c_{\iota\kappa})$, performing the same transformation on the identity we obtain two matrices,

$$D := (d_{\iota\ell}) \qquad \text{and} \qquad E := (e_{\kappa\ell}),$$

which satisfy:

    (m) $D = CE$;

    (n) E is invertible;

    (o) D is lower triangular, so that $\ell > \iota \implies d_{\iota\ell} = 0$;

    (p) $\ell > S = \deg(\mathsf{J}_0) \implies d_{\iota\ell} = 0$ for each ι;

    (q) $\mathbf{q}' := \{q'_1, \ldots, q'_S\}, q'_\ell := \sum_{\iota=1}^R b_\iota d_{\iota\ell}, 1 \leq \ell \leq S$, is a linear basis both of $\mathrm{Im}(\phi)$ and $\pi(\mathsf{J}_0) \subset \mathcal{Q}/\mathsf{J}$;

    (r) for each $\ell, 1 \leq \ell \leq S, \chi_\ell := \sum_{\kappa=1}^R q_\kappa e_{\kappa\ell} \in \mathrm{Span}_K(\mathbf{q})$ satisfies

$$p^\delta \chi_\ell = \phi(\chi_\ell) = q'_\ell;$$

(s) for each ℓ, $1 \le \ell \le S$, $p^{\delta-1}\chi_\ell$ has the representation[5]

$$p^{\delta-1}\chi_\ell = \sum_{\lambda=1}^{R} q_\lambda \left(\sum_{\kappa=1}^{R} \gamma(p^{\delta-1}q_\kappa, q_\lambda, \mathbf{q})e_{\kappa\ell} \right);$$

(t) $\mathbf{q}'' := \{\chi_{S+1}, \ldots, \chi_R\}$, $\chi_\ell := \sum_{\kappa=1}^{R} q_\kappa e_{\kappa\ell}$, $S+1 \le \ell \le R$, is a linear basis of both $\ker(\phi)$ and $\pi(\mathsf{J}_1) \subset \mathcal{Q}/\mathsf{J}$;

(u) $\{q'_1, \ldots, q'_S\} \cup \{\chi_{S+1}, \ldots, \chi_R\}$ is a K-basis of $\mathcal{Q}/\mathsf{J}$;

(v) there is a linear relation (mod J)

$$\begin{aligned}
1 &= \sum_{\ell=1}^{S} c_\ell q'_\ell + \sum_{\ell=S+1}^{R} c_\ell \chi_\ell \\
&= \sum_{\ell=1}^{S} c_\ell p^\delta \chi_\ell + \sum_{\ell=S+1}^{R} c_\ell \chi_\ell \\
&= p \sum_{\ell=1}^{S} c_\ell \left(\sum_{\lambda=1}^{R} q_\lambda \left(\sum_{\kappa=1}^{R} \gamma(p^{\delta-1}q_\kappa, q_\lambda, \mathbf{q})e_{\kappa\ell} \right) \right) + \sum_{\ell=S+1}^{R} c_\ell \chi_\ell \\
&= p \left(\sum_{\lambda=1}^{R} q_\lambda \left(\sum_{\ell=1}^{S}\sum_{\kappa=1}^{R} \gamma(p^{\delta-1}q_\kappa, q_\lambda, \mathbf{q})e_{\kappa\ell}c_\ell \right) \right) + \sum_{\ell=S+1}^{R} c_\ell \chi_\ell;
\end{aligned}$$

(w) setting

$$\eta_\lambda := \sum_{\ell=1}^{S}\sum_{\kappa=1}^{R} \gamma(p^{\delta-1}q_\kappa, q_\lambda, \mathbf{q})e_{\kappa\ell}c_\kappa, \quad 1 \le \lambda \le R,$$

and $q := \sum_{\lambda=1}^{R} q_\lambda \eta_\lambda$, we have $1 \equiv qp \bmod \mathsf{J}_1$;

(x) for each κ, $S < \kappa \le R$, there is a value[6] $\rho(\kappa)$ for which $e_{\rho(\kappa)\kappa} = 1$ and $e_{\rho(\kappa)\ell} = 0$ for each $\ell \ne \kappa$.

As a consequence we have the following

Corollary 45.4.4. *With the present notation it holds that:*

o *if $S = 0$ then C is the null matrix, $\mathsf{J}_1 = (1)$, $\mathsf{J}_0 = \mathsf{J}$ and $p(\mathbf{b}_{ij}) = 0$ for each* $\mathbf{b}_{ij} \in \mathcal{Z}(\mathsf{J})$;

o *if $S = R$ then C is invertible, $\mathsf{J}_0 = (1)$, $\mathsf{J}_1 = \mathsf{J}$, $p(\mathbf{b}_{ij}) \ne 0$ and $p^{-1}(\mathbf{b}_{ij}) = q(\mathbf{b}_{ij})$ for each $\mathbf{b}_{ij} \in \mathcal{Z}(\mathsf{J})$;*

o *otherwise, for each $\mathbf{b}_{ij} \in \mathcal{Z}(\mathsf{J})$, we have*

---

[5] We have

$$p^{\delta-1}\chi_\ell = \sum_{\kappa=1}^{R} p^{\delta-1}q_\kappa e_{\kappa\ell} = \sum_{\kappa=1}^{R}\sum_{\lambda=1}^{R} q_\lambda \gamma(p^{\delta-1}q_\kappa, q_\lambda, \mathbf{q})e_{\kappa\ell} = \sum_{\lambda=1}^{R} q_\lambda \left(\sum_{\kappa=1}^{R} \gamma(p^{\delta-1}q_\kappa, q_\lambda, \mathbf{q})e_{\kappa\ell} \right).$$

[6] Note that the κth column of D, which is null, corresponds to a repeated transformation of the $\rho(\kappa)$th column of C, which is never chosen as a *pivot* element.

$$p(\mathbf{b}_{ij}) = 0 \iff \mathbf{b}_{ij} \in \mathcal{Z}(\mathsf{J}_0),$$
$$p(\mathbf{b}_{ij}) \neq 0 \iff \mathbf{b}_{ij} \in \mathcal{Z}(\mathsf{J}_1), \text{ in which case } p^{-1}(\mathbf{b}_{ij}) = q(\mathbf{b}_{ij}).$$

In conclusion, Gaussian reduction is all we need in order to obtain the Duval splitting $\mathsf{J} = \mathsf{J}_0 \cap \mathsf{J}_1$. What we have to do is:

- compute the canonical representation of $p^{\delta-1}$;
- compute the matrix A representing the canonical representation of $p^{\delta-1}q_\kappa$, $1 \leq \kappa \leq R$;
- compute the canonical representation of

$$p^\delta = pp^{\delta-1} \qquad \text{and} \qquad p(p^{\delta-1}q_\kappa) \qquad 1 \leq \kappa \leq R,$$

  thus obtaining the matrix C;
- perform Gaussian column-reduction on C, deducing D and E;
- extract the K-linearly independent bases $\mathbf{q}'$ and $\mathbf{q}''$ of, respectively, $\pi(\mathsf{J}_0)$ and $\pi(\mathsf{J}_1)$;
- apply the solution, discussed in Algorithm 45.4.8 below, of Problem 45.4.2 in order to obtain a Gröbner representation of both J_0 and J_1;
- compute, for each κ, λ, $1 \leq \kappa \leq S$, $1 \leq \lambda \leq R$, the values

$$\epsilon_{\kappa,\lambda} := \sum_{\iota=1}^{R} e_{\iota\kappa} \gamma(p^{\delta-1}q_\iota, q_\lambda, \mathbf{q});$$

- compute the unique solution $(c_1, \ldots, c_R)$ of the linear equations

$$\begin{cases} 1 &= \sum_{\ell=1}^{S} c_\ell \left(\sum_{\kappa=1}^{R} \gamma(p^{\delta-1}q_\kappa, q_1, \mathbf{q})e_{\kappa\ell} \right) + \sum_{\ell=S+1}^{R} c_\ell e_{1\ell}, \\ 0 &= \sum_{\ell=1}^{S} c_\ell \left(\sum_{\kappa=1}^{R} \gamma(p^{\delta-1}q_\kappa, q_2, \mathbf{q})e_{\kappa\ell} \right) + \sum_{\ell=S+1}^{R} c_\ell e_{2\ell}, \\ &\quad \ldots \\ 0 &= \sum_{\ell=1}^{S} c_\ell \left(\sum_{\kappa=1}^{R} \gamma(p^{\delta-1}q_\kappa, q_R, \mathbf{q})e_{\kappa\ell} \right) + \sum_{\ell=S+1}^{R} c_\ell e_{R\ell}; \end{cases}$$

- compute $\eta_\lambda := \sum_{\ell=1}^{S} \sum_{\kappa=1}^{R} \gamma(p^{\delta-1}q_\kappa, q_\lambda, \mathbf{q})e_{\kappa\ell}c_\kappa$, $1 \leq \lambda \leq R$;
- return $q := \sum_{\lambda=1}^{R} q_\lambda \eta_\lambda$.

Remark 45.4.5.   It is clear that the above algorithm is essentially a refinement of Traverso's Algorithm 29.3.8 for computing J_0, extended and adapted in order to obtain also $\mathsf{J}_1 := \pi^{-1}(\ker(\psi))$ and q.

Example 45.4.6.   Let us consider

- the ideal

$$\mathsf{J} := \mathbb{I}(Z_1^3 - Z_1^2, Z_1 Z_2, Z_2^2 - Z_2) \subset K[Z_1, Z_2] = \mathcal{Q}$$

whose roots are $\{(0, 0), (1, 0), (0, 1)\}$, $(0, 0)$ having multiplicity 2 and whose primary component is (X_1^2, X_2),

- the Lagrange basis

$$q_1 := 1 - Z_1^2 - Z_2, \quad q_2 := Z_1 - Z_1^2, \quad q_3 := Z_1^2, \quad q_4 := Z_2$$

- and the polynomial $p := 1 - Z_1 + Z_2 = q_1 - q_2 + 2q_4$.

We thus have

$$A := \begin{pmatrix} & q_1 & q_2 & q_3 & q_4 \\ \hline q_1 & 1 & 0 & 0 & 0 \\ q_2 & -3 & 1 & 0 & 0 \\ q_3 & 0 & 0 & 0 & 0 \\ q_4 & 0 & 0 & 0 & 8 \end{pmatrix} \quad \text{and} \quad C := \begin{pmatrix} & q_1 & q_2 & q_3 & q_4 \\ \hline Z_1^2 & 3 & -1 & 0 & 0 \\ Z_2 & -1 & 0 & 0 & 16 \\ Z_1 & -4 & 1 & 0 & 0 \\ 1 & 1 & 0 & 0 & 0 \end{pmatrix},$$

and we deduce that

$$D := \begin{pmatrix} Z_1^2 & 1 & 0 & 0 & 0 \\ Z_2 & 0 & 1 & 0 & 0 \\ Z_1 & 0 & 0 & 1 & 0 \\ 1 & -1 & 0 & -1 & 0 \end{pmatrix} \quad \text{and} \quad E := \begin{pmatrix} q_1 & -1 & 0 & -1 & 0 \\ q_2 & -4 & 0 & -3 & 0 \\ q_3 & 0 & 0 & 0 & 1 \\ q_4 & -\frac{1}{16} & \frac{1}{16} & -\frac{1}{16} & 0 \end{pmatrix}.$$

Therefore we have

$$J_0 = J + \mathbb{I}(p^3) = J + \{q_1', q_2', q_3'\} = J + \{Z_1^2 - 1, Z_2, Z_1 - 1\} = \mathbb{I}(Z_2, Z_1 - 1),$$
$$\mathcal{Z}(J_0) = \{(1, 0)\} = \mathbf{Z}_0,$$
$$\phi(\chi_1) = \phi(-q_1 - 4q_2 - \tfrac{1}{16}q_4) = \phi(5Z_1^2 + \tfrac{15}{16}Z_2 - 4Z_1 - 1) = q_1',$$
$$\phi(\chi_2) = \phi(\tfrac{1}{16}q_4) = \phi(\tfrac{1}{16}Z_2) = q_2',$$
$$\phi(\chi_3) = \phi(-q_1 - 3q_2 - \tfrac{1}{16}q_4) = \phi(4Z_1^2 + \tfrac{15}{16}Z_2 - 3Z_1 - 1) = q_3',$$

$$AE = \begin{pmatrix} q_1 & -1 & 0 & -1 & 0 \\ q_2 & -1 & 0 & 0 & 0 \\ q_3 & 0 & 0 & 0 & 0 \\ q_4 & -\frac{1}{2} & \frac{1}{2} & -\frac{1}{2} & 0 \end{pmatrix},$$

$$p^3\chi_1 = -q_1 - q_2 - \tfrac{1}{2}q_4, \quad p^3\chi_2 = \tfrac{1}{2}q_4, \quad p^3\chi_3 = -q_1 - \tfrac{1}{2}q_4;$$
$$J_1 = J : p^3 = J + \{\chi_4\} = J + \{q_3\} = J + \{Z_1^2\} = \mathbb{I}(Z_1^2, Z_1 Z_2, Z_2^2 - Z_2),$$
$$\mathcal{Z}(J_1) = \{(0, 0), (0, 1)\} = \mathbf{Z}_1,$$
$$\{q_1', q_2', q_3', \chi_4\} = \{Z_1^2 - 1, Z_2, Z_1 - 1, Z_1^2\} \text{ is a } K\text{-basis of } \mathcal{Q}/J;$$
$$1 = -q_1' + \chi_4;$$
$$(\eta_1, \ldots, \eta_R) = AE(-1, 0, 0, 1)^T = (1, 1, 0, \tfrac{1}{2});$$
$$q := q_1 + q_2 + \tfrac{1}{2}q_4 = -2Z_2^2 - \tfrac{1}{2}Z_2 + Z_1 + 1 \text{ satisfies } 1 \equiv qp \bmod J_1;$$
$$\rho(4) = 3, \ e_{3\ell} = 0, \ell \neq 4.$$

⊡

Example 45.4.7. Let us consider the *radical* ideal $J \subset K[Z_1, Z_2, Z_3]$ discussed in Examples 39.2.3, 40.3.3 and 42.8.8, for which we choose

$$\mathbf{q} := \mathbf{N}(J) := \{1, Z_1, Z_2, Z_3, Z_1^2, Z_1 Z_2, Z_2^2, Z_1 Z_3, Z_3^2\}$$

and the algebraic expression $p = Z_1 Z_3$, so that we have

$$\mathbf{Z}_0 = \mathcal{Z}(J_0) = \{\mathbf{b}_i : i \in \{1, 2, 4, 9\}\}, \qquad \mathbf{Z}_1 = \mathcal{Z}(J_1) = \{\mathbf{b}_i : i \in \{3, 5, 6, 7, 8\}\}.$$

We thus obtain

$$
C := \begin{pmatrix}
\begin{array}{c|ccccccccc}
 & 1 & Z_1 & Z_2 & Z_3 & Z_1^2 & Z_1Z_2 & Z_2^2 & Z_1Z_3 & Z_3^2 \\
\hline
Z_3^2 & 0 & 0 & 0 & 2 & 0 & 0 & 0 & 4 & 6 \\
Z_1Z_3 & 1 & 3 & -1 & 4 & 7 & -1 & -1 & 12 & 7 \\
Z_2^2 & 0 & 3 & -3 & 15 & 9 & -3 & -3 & 42 & 36 \\
Z_1Z_2 & 0 & 6 & -3 & 30 & 18 & -3 & -3 & 84 & 72 \\
Z_1^2 & 0 & 1 & 2 & 0 & 3 & 2 & 2 & 1 & 0 \\
Z_3 & 0 & -2 & 2 & -8 & -6 & 2 & 2 & -24 & -18 \\
Z_2 & 0 & -9 & 9 & -45 & -27 & 9 & 9 & -126 & -108 \\
Z_1 & 0 & -3 & -3 & -3 & -9 & -3 & -3 & -12 & -6 \\
1 & 0 & 2 & -2 & 6 & 6 & -2 & -2 & 20 & 12
\end{array}
\end{pmatrix},
$$

whence we compute

$$
D := \begin{pmatrix}
\begin{array}{c|ccccc|cccc}
Z_3^2 & 1 & 0 & 0 & 0 & 0 & 0 & 0 & 0 & 0 \\
Z_1Z_3 & 0 & 1 & 0 & 0 & 0 & 0 & 0 & 0 & 0 \\
Z_2^2 & 0 & 0 & 1 & 0 & 0 & 0 & 0 & 0 & 0 \\
Z_1Z_2 & 0 & 0 & 0 & 1 & 0 & 0 & 0 & 0 & 0 \\
Z_1^2 & 0 & 0 & 0 & 0 & 1 & 0 & 0 & 0 & 0 \\
\hline
Z_3 & 1 & 0 & -\frac{2}{3} & 0 & 0 & 0 & 0 & 0 & 0 \\
Z_2 & 0 & 0 & -3 & 0 & 0 & 0 & 0 & 0 & 0 \\
Z_1 & 1 & 0 & -\frac{1}{3} & 0 & -2 & 0 & 0 & 0 & 0 \\
1 & -2 & 0 & \frac{2}{3} & 0 & 0 & 0 & 0 & 0 & 0
\end{array}
\end{pmatrix}
$$

and

$$
E := \begin{pmatrix}
\begin{array}{c|ccccc|cccc}
1 & -\frac{7}{6} & 1 & -\frac{37}{9} & 2 & -\frac{8}{3} & 2 & 0 & 0 & 4 \\
Z_1 & \frac{5}{6} & 0 & \frac{17}{9} & -1 & \frac{4}{3} & -3 & 0 & 0 & -1 \\
Z_2 & 0 & 0 & -\frac{2}{3} & \frac{1}{3} & 0 & 0 & -1 & -1 & 0 \\
Z_3 & \frac{13}{6} & 0 & \frac{10}{9} & -\frac{2}{3} & \frac{2}{3} & 0 & 0 & 0 & -5 \\
Z_1^2 & 0 & 0 & 0 & 0 & 0 & 1 & 0 & 0 & 0 \\
Z_1Z_2 & 0 & 0 & 0 & 0 & 0 & 0 & 1 & 0 & 0 \\
Z_2^2 & 0 & 0 & 0 & 0 & 0 & 0 & 0 & 1 & 0 \\
Z_1Z_3 & -\frac{5}{6} & 0 & -\frac{5}{9} & \frac{1}{3} & -\frac{1}{3} & 0 & 0 & 0 & 1 \\
Z_3^2 & 0 & 0 & 0 & 0 & 0 & 0 & 0 & 0 & 1
\end{array}
\end{pmatrix}.
$$

Moreover,

$$
\begin{aligned}
1 = {} & -\tfrac{1}{2}(Z_3^2 + Z_3 + Z_1 - 2) - \tfrac{1}{2}Z_1Z_3 - \tfrac{3}{2}(3Z_2^2 - 2Z_3 - 9Z_2 - Z_1 + 2) \\
& - 9Z_1Z_2 - \tfrac{1}{2}(Z_1^2 - 2Z_1) \\
& + \tfrac{1}{2}(Z_1^2 - 3Z_1 + 2) + 9(Z_1Z_2 - Z_2) \\
& + \tfrac{9}{2}(Z_2^2 - Z_2) + \tfrac{1}{2}(Z_3^2 + Z_1Z_3 - 5Z_3 - Z_1 + 4),
\end{aligned}
$$

so that

$$J_0 = \mathbb{I}\left(Z_3^2 + Z_3 + Z_1 - 2, \ Z_1 Z_3, \ 3Z_2^2 - 2Z_3 - 9Z_2 - Z_1 + 2, \ Z_1 Z_2, \ Z_1^2 - 2Z_1\right),$$

$$J_1 = \mathbb{I}\left(Z_1^2 - 3Z_1 + 2, \ Z_1 Z_2 - Z_2, \ Z_2^2 - Z_2, \ Z_3^2 + Z_1 Z_3 - 5Z_3 - Z_1 + 4\right),$$

$$q = \tfrac{1}{24}\left(2Z_3^2 + 4Z_1 Z_3 + 33Z_2^2 + 66Z_1 Z_2 - 18Z_1^2 - 20Z_3 - 99Z_2 + 38Z_1 + 18\right).$$

$$\boxed{\odot}$$

Algorithm 45.4.8. Let us now finally discuss a solution of Problem 45.4.2:

- for each l, λ, h compute the values

$$b_{l\lambda}^{(h)} := \sum_{i=1}^{R} \gamma(q_l', q_i, \mathbf{q}) a_{i\lambda}^{(h)}$$

which satisfy, for each l, h (mod $\mathbf{J}'$)

$$\sum_{\lambda=1}^{R} b_{l\lambda}^{(h)} q_\lambda = \sum_{\lambda=1}^{R} \left(\sum_{i=1}^{R} \gamma(q_l', q_i, \mathbf{q}) a_{i\lambda}^{(h)}\right) q_\lambda \equiv \sum_{i=1}^{R} \gamma(q_l', q_i, \mathbf{q}) Z_h q_i \equiv Z_h q_l' \in \mathbf{J}';$$

- by linear algebra compute, for each l, h, the unique values $a'^{(h)}_{lj}, 1 \leq j \leq S$, satisfying

$$\sum_{j=1}^{S} a'^{(h)}_{lj} q_j' = \sum_{\lambda=1}^{R} b_{l\lambda}^{(h)} q_\lambda \equiv Z_h q_l' \ (\text{mod } \mathbf{J}');$$

- for each i, j, λ compute the values

$$\delta_{ij}^{(\lambda)} := \left(\sum_\iota \sum_\kappa \gamma(q_i', q_\iota, \mathbf{q}) \gamma_{\iota\kappa}^{(\lambda)} \gamma(q_j', q_\kappa, \mathbf{q})\right)$$

which satisfy, for each i, j (mod $\mathbf{J}'$),

$$\sum_{\lambda=1}^{R} \delta_{ij}^{(\lambda)} q_\lambda = \sum_{\lambda=1}^{R} \left(\sum_{\iota=1}^{R} \sum_{\kappa=1}^{R} \gamma(q_i', q_\iota, \mathbf{q}) \gamma_{\iota\kappa}^{(\lambda)} \gamma(q_j', q_\kappa, \mathbf{q})\right) q_\lambda$$

$$\equiv \sum_{\iota=1}^{R} \sum_{\kappa=1}^{R} \gamma(q_i', q_\iota, \mathbf{q}) \gamma(q_j', q_\kappa, \mathbf{q}) \left(\sum_{\lambda=1}^{R} \gamma_{\iota\kappa}^{(\lambda)} q_\lambda\right)$$

$$\equiv \sum_{\iota=1}^{R} \sum_{\kappa=1}^{R} \gamma(q_i', q_\iota, \mathbf{q}) \gamma(q_j', q_\kappa, \mathbf{q}) q_\iota q_\kappa$$

$$= q_i' q_j' \in \mathbf{J}';$$

- by linear algebra, for each i, j compute the unique values $\gamma'^{(l)}_{ij}, 1 \leq l \leq S$, satisfying

$$\sum_{l=1}^{S} \gamma'^{(l)}_{ij} q_l' = \sum_{\lambda=1}^{R} \delta_{ij}^{(\lambda)} q_\lambda \equiv q_i' q_j' \ \text{mod } \mathbf{J}'.$$

Thus the data

(a) $\mathbf{q}' = \{q'_1, \ldots, q'_S\}$,

(b) $\mathcal{M}(\mathbf{q}') := \left\{ \left(a'^{(h)}_{lj} \right) \right\}$,

(c) the structure constants $\gamma'^{(l)}_{ij}$,

give the required Gröbner representation of J'.

45.5 Linear Representation

Let us now specialize the results of the previous section to the case in which we have:

(A) $\mathbf{q} = \mathbf{N}_\prec(J)$, ordered in such a way that $1 = q_1 \prec q_2 \prec \cdots \prec q_R$;

(B) at each step of the Gaussian algorithm, as a *pivot* element among all possible choices we systematically choose the leftmost available column;

(C) $\mathbf{b} = \mathbf{N}_\prec(J)$, ordered in such a way that $b_1 \succ \cdots \succ b_{R-1} \succ b_R = 1$, so that $q_i = b_{R-i+1}$ for each i.

Remark 45.5.1. Given any Gröbner representation of J and a term ordering $\prec$, a direct application of Möller's Algorithm 28.2.7 to the functionals $\gamma(\cdot, q_i, \mathbf{q})$ returns $\mathbf{N}_\prec(J)$. ◉

Example 45.5.2. Conditions (A) and (B) are satisfied by both Examples 45.4.6 and 45.4.7; the latter satisfies condition (C) also. Therefore the reader can easily verify our claims using these examples. ◉

As a consequence of (B), in the transformation of C into D, each column which is not a *pivot* element is modified only by the columns to its left. Therefore, since in (x) (see above Corollary 45.4.4) the set $J := \{\rho(\kappa) : S < \kappa \leq R\}$ denotes the indices corresponding to the columns of C which have not been used as a *pivot* element, assumption (B) allows us to reformulate (x) as follows:

(x)′ writing $\rho(\kappa) := \max\{\ell : e_{\ell\kappa} \neq 0\}$ for each κ, $S < \kappa \leq R$, and $J := \{\rho(\kappa) : S < \kappa \leq R\}$, we have $e_{\ell\kappa} = 0$ for each $\ell \neq \kappa$.

Thus we trivially have

Corollary 45.5.3. *If assumptions (A) and (B) are satisfied then* $\mathbf{N}(J_1) = \{b_i, 1 \leq i \leq R, i \notin J\}$. ◉

In the same vein, as a direct consequence of the order imposed on $\mathbf{b}$ by condition (C) we also obtain

Corollary 45.5.4. *If assumptions (A)–(C) are satisfied then* $\mathbf{N}(J_0) = \{b_{S+1}, \ldots, b_R\}$. ◉

Bibliography

Aho A.V., Hopcroft J.E., Ullman J.D., *The Design and Analysis of Computer Algorithms*, Addison–Wesley (1974).

Alonso M.E., Becker E., Roy M.-F., Wörmann T., Zeroes, multiplicities and idempotents for zero dimensional systems, in *Prog. Math.* **143** (1996), pp. 1–16, Birkhäuser.

Ampère A.-M., *Fonctions Interpolaires*, Annales de M. Gergonne (1826).

Arnaudiès J.M., Valibouze A., Résolventes de Lagrange, Report LIPT **93.61** (1993),

Arnaudiès J.M., Valibouze A., Lagrange resolvents, *J. Pure Appl. Algebra* **117–118** (1996), 23–40.

Aubry P., Moreno Maza M., Triangular set for solving polynomial systems: a comparative implementation of four methods, *J. Symb. Comp.* **28** (1999), 125–154.

Aubry P., Valibouze A., Using Galois ideals for computing relative resolvents, *J. Symb. Comp.* **30** (2000), 635–651.

Aubry P., Lazard D., Moreno Maza M., On the theories of triangular sets, *J. Symb. Comp.* **28** (1999), 105–124.

Auzinger W., Stetter H.J., An elimination algorithm for the computation of all zeros of a system of multivariate polynomial equations, in *I.S.N.M.* **86** (1988), pp. 11–30, Birkhäuser.

Becker E., Cardinal J.-P., Roy M.-F., Szafraniec Z., Multivariate Bezoutians, Kronecker symbol and Eisenbud–Levin formula, in *Prog. Math.* **143** (1996), pp. 79–104, Birkhäuser.

Bézout E., Recherches sur le degré des équations résultantes de l'évanouissement des inconnues, et sur les moyens qu'il convient d'employer pour trouver ses équations, *Mém. Acad. Roy. Sci. Paris* (1764), 288–233.

Bézout E., *Théorie Generale des Èquations Algébriques*, Pierres, Paris (1771).

Bini D., Pan V., *Polynomial and Matrix Computations*, Birkhäuser (1994).

Bostajn A., Salvy B., Schost E., Fast algorithm for zero-dimensional polynomial systems using duality, *J. AAECC* **14** (2003), 239–272.

Bürgisser P., Clausen M., Shorolahi M.A., *Algebraic Complexity Theory*, Springer (1997).

Burnside W., *Theory of Groups of Finite Order*, Cambridge University Press (1911).

Canny J., Generalized characteristic polynomials, in *L. N. Comp. Sci.* **358** (1988), pp. 293–299, Springer.

Canny J., An effective algorithm for the sparse mixed resultant, in *L. N. Comp. Sci.* **673** (1993), pp. 89–104, Springer.

Cardinal J.P., *Dualité et algorithms itératifs pour la résolution de systémes polynomiaux*, Ph.D. thesis, University of Rennes I (1993).

Cardinal J.P., Mourrain B., Algebraic approach of residues and applications, in *L. N. Appl. Math.* **32** (1999), American Mathematical Society Press.

Cauchy A., Usage des fonctions interpolaires dans ls determination des fonctions symmetriques des racines d'une équation algébrique donnée, *C.R. Acad. Sci. Paris* **11** (1840), 933.

Cauchy A., *Oeuvres* t. V, Gauthier–Villars, Paris, (1882).

Cayley A., On the theory of elimination, *Camb. Dublin Math. J.* **III** (1848), 116–120.

Cayley A., Note sur la méthode d'élimination de Bezout, *J. Reine Ang. Math.* **LIII** (1857), 366–367

Cayley, A., A fourth memoir upon quantics, *Phil. Trans. Royal Soc. London* **CXLVIII** (1858), 415–427.

Charden M., Un algorithm pour les calcul des resultants, in *Prog. Math.* **94** (1990), pp. 47–62, Birkhäuser.

Charden M., The resultant via a Koszul complex, *Prog. Math.* **109** (1993), pp. 29–40, Birkhäuser.

Chen C., Golubitsky O., Lemaire F., Moreno Maza M., Comprehensive triangular decomposition, in *Proc. CASC 2007* (2007), pp. 73–101.

Dahan X., *Sur la complexité des représentations des systèmes polynomiaux: triangulation, méthodes modulaires, évaluation dynamique*, Ph.D. thesis, École Polytechnique (2006).

Dahan X., Schost E., Sharp estimates for triangular sets, in *Proc. ISSAC'04* (2004), pp. 103–110, Association for Computing Machinery.

Dahan X., Moreno Maza M., Schost E., Wu W., Xie Y., Lifting techniques for triangular decomposition, in *Proc. ISSAC'05* (2005), pp. 108–115.

Delassus E., Sur les systèmes algébriques et leurs relations avec certains systèmes d'equations aux dérivées partielles. *Ann. Éc. Norm. 3e série* **14** (1897), 21–44.

Dixon, A.L., On a form of the eliminant of two quantics, *Proc. London Math. Soc.* **6** (1908a), 468–478.

Dixon, A.L., The eliminant of three quantics in two independent variables, *Proc. London Math. Soc.* **7** (1908b), 49–69.

Dixon, A.L., Some results in the theory of elimination, *Proc. Roy. Soc. London* **82** (1909), 468–478.

Euler, L., *Introductio in Analysin Infinitorum*, Tom. 2, Lausanne, (1748).

Felszeghy B., Ráth B., Rónyai L., The lex game and some applications, *J. Symb. Comp.* **41** (2006), 663–681.

Gallo G., Mishra B., Effective algorithms and bounds for Wu–Ritt characteristic sets, in *Prog. Math.* **94** (1990), pp. 119–142, Birkhäuser.

Gallo G., Mishra, B., A solution to Kronecker's problem, *J. AAECC* **5** (1994), 343–370.

Gallo G., Mishra, B., Olivier F., Some constructions in rings of differential polynomials, in *L. N. Comp. Sci.* **539** (1991), pp. 171–182, Springer.

Gianni P., Properties of Gröbner bases under specialization, in *L. N. Comp. Sci.* **378** (1987), pp. 293–297, Springer.

Giusti M., Heintz J., Algorithmes – disons rapides – pour la décomposition d'une variété algébrique en composantes irréductibles, in *Prog. Math.* **94** (1990), pp. 169–194, Birkhäuser.

Giusti M., Heintz J., La detérmination des point isolés et de la dimension d'une variété algébrique peut se faire en temps polynomial, in *Symp. Math.* **34** (1993), pp. 216–256, Cambridge University Press.

Giusti M., Schost E., Solving some overdetermined polynomial systems, in *Proc. ISSAC'99* (1999), pp. 1–8, Association for Computing Machinery.

Giusti M., Heintz J., Morais J.E., Pardo L.M., When polynomial equation systems can be "solved" fast?, in *L. N. Comp. Sci.* **948** (1995), pp. 205–231, Springer.

Giusti M., Heintz J., Hägele K., Morais J.E., Pardo L.M., Montaña J.M., Lower bounds for diophantine approximation, *J. Pure Appl. Algebra* **117–118** (1997a), 277–311.

Giusti M., Heintz J., Morais J.E., Pardo L.M., Le rôle des structures de données dans les problèmes d'élimination, *C.R. Acad. Sci. Paris* **325** (1997b), 1223–1228.

Giusti M., Heintz J., Morais J.E., Morgensten J., Pardo L.M., Straight-line programs in geometric elimination theory, *J. Pure Appl. Algebra* **124** (1998), 101–146.

Giusti M., Hägele K., Lecerf G., Marchand J., Salvy B., The projective Noether maple package: computing the dimension of a projective variety, *J. Symb. Comp.* **30** (2000), 291–307.

Giusti M., Lecerf G., Salvy B., A Gröbner free alternative for polynomial system solving, *J. Complexity* **17** (2001), 154–211.

Gonzalez-Vega L., Rouiller F., Roy M.-F., Symbolic recipes for polynomial system solving, in *Some Tapas of Computer Algebra*, ed. A. Cohen, (1997), Springer.

Gunther N., Sur la forme canonique des systèmes équations homogènes (in Russian) (*Journal de l'Institut des Ponts et Chaussées de Russie*), *Izdanie Inst. Inž. Putej Soobščenija Imp. Al. I.* **84** (1913).

Gunther N., Sur les caractéristiques des systémes d'équations aux dérivées partialles, *C.R. Acad. Sci. Paris* **156** (1913), 1147–1150.

Gunther N., Sur la forme canonique des equations algébriques, *C.R. Acad. Sci. Paris* **157** (1913), 577–80.

Hägele K., Morais J.E., Pardo L.M., Sombra M., On the intrinsic complexity of the arithmetic Nullstellensatz, *J. Pure Appl. Algebra* **146** (2000), 103–183.

Hashemi A., Structure et Complexité des bases de Gröbner, *J. Symb. Comp.* **45** (2010), 1330–1340.

Hashemi A., Lazard D., Sharper complexity bounds for zero-dimensional Gröbner bases and polynomial system solving, *Int. J. Algebra Comp.* **21** (2011), 705–713.

Jacobi C.G.I., De eliminatione variabilis e duabus aequationibus algebraicas, *J. Reine Ang. Math.* **XV** (1836) 101–124.

Jouanoulou J.-P., Le formalisme du résultant, *Adv. Math.* **90** 117–263.

Kalkbrener M., Solving systems of algebraic equations by using Gröbner bases, in *L. N. Comp. Sci.* **378** (1987), pp. 282–292, Springer.

Kalkbrener M., *Three contributions to elimination theory*, Ph.D. thesis, Linz University (1991).

Kalkbrener M., A generic euclidean algorithm for computing triangular representations of algebraic varieties, *J. Symb. Comp.* **15** (1993), 153–167.

Kalkbrener M., On the stability of Gröbner bases under specialization, *J. Symb. Comp.* **24** (1997), 51–58.

Kapur D., Cai Y., An algorithm for computing a Gröbner basis of a polynomial ideal over a ring with zero divisors, *Math. Comput. Sci.* **2** (2009), 601–634.

Kapur D., Chtcherba A.D., Conditions for exact resultants using the Dixon resultant formulation, in *Proc. ISSAC 2000* (2000), pp. 62–70, Association for Computing Machinery.

Kapur D., Chtcherba A.D., On the efficiency and optimality of Dixon-based resultant method, in *Proc. ISSAC 2002* (2002), pp. 29–36, Association for Computing Machinery.

Kapur D., Saxena T., Yang, L., Algebraic and geometric reasoning using Dixon resultants, in *Proc. ISSAC 94* (1994), pp. 99–136, Association for Computing Machinery.

Kapur D., Saxena T., Extraneous factors in the Dixon resultant formulation, in *Proc. ISSAC 97* (1997), pp. 141–148, Association for Computing Machinery.

Kobayashi H., Moritsugu S., Hogan R.W., On radical zero-dimensional ideals, *J. Symb. Comp.* **8** (1989), 545–552.

Kratzer M., *Computing the dimension of a polynomial ideal and membership in low-dimensional ideals*. Master's thesis, Technische Universität München (2008).

Krick T., Pardo L.M., Une approache informatique pour l'approximation diophantienne, *C.R. Acad. Sci. Paris* **318** (1994), 407–412.

Krick T., Pardo L.M., A computational method for Diphantine approximation, in *Prog. Math.* **143** (1996), pp. 193–254, Birkhäuser.

Lakshman Y.N., Lazard D., On the complexity of zero-dimensional algebraic systems, in *Prog. Math.* **94** (1990), pp. 217–226, Birkhäuser.

Lazard D., Algèbre linéaire sur $K[X_1, \ldots, X_n]$ et élimination, *Bull. Soc. Math. France* **105** (1977), 165–190.

Lazard D., Systems of algebraic equations, in *L. N. Comp. Sci.* **72** (1979), pp. 88–94, Springer.

Lazard, D., Resolution des systemes d'equations algebriques, *Theoret. Comput. Sci.* **15** (1981), 77–110.

Lazard, D., A new method for solving algebraic systems of positive dimension, *Disc. Appl. Math.* **33** (1991), 147–160.

Lazard D., Solving zero-dimensional algebraic systems, *J. Symb. Comp.* **15** (1992), 117–132.

Lazard D., Systems of algebraic equations (algorithms and complexity), *Symp. Math.* **34** (1993), 84–106, Cambridge University Press.

Lazard D., Resolution of polynomial systems, in *Proc. ASCM 2000* (2000), pp. 1–8, World Scientific.

Lazard D., On the specification for solvers of polynomial systems, in *Proc. ASCM 2001* (2001), pp. 1–10, World Scientific.

Lazard D., Thirty years of polynomial system solving, and now?, *J. Symb. Comp.* **44** (2009), 222–239.

Lazard D., Rouillier F., Solving parametric polynomial systems, *J. Symb. Comp.* **42** (2007), 636–667.

Lecerf G., *Une alternative aux méthodes de réécriture pour résolution des systémes algébriques*, Ph.D. thesis, École Polytechnique (2001).

Lundqvist S., Complexity of comparing monomials and two improvements of the BM-algorithm, in *L. N. Comp. Sci.* **5393** (2008), pp. 105–125, Springer.

Lundqvist S., Vector space bases associated to vanishing ideals of points, *J. Pure Appl. Algebra* **214** (2010), 309–321.

Macaulay F. S., Some formulae in elimination, *Proc. London Math. Soc.* (1) **35** (1903), 3–27.

Macaulay F. S., *The Algebraic Theory of Modular Systems*, Cambridge University Press (1916).

Malle G., Trinks W., *Zur Behandlung algebraischer Gleichungssysteme mit dem Computer*, preprint (1985).

Manocha D., Cannon J.F., Multipolynomial resultant algorithms, *J. Symb. Comp.* **15** (1993), 99–122.

Möller H.M., Systems of algebraic equations solved by means of endomorphisms, in *L. N. Comp. Sci.* **673** (1993a), pp. 43–56, Springer.

Möller H.M., On decomposing systems of polynomial equations with finitely many solutions, *J. AAECC* **4** (1993b), 217–230.

Möller M., Stetter H., Multivariate polynomial equations with multiple zeros solved by matrix eigenproblems, *Num. Math.* **70** (1995), 311–325.

Monico C., Computing the primary decomposition of zero-dimensional ideals, *J. Symb. Comp.* **34** (2002), 451–459.

Morais J.E., *Resolución eficaz de systemas de ecuaciones polinomiales*, Ph. D. thesis, University of Cantabria, Santander (1997).

Moreno Maza M., Rioboo R., Polynomial gcd computation over tower of algebraic extension, in *L. N. Comp. Sci.* **948** (1995), pp. 365–382, Springer.

Moritzugu S., Kuriyama K., On multiple zeros of systems of algebraic equations, in *Proc. ISSAC'99* (1999), pp. 23–30, Association for Computing Machinery.

Mourrain B., Computing the isolated roots by matrix methods, *J. Symb. Comp.* **26** (1998), 715–738.

Mourrain B., A new criterion for normal form algorithms, in *L. N. Comp. Sci.* **1719** (1999), pp. 430–443, Springer.

Mourrain B., Bezoutian and quotient ring structure, *J. Symb. Comp.* **39** (2005), 397–415.

Mourrain B., Pan Y.V., Multivariate polynomials, duality and structured matrices, *J. Complexity* **16** (2000), 110–180.

Mourrain B., Ruatta O., Relation between roots and coefficients, interpolation and application to system solving, *J. Symb. Comp.* **33** (2002), 679–699.

Mourrain B., Trebuchet P., Solving projective complete intersection faster, in *Proc. ISSAC'00* (2000), pp. 234–241, Association for Computing Machinery.

Muir T., *The Theory of Determinants in the Historical Order of Development*, MacMillan (1906).

Netto E., *Vorlesungen über Algebra*, Zweiter Band Teubner (1900).

Pardo L.M., How lower and upper complexity bounds meet in elimination, in *L. N. Comp. Sci.* **948** (1995), pp. 33–69, Springer.

Pierce R.S., Modules over commutaive regular rings, *Mem. A.M.S.* **70** (1967).

Pohst M., Yun D., On solving systems of algebraic equations via ideal bases and elimination, in *Proc. 1981 SymSAC* (1981), pp. 206–211, Association for Computing Machinery.

Poisson, S.D. Mémoire sur l'élimination dans les équations algébriques, *J. École Polytechnique t.* IV (1802), 199–203.

Rennert N., Valibouze A., Calcule de résolventes avec les modules de Cauchy, *Exp. Math.* **8** (1999), 351–366.

Riordan J., *Combinatorial Identities*, Wiley, (1968).

Ritt J.F., Prime and composite polynomials, *Trans. A.M.S.* **23** (1922), 51–66.

Ritt J.F., *Differential Equations from the Algebraic Standpoint*, A.M.S. Colloquium Publications **14** (1932).

Ritt J.F., *Differential Algebra*, A.M.S. Colloquium Publications **33** (1950).

Robinson L.B., *Sur les systémes d'équations aux dérivées partialles*, C.R. Acad. Sci. Paris **157** (1913), 106–108.

Rouillier F., *Algorithmes efficaces pour l'étude des zéros réels des systèmes polynomiaux*, Ph.D. thesis, University of Rennes I (1996).

Rouillier F., Solving zero-dimensional systems through the Rational Univariate Representation, *J. AAECC* **9** (1999), 433–461.

Salmon G., *Lessons Introductory to the Modern Higher Algebra*, Fifth Edn., Chelsea (1885).

Sims C., *Computation with Finitely Presented Groups*, Cambridge University Press (1994).

Stetter H., Matrix eigenprobelms are at the heart of polynomial system solving, *SIGSAM Bull.* **30** (1996), 22–25.

Stetter H., *Numerical Polynomial Algebra*, Tutorial Notes at ISSAC'98, Rostock (1998).

Stetter H., *Numerical Polynomial Algebra*, SIAM (2004).

Sylvester J.J., A method of determining by mere inspection the derivatives from two equations of any degree, *Phil. Mag.* **XVI** (1840), 132–135.

Sylvester J.J., Memoir on the dialytic method of elimination. Part I. *Phil. Mag.* **XXXI** (1842), 534–539.

Sylvester J.J., On a theory of the syzygietic relations of two rational integral functions, comprising an application to the theory of Sturm's functions, and that of the greatest algebraic common measure, *Phil. Trans. Royal Soc. London* **CXLIII** (1853), 407–548.

Trinks W., Über B. Buchberger Verfahren, Systeme algebraischer Gleichungen zu lösen, *J. Numb. Theory* **10** (1978), 475–488.

Valibouze A., Resolutions et functions symmetriques, in *Proc. ISSAC'89* (1989), pp. 390–399, Association for Computing Machinery.

Valibouze A., Computation of the Galois groups of the resolvent factors for the direct and inverse Galois problems, in *L. N. Comp. Sci.* **948** (1995), pp. 456–468, Springer.

Valibouze A., Étude des relations algébriques entre les racines d'un polynôme d'une variable, *Bull. Belg. Math. Soc. Simon Stevin* **6** (1999), 507–535.

Valibouze A., *Théorie de Galois constructive*, Mémoir d'Habilitation, Paris (1998).

Wang D.-M., An elimination method for polynomial systems, *J. Symb. Comp.* **16** (1993), 83–114.

Wang D.-M., Decomposing polynomial systems into simple systems, *J. Symb. Comp.* **25** (1998), 295–314.

Wimmer H.K., On the history of the Bezoutian and the resultant matrix, *Linear Algebra Appl.* **128** (1990), 27–34.

Wu W.-T., On the decision problem and the mechanization of the theorem-proving in elementary geometry, *Scinetia Sinica* **21** (1978), 159–172.

Wu W.-T., Basic principles of mechanical theorem proving in elementary geometry, *J. Sys. Sci. & Math. Sci.* **4** (1984), 207–235.

Wu W.-T., On the decision problem and the mechanization of the theorem-proving in elementary geometry, in *Contemp. Math.* **29** (1984), pp. 213–234, American Mathematical Society.

Wu W.-T., Some recent advances in mechanical theorem-proving of geometry, (reprinted) in *Contemp. Math.* **29** (1984), pp. 235–241, American Mathematical Society.

Wu W.-T., A zero structure theorem for polynomial equations solving, *M.M. Research Preprints* **1** (1987), 2–12.

Wißmann D., *Anwendung von Rewriting-Techniken in polyzyklischen Gruppen*, Dissertation, Kaiserslautern (1989).

Yokoyama K., Noro M., Takeshima T., Solutions of systems of algebraic equations and linear maps on residue class rings, *J. Symb. Comp.* **14** (1992), 399–417.

Zariski O., Samuel P., *Commutative Algebra*, Van Nostrand (1958).

Index

CPSIA information can be obtained at www.ICGtesting.com
Printed in the USA
LVOW11*1441031115

460922LV00011B/92/P